Electronics and Microcomputer Circuits

146 PRACTICAL PROJECTS

Roger L. Tokheim

McGraw-Hill Book Company

New York St. Louis San Francisco Auckland Bogotá Guatemala Hamburg
Johannesburg Lisbon London Madrid Mexico Montreal New Delhi
Panama Paris San Juan São Paulo Singapore Sydney Tokyo Toronto

The author has carefully reviewed the programs in this book to ensure their performance in accordance with the specifications described. Neither the author nor McGraw-Hill, however, makes any warranties whatever concerning the programs. They assume no responsibility or liability of any kind for errors in the programs or for the consequences of any such errors.

ELECTRONICS AND MICROCOMPUTER CIRCUITS

A Byte Book

1 2 3 4 5 6 7 8 9 0 SEM SEM 0 9 8 7 6 5

ISBN 0-07-064984-7

Library of Congress Cataloging in Publication Data

Tokheim, Roger L.
Electronics and microcomputer circuits.

(A Byte book)
Includes index.
1.Electronics—Amateurs' manuals. 2.Microcomputers —Amateurs' manuals. I.Title. II.Title: Electronics and microcomputer circuits. III.Series: Byte books.
TK9965.T65 1986 621.381 85-116
ISBN 0-07-064984-7

Editor: **Elizabeth Zayatz**
Editing Supervisor: **Marthe Grice**
Book design by **M. R. P. Design.**

Contents

Preface

This is a sourcebook of modern electronic circuits and projects. The 146 projects range in level of difficulty from very simple to quite complex. Grade school students can wire the most simple circuits while only veteran hobbyists, technicians, or engineers should attempt the most complex. These projects were selected after researching more than 20,000 pages of books, magazines, and manufacturer's literature. In my electronics teaching career, I have supervised the construction of thousands of electronic projects. Two-thirds of the projects in this book were actually wired and tested under my supervision to make certain that they operate adequately. More than 50 possible circuits were rejected because they did not work properly.

Microcomputers and robotics have fueled the imagination of the general public. Many principles of computer control of machines are illustrated in the microcomputer interface circuits in Chapters 2, 3, and 11. Fifteen circuits are included that interface the outside world to the Apple II/IIe microcomputer. Chapter 11 contains another eight microcomputer-related circuits, including a complete microcomputer system.

Electronic circuits that use household current from standard wall receptacles can be dangerous. Safe practices must be used to prevent electrical shock, fires, and injuries. Electrical shocks can be *fatal*. Shock hazards are greatest when working with household ac current. *It is extremely important that all high-voltage wires be covered with adequate insulating materials* (such as shrinkable tubing). Any project using household current must be placed in a proper enclosure so no one can possibly come in contact with the high-voltage wiring. Finally, use extreme care to provide mechanical strain relief where high voltages enter and exit an enclosure. For safety, beginners and younger students should start with projects that require only batteries.

I am grateful to the many authors, companies, and manufacturers who furnished materials and granted permission for their circuits to be included in this sourcebook. My thanks to and admiration for the creative individuals who designed the circuits. I give special thanks to engineering student Don Hallgren for his design work on the microcomputer interfacing circuits and for testing hundreds of circuits. Finally, I would like to express my appreciation to my family, to Dan for testing many circuits, and to Marshall and Carrie for their help and patience.

Roger L. Tokheim

chapter 1

Alarm Circuits

BREAK-BEAM ALARM

The break-beam alarm will emit a momentary sound as customers, students, friends, pets, or family members move past the unit. Almost any ambient level of light can be used. The trigger adjust R_3 shown in Fig. 1-1 adjusts the sensitivity of the alarm. The volume control R_9 adjusts the tone and volume of sound emitted from the buzzer.

The break-beam alarm circuit shown in Fig. 1-1 is based on the 339 voltage comparator IC. The voltages at pins 4 and 5 of the IC are compared. The cadmium sulfide photo cell R_2 and resistor R_1 form a voltage divider across the 12-volt (12-V) power supply. When less light strikes photo resistor R_2, its resistance increases causing the voltage at pin 5 of the comparator to increase above pin 4. This causes the voltage at pin 2 of the comparator to increase, thus activating transistor Q_1. When the transistor is activated, the ground path is completed for the piezo buzzer, causing it to sound.

COMBINATION-LOCK/ALARM CONTROL

The schematic diagram and parts list for the combination-lock/alarm control circuit are shown in Fig. 1-2(*a*). The combination-lock/alarm control circuit is typically used as a burglar alarm. Typical wiring of the "jumper matrix" area is shown in Fig. 1-2(*b*). When the header is wired as in Fig. 1-2(*b*), the combination to disarm the lock/alarm will be the number sequence 4, 1, 6, and 3. The # button is used to arm the alarm (connected to "lock" input).

First, *arm* the lock by pushing "#" on the keypad. When the "lock" input to Q_1 in Fig. 1-2(*a*) goes high, pin 1 of the 7220 digital lock IC goes low. This causes pin 8 of the IC to go high and the LED D_1 to light. Also, pin 13 of the IC goes low. Transistor Q_2 will conduct, thus triggering the SCR and the alarm. Transistor Q_2 will conduct only when pin 13 is high *and* the "switch on the door" is closed. Once the

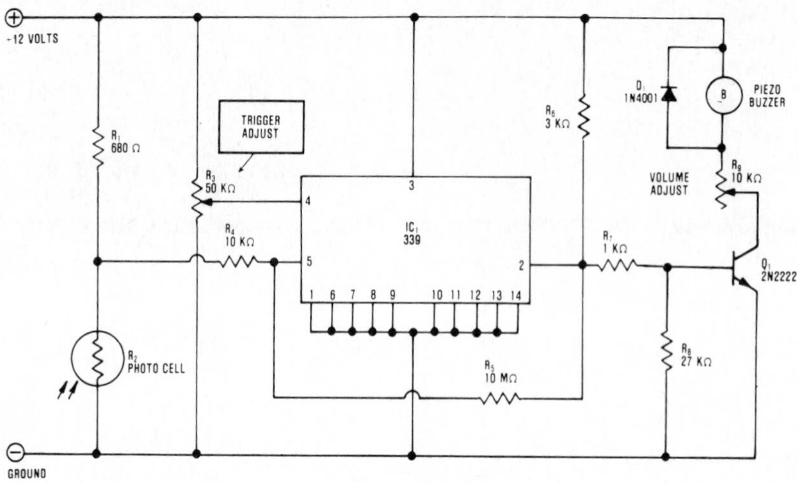

PARTS LIST

B_1	Piezo buzzer (Radio Shack 273-060)
D_1	1N4001 silicon diode 1 A, 50 PIV
IC_1	339 voltage comparator IC
Q_1	2N2222 NPN transistor (or similar NPN)
R_1	680-Ω, ½-W resistor
R_2	Cadmium sulfide photocell (Radio Shack 276-116)
R_3	50-kΩ linear taper potentiometer
R_4	10-kΩ, ¼-W resistor
R_5	10-MΩ, ¼-W resistor
R_6	3-kΩ, ¼-W resistor
R_7	1-kΩ, ¼-W resistor
R_8	27-kΩ, ¼-W resistor
R_9	10-kΩ, audio taper potentiometer

FIG. 1-1 Break-beam alarm. (Michael Gannon, *Workbench Guide to Semiconductor Circuits and Projects,* Prentice-Hall, New Jersey, 1982, pp. 172–173. Used by permission of Prentice-Hall, Inc.)

SCR is activated, it will continue to sound the alarm until the reset switch S_{14} is pressed and the alarm is disarmed.

Second, *disarm* the lock. The lock can be disarmed by activating pins 3, 4, 5, and then 6 on the 7220 digital lock IC in order. This causes pin 8 to go low, extinguishing the LED. Also, pin 13 goes high, thereby disarming the alarm. The reset switch S_{14} also must be opened to stop the alarm from sounding.

A typical use for the combination-lock/alarm control circuit is shown in Fig. 1-3. The keypad and LED are mounted on the outside of the door while the combination-lock printed-circuit (pc) board is attached to the inside. In this case, the combination-lock/alarm control circuit is driving a second circuit board, the whooper alarm. The whooper alarm is a product of Electronic Kits International, Inc., but any alarm or siren may be used. The home installation shown in Fig. 1-3 is used as follows:

1. When switch S_{13} is turned on, the alarm will sound. To avoid this, hold the reset button while turning S_{13} on *and* enter the combination (4, 1, 6, 3 in this example). The LED will go out.
2. To activate alarm when leaving a room, close door and press *lock* or *arm* button on the keypad (press # in this example). The LED will light, and the alarm will be armed. The whooper alarm will sound if the correct four-digit combination is not entered before the door is opened.

3. The four-digit code (4, 1, 6, 3 in this example) must be entered to disarm the alarm before the door is opened. The LED will go out when the proper code is entered, and the whooper alarm will not sound as the door is opened.

The combination-lock/alarm control circuit will operate on 9 to 12 V. An alkaline or auto storage battery is recommended. A parts kit, a pc board, and full instructions are currently available for both the combination lock and the whooper alarm from Electronic Kits Interna-

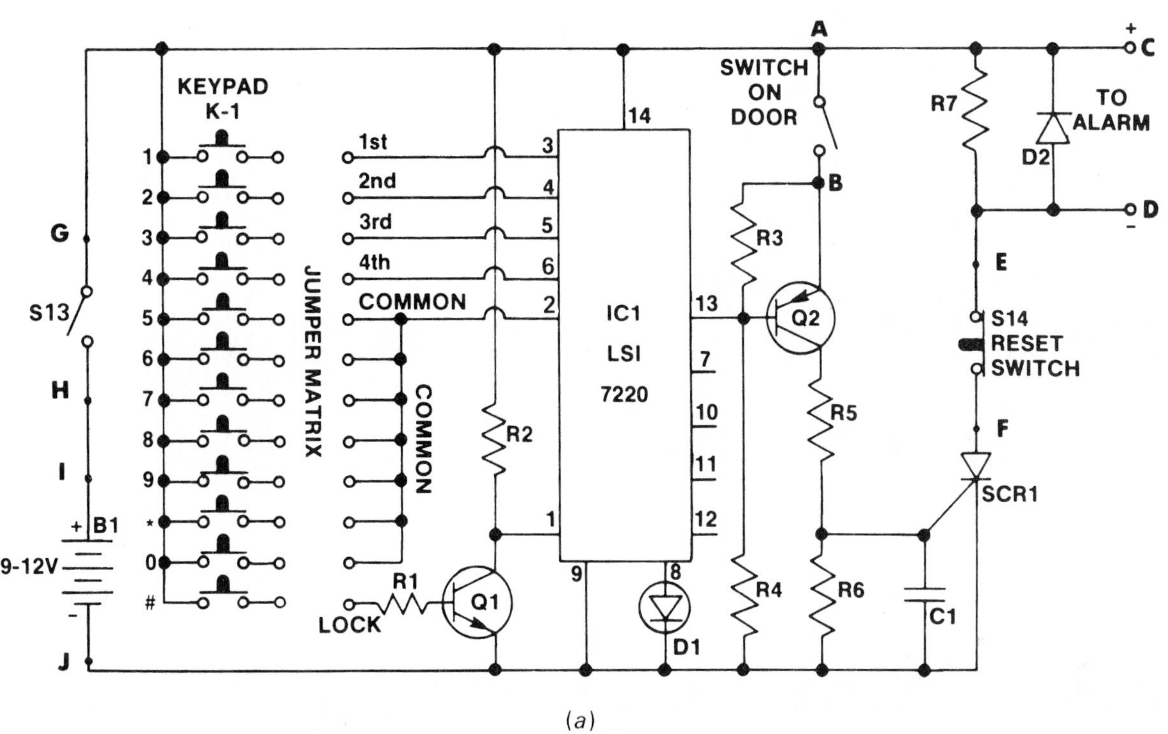

(*a*)

PARTS LIST

B_1	9-V battery
C_1	0.1-μF capacitor
D_1	Light-emitting diode
D_2	1N4003 silicon diode, 1 A, 200 PIV
IC_1	7220 digital lock IC (LSI Computer Systems, Inc.)
Q_1	2N3904 NPN transistor
Q_2	2N3638A PNP transistor
R_1, R_4, R_5	10-kΩ, ½-W resistor
R_2, R_6, R_7	1-kΩ, ½-W resistor
R_3	2.2-kΩ, ½-W resistor
K_1	Keypad
S_{13}	SPST switch
S_{14}	Normally closed push-button switch
SCR_1	4-A, 200-PIV silicon controlled rectifier (such as GE C106B1)
Miscellaneous	24-pin jumper header
	14-pin DIP IC socket
	24-pin DIP IC socket

FIG. 1-2 Combination-lock/alarm control. (Used by permission of Electronic Kits International, Inc.)

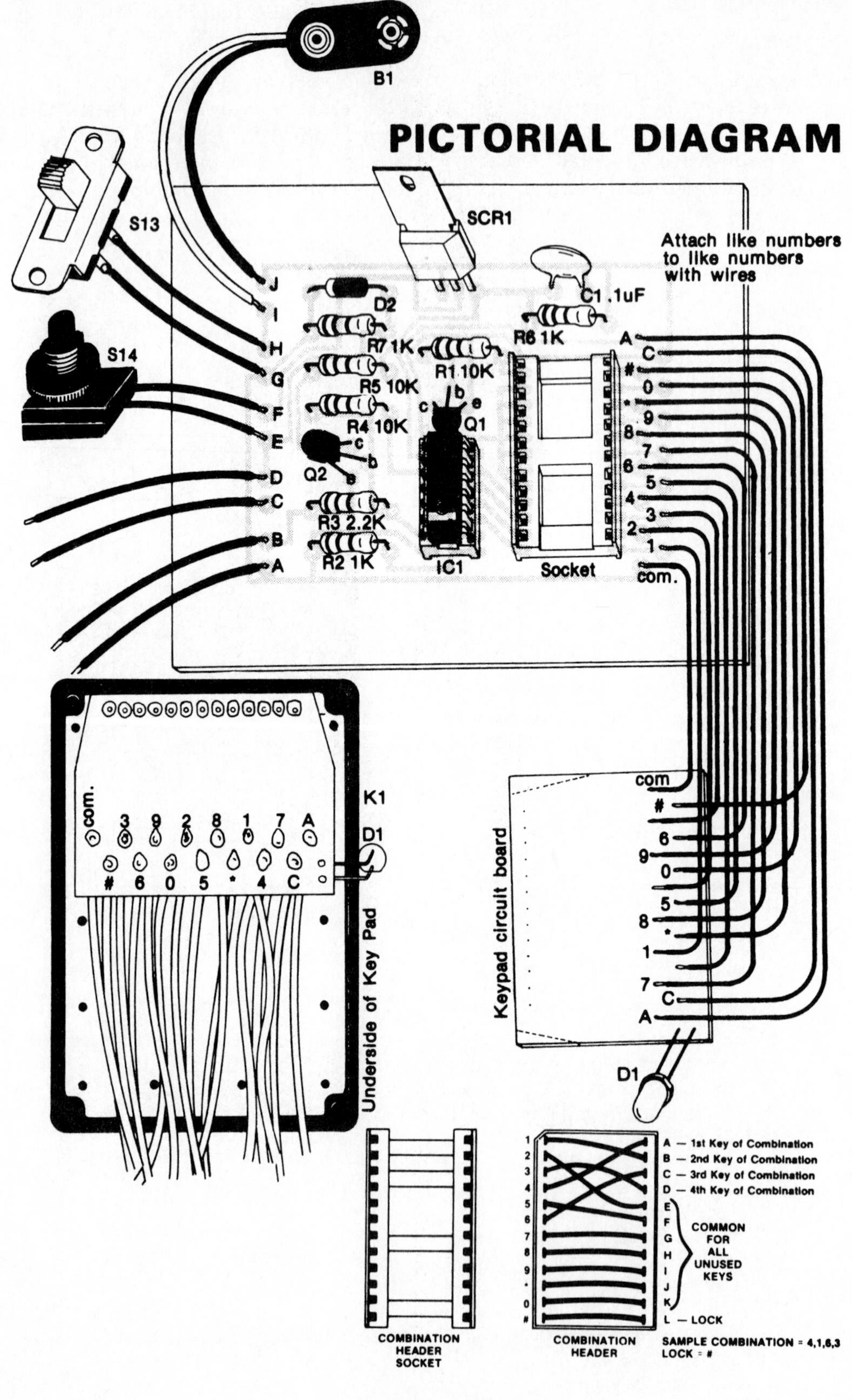

(*b*)

FIG. 1-2 (*Continued*)

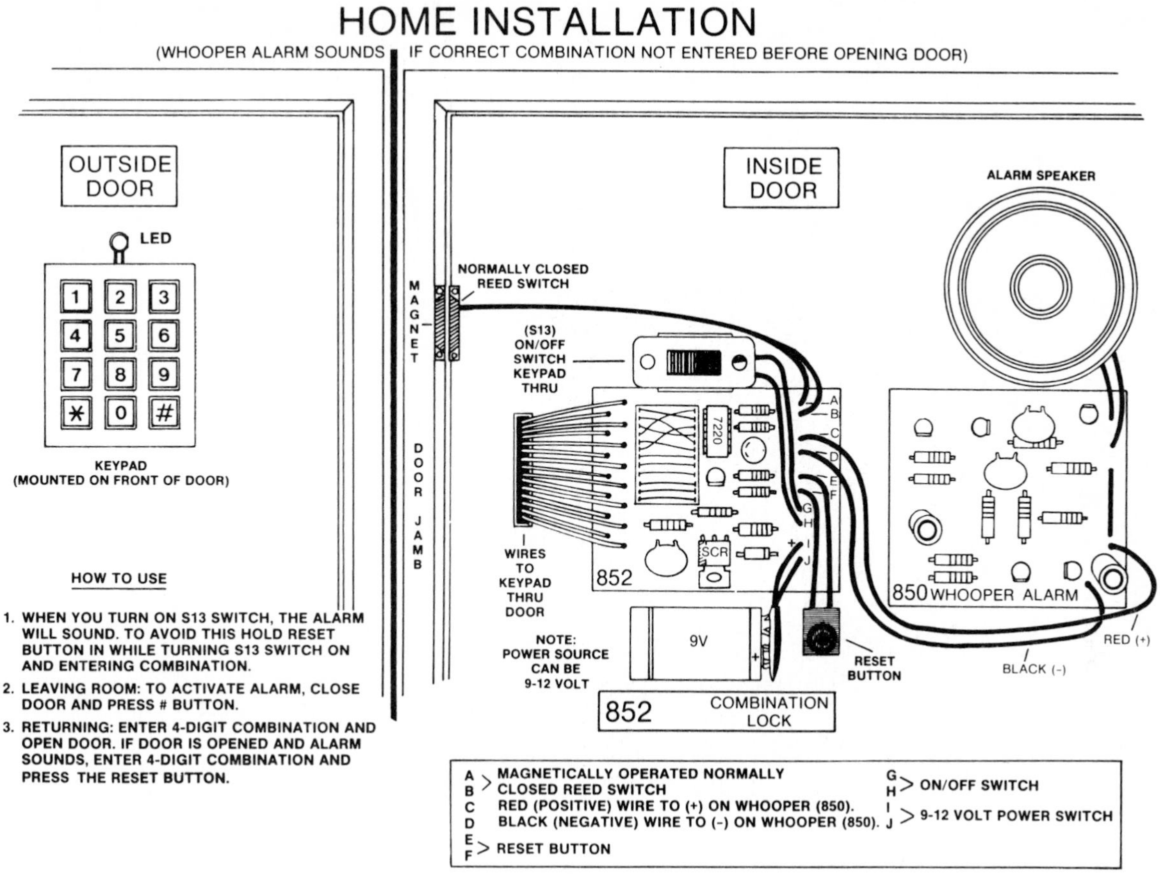

FIG. 1-3 Typical installation of combination lock/alarm control on home door. A whooper alarm is the sounding device in this system. (Used by permission of Electronic Kits International, Inc.)

tional, Inc. This company also includes instructions for installing the combination-lock/alarm control in an automobile.

HIGH-TEMPERATURE ALARM

The high-temperature alarm circuit shown in Fig. 1-4 will monitor the temperature surrounding the thermistor R_1 and sound an alarm if it becomes too warm. To calibrate, heat a container of water to the temperature at which the alarm should sound. Place the thermistor in the water and rotate the temperature adjust control R_3 until the alarm just sounds. To use, place the thermistor R_1 in the remote area where the temperature is to be monitored.

The high-temperature alarm circuit uses the 339 voltage comparator IC to compare the input voltage at pin 5 to the reference voltage at pin 4. The temperature adjust control sets the voltage reference at pin 4 slightly more positive than that at pin 5. As the temperature surrounding the thermistor R_1 increases, its resistance decreases, causing the voltage at pin 5 to become more positive. When pin 5 becomes more positive than the reference voltage at pin 4, the comparator's output (pin 2) goes positive, causing the NPN transistor Q_1 to turn on. With the transistor on, the piezo buzzer sounds and the LED indicator D_2 lights.

The thermistor in the high-temperature alarm should have a cold resistance of 100 to 300 ohms (Ω) and have a negative temperature coefficient. If the cold resistance of the thermistor is either higher or lower than 100 to 300 Ω, simply change the value of voltage divider resistor R_2 by a proportionate amount.

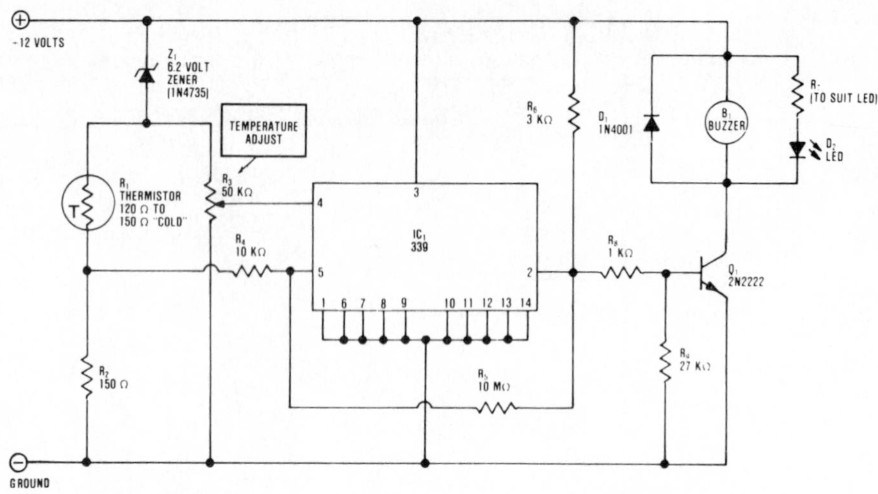

PARTS LIST

B_1	Piezo buzzer (Radio Shack 273-060)
D_1	1N4001 silicon diode, 1 A, 50 PIV
D_2	Light-emitting diode
IC_1	339 voltage comparator IC
Q_1	2N2222 NPN transistor (or similar NPN)
R_1	Thermistor cold resistance of 100 to 300 Ω negative temperature coefficient
R_2	150-Ω, ¼-W resistor
R_3	50-kΩ linear taper potentiometer
R_4	10-kΩ, ¼-W resistor
R_5	10-MΩ, ¼-W resistor
R_6	3-kΩ, ¼-W resistor
R_7	470-Ω, ¼-W resistor
R_8	1-kΩ, ¼-W resistor
R_9	27-kΩ, ¼-W resistor
Z_1	1N4735 zener diode (6.2 V)

FIG. 1-4. High-temperature alarm. (Michael Gannon, *Workbench Guide to Semiconductor Circuits and Projects,* Prentice-Hall, New Jersey, 1982, pp. 167–169. Used by permission of Prentice-Hall, Inc.)

HOME BURGLAR ALARM

The burglar alarm circuit shown in Fig. 1-5 includes many of the popular features of commercial units. These features include an exit and entrance delay, an automatic bell shutoff, and relay control to handle any signaling device. The schematic and parts list are shown in Fig. 1-5(a). External connections to the burglar alarm are sketched in Fig. 1-5(b). The numbers 1 through 6 in Fig. 1-5(b) correspond to those on the schematic diagram. Note that either normally open or normally closed sensors may be used.

Operational details of the home burglar alarm are as follows:

1. *Power source*—1 microampere (μA) of current is needed during the "ready" state, while approximately 1 ampere (A) is needed to drive most signaling devices.
2. *Exit delay*—30-second delay; the value of C_1 changes exit time delay. (Note: Increasing the size of the capacitor increases the delay.)
3. *Entrance delay*—30-second delay; the value of C_2 changes entrance time delay.

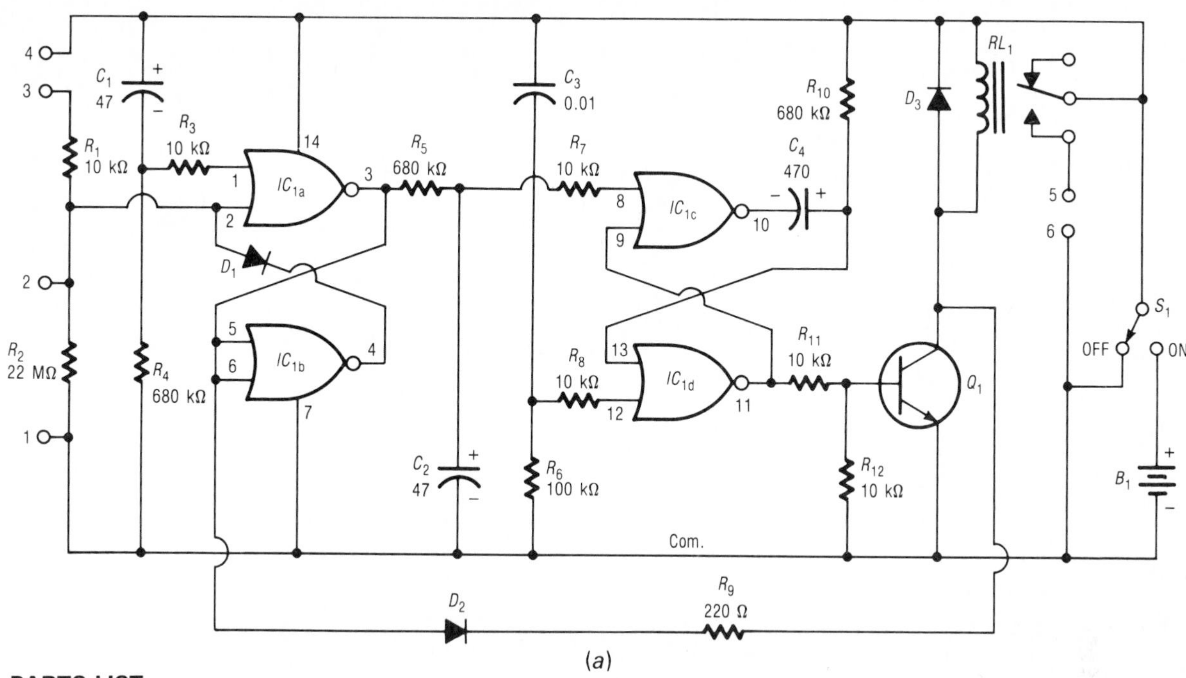

PARTS LIST

B_1	12-V lantern battery
C_1, C_2	47-µF, 16-V electrolytic capacitor
C_3	0.01-µF disk capacitor
C_4	470-µF, 16-V electrolytic capacitor
D_1, D_2	1N914 or 1N4148 diode
D_3	1N4001 silicon diode, 1 A, 50 PIV
IC_1	4001 CMOS quad NOR gate IC
Q_1	2N2222 NPN transistor
$R_1, R_3, R_7, R_8, R_{11}, R_{12}$	10-kΩ, ½-W resistor
R_2	22-MΩ, ½-W resistor
R_4, R_5, R_{10}	680-kΩ, ½-W resistor
R_6	100-kΩ, ½-W resistor
R_9	220-Ω, ½-W resistor
RL_1	Relay, 12-V coil
S_1	SPDT switch

6 5 4 3 2 1
Audible signal* (12 V)
NC switch
NO switch
or switches
*Load device; terminals 5 + 6 energize the load
(b)

FIG. 1-5 Home burglar alarm. (Used by permission of Robert Delp Electronics.)

4. *Automatic bell cutoff*—audible signal cuts off after 6 minutes; the value of C_4 changes bell cutoff time.
5. *Automatic reset*—after automatic bell cutoff, the alarm will reset to the "ready" state until activated again.

LEVEL-SENSING SWITCH FOR SUMP PUMPS

The schematic diagram and parts list for a solid-state level-sensing switch for a sump pump is shown in Fig. 1-6. As shown in Fig. 1-6, the common probe (metal wire or plate) is always touching the water in the sump. As the water rises in the sump, it first contacts the keep-alive probe. The keep-alive probe will not energize the sump pump. As the water rises further, it contacts the pump-trigger probe, which activates both relays (K_1 and K_2), thus turning on the sump pump. The sump pump continues to run until the water drops below the keep-alive probe.

The level-sensing switch also has an alarm feature in case either the pump or the switch

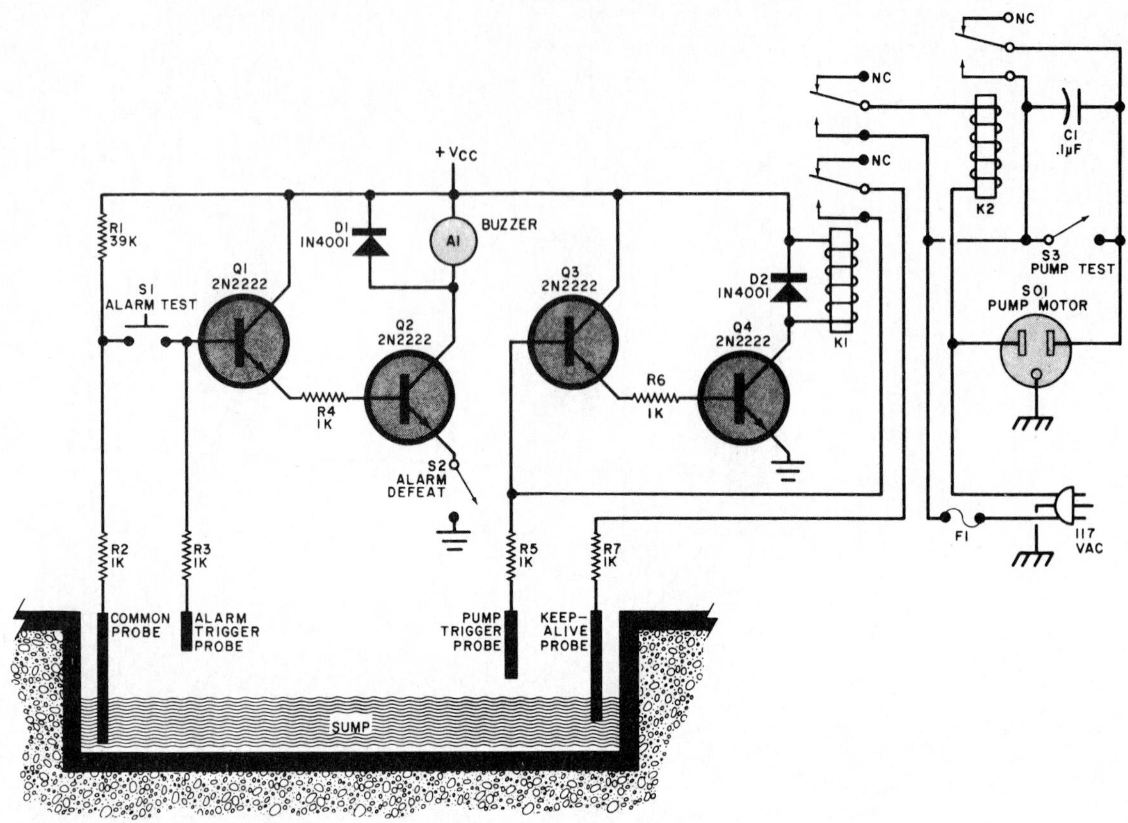

PARTS LIST

A_1 — DC buzzer (match to battery voltage)
C_1 — 0.1-μF, 1000-V disk capacitor
D_1, D_2 — 1N4001 silicon diode, 1 A, 50 PIV
F_1 — Fast-blow fuse (match to pump motor)
K_1 — DC energized relay (match to battery voltage)
K_2 — 117-V ac relay
Q_1, Q_2, Q_3, Q_4 — 2N2222 NPN transistor (or similar)
R_1 — 39-kΩ, ¼-W resistor
R_2 through R_7 — 1-kΩ, ¼-W resistor
S_1 — Normally open push-button switch
S_2 — SPST switch
S_3 — SPST switch (match to pump motor)
SO_1 — Grounded ac power socket

FIG. 1-6 Level-sensing switch for sump pumps. (Reprinted from *Popular Electronics*. Copyright © August 1979, Ziff-Davis Publishing Company.)

fail, causing the water to rise too high in the sump. The alarm is activated by either the alarm test push-button switch S_1 or if water were to touch the highest probe (alarm trigger probe) in the sump pump. When the base of transistor Q_1 is activated with a positive voltage, it turns on transistors Q_1 and Q_2. When the emitter-to-collector resistance of transistor Q_2 drops, greater current flows through the buzzer, causing the alarm to signal that the water is too high. The alarm feature may be turned off by opening the alarm defeat switch S_2.

The dc power may be supplied from either a dc power supply or battery with a voltage of 6 to 15 V. The coil voltage of relay K_1 should match the dc power supply voltage. Relay K_2 must be a 117-V ac relay with contacts heavy enough to handle the sump pump motor current.

Great care must be used in wiring the high-voltage ac sections of the sump pump switch. Be sure the high voltage wiring is insulated and housed in a safe enclosure. Be sure to use proper strain relief techniques where cords enter and exit the enclosure.

LOW-TEMPERATURE ALARM

The low-temperature alarm circuit shown in Fig. 1-7 will monitor the temperature surrounding the thermistor R_2 and sound an alarm if it becomes too cold. To calibrate, cool a container of water to the temperature at which the alarm should sound. Place the thermistor in the water and rotate the temperature adjust control R_3 until the alarm just sounds. To use,

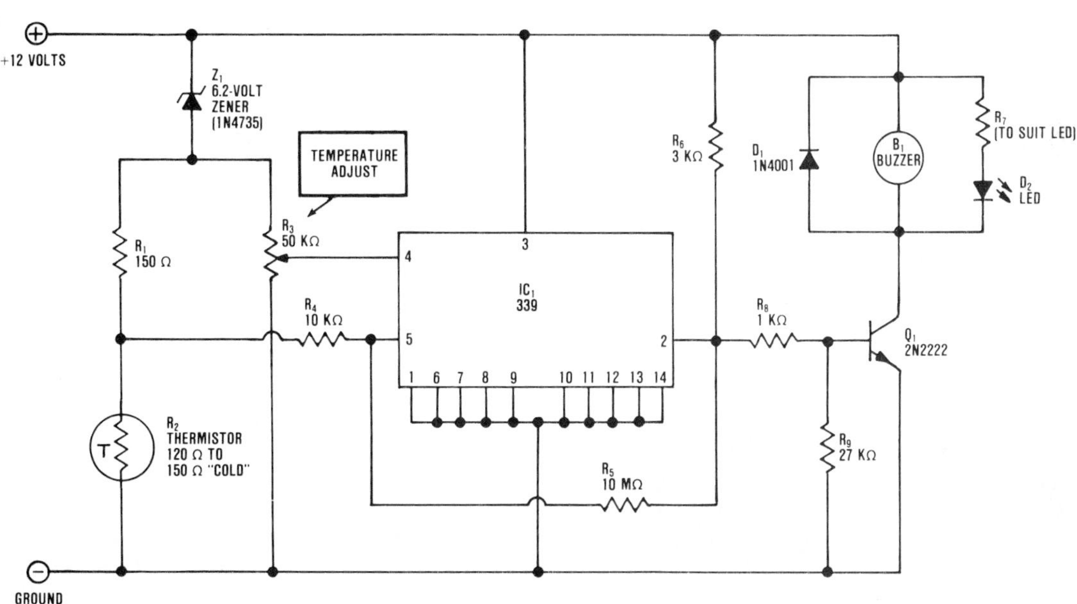

PARTS LIST

B_1 Piezo buzzer (Radio Shack 273-060)
D_1 1N4001 silicon diode, 1 A, 50 PIV
D_2 Light-emitting diode
IC_1 339 voltage comparator IC
Q_1 2N2222 NPN transistor (or similar NPN)
R_1 150-Ω, ¼-W resistor
R_2 Thermistor cold resistance of 100 to 300 Ω negative temperature coefficient
R_3 50-kΩ linear taper potentiometer
R_4 10-kΩ, ¼-W resistor
R_5 10-MΩ, ¼-W resistor
R_6 3-kΩ, ¼-W resistor
R_7 470-Ω, ¼-W resistor
R_8 1-kΩ, ¼-W resistor
R_9 27-kΩ, ¼-W resistor
Z_1 1N4735 zener diode (6.2 V)

FIG. 1-7 Low-temperature alarm. (Michael Gannon, *Workbench Guide to Semiconductor Circuits and Projects,* Prentice-Hall, New Jersey, 1982, pp. 169–171. Used by permission of Prentice-Hall, Inc.)

place the thermistor R_2 in the remote area where the temperature is to be monitored.

The low-temperature alarm circuit in Fig. 1-7 uses the 339 voltage comparator IC to compare the input voltage at pin 5 to the reference voltage at pin 4. The temperature adjust control sets the voltage reference at pin 4 slightly more positive than that at pin 5. As the temperature surrounding the thermistor R_2 decreases, its resistance increases, causing the voltage at pin 5 to become more positive. When pin 5 becomes more positive than the reference voltage at pin 4, the comparator's output (pin 2) goes positive, causing the NPN transistor Q_1 to turn on. With the transistor on, the piezo buzzer sounds and the LED indicator D_2 lights.

The thermistor in the low-temperature alarm should have a cold resistance of 100 to 300 Ω and have a negative temperature coefficient. If the cold resistance of the thermistor is either higher or lower than 100 to 300 Ω, change the value of voltage divider resistor R_1 by a proportionate amount.

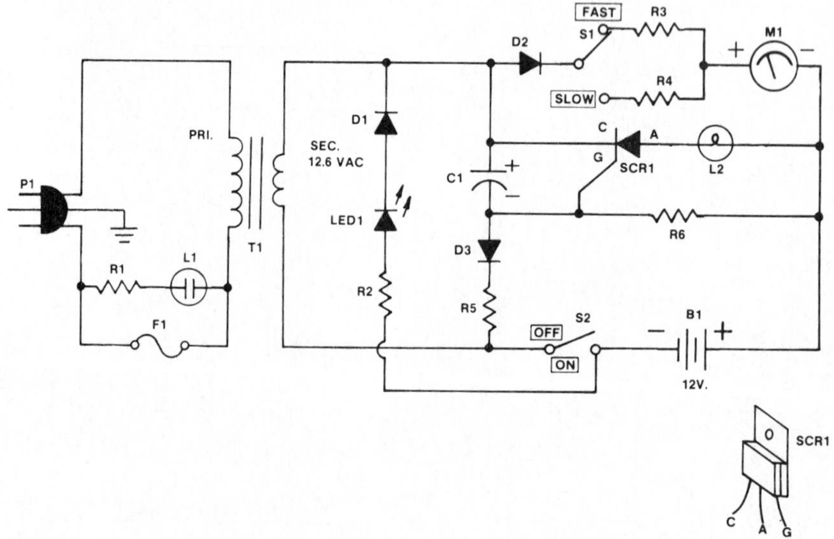

PARTS LIST

B_1	12-V rechargeable battery, nickel-cadmium or gel-cell [4 ampere-hours (4 A-hours)]
C_1	200-μF, 25-V electrolytic capacitor
D_1, D_2, D_3	Silicon diode, 1 A, 50 PIV
F_1	½-A fuse
L_1	NE-2 neon lamp
L_2	12-V lamp; match current rating of lamp with current rating of SCR
LED_1	Light-emitting diode
M_1	0–1-A ammeter
P_1	117-V ac three-pronged plug
R_1	56-kΩ, ½-W resistor
R_2	470-Ω, ½-W resistor
R_3	5-Ω, 5-W resistor
R_4	27-Ω, 2-W resistor
R_5	47-Ω, 1-W resistor
R_6	1.5-kΩ, ½-W resistor
S_1	SPDT switch
S_2	SPST switch
SCR_1	4-A, 50-PIV silicon controlled rectifier (such as GE C106F1 or any GE C106 series); match to lamp L_2
T_1	12.6-V, 1.2-A power transformer

FIG. 1-8 Power failure lantern. (Carl G. Grolle and Michael B. Girosky, *Workbench Guide to Electronic Projects You Can Build in Your Spare Time*, Parker, New York, 1981, pp. 149–154. Used by permission of Parker Publishing Company, Inc.)

POWER FAILURE LANTERN

The schematic diagram and parts list for the power failure lantern are shown in Fig. 1-8. This inexpensive unit will provide emergency lighting in the event of power failure. The unit can also double as a rechargeable 12-V lantern or flashlight.

When used to provide automatic emergency standby lighting, the unit is powered by 117 V ac. When used for standby lighting, switch S_2 must be closed or in the ON position. Switch S_1 is the charge rate switch (slow or fast charge) which continuously maintains the charge on the rechargeable batteries B_1. If the power fails, the emergency lamp L_2 lights automatically.

When used as a portable rechargeable lantern, lamp L_2 is powered by the rechargeable batteries B_1 and S_2 serves as the ON-OFF switch. Plugging the lantern into 117 V ac will recharge the batteries if S_2 is closed or in the ON position. Diode LED_1 will light when S_2 is closed and ac is applied.

SIREN

The siren shown in Fig. 1-9 produces a loud wailing sound. The circuit is based on the 555 timer (IC_1). The timer is wired as a free-running multivibrator. The output of the 555 timer drives the output amplifier Q_1, which in turn drives the speaker. The basic tone of the 555 timer is set by the RC circuit containing R_1, R_4, and C_1. The rising pitch of the siren is caused by the relaxation oscillator formed by the unijunction transistor Q_2 and timing section R_5 and C_2. The signal from the low-frequency relaxation oscillator is coupled to the

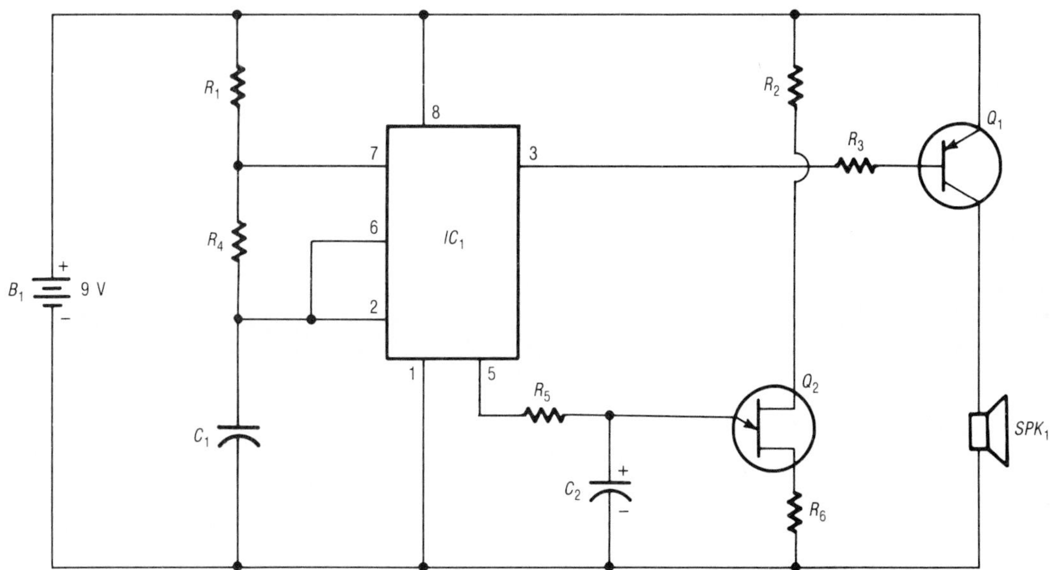

PARTS LIST

B_1	9-V battery (alkaline or ni-cad)
C_1	0.2-μF, 50-V capacitor
C_2	33-μF, 25-V electrolytic capacitor
IC_1	555 timer IC
Q_1	PNP transistor (Radio Shack 276-2027)
Q_2	Unijunction transistor (ECG 6410)
R_1,R_4	2.4-kΩ, ½-W resistor
R_2	1-kΩ, ½-W resistor
R_3	150-Ω, ½-W resistor
R_5	15-kΩ, ½-W resistor
R_6	56-Ω, ½-W resistor
SPK_1	8-Ω speaker

FIG. 1-9 Siren.

timer and modulates the basic tone of the siren.

A larger speaker will give a louder sound with this circuit. An alkaline or ni-cad (nickel-cadmium) battery is recommended for use with the siren.

WATER-LEVEL ALARM

The water-level alarm shown in Fig. 1-10 may be used as a warning of high water. It is typically used in basement or well sumps, pools, or greenhouses. It should be used with water only and should *not be used with flammable liquids.*

The 555 timer IC is wired as an astable multivibrator in Fig. 1-10. It generates a tone of about 800 Hz. If the value of C_1 is changed, the frequency will also change. When water touches the probes, Q_1 is activated, causing the alarm to sound. The circuit can be tested by touching the probes together. A 9-V transistor-type battery is used as the power source.

TWO-TONE SIREN

The circuit in Fig. 1-11 mimics the two-tone sirens on some emergency vehicles. The siren consists of two basic sections: the basic tone

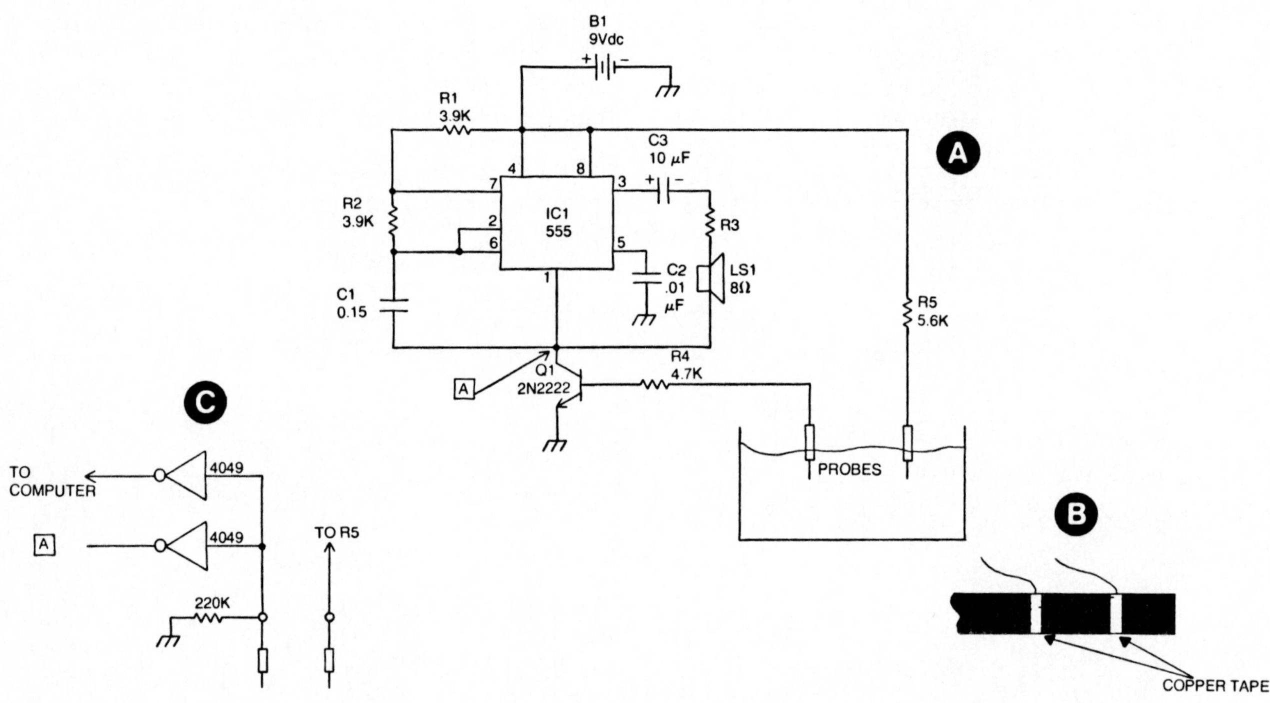

PARTS LIST

B_1	9-V battery
C_1	0.15-μF, 50-V disk capacitor
C_2	0.01-μF, 50-V disk capacitor
C_3	10-μF, 16-V electrolytic capacitor
IC_1	555 timer IC
LS_1	8-Ω speaker
Q_1	2N2222 NPN transistor (or similar NPN)
R_1,R_2	3.9-kΩ, ¼-W resistor
R_3	22-Ω, ¼-W resistor
R_4	4.7-kΩ, ¼-W resistor
R_5	5.6-kΩ, ¼-W resistor

FIG. 1-10 Water-level alarm. (Joseph J. Carr, *104 Weekend Electronics Projects,* TAB Books, Pennsylvania, 1982, pp. 149–153. Used by permission of TAB Books, Inc.)

oscillator/driver and the very low frequency oscillator. The LM13080 chip (IC_2) is wired as an oscillator with feedback components R_9 and C_2 determining the frequency. The basic tone [about 200 hertz (Hz)] generated by the oscillator is applied directly to the speaker through capacitor C_3. Unlike most operational amplifiers (op-amps), the LM13080 can deliver enough power to drive the speaker.

The basic tone of the oscillator IC_2 in Fig. 1-11 is made to switch to a higher tone (250 to 300 Hz) for some time and then drop back to its basic tone. This process is repeated. The very low frequency oscillator is based on the LM393 voltage comparator (IC_1). Feedback components R_3 and C_1 determine the frequency (1 to 2 Hz) of the oscillator. The output of the oscillator is fed through R_6 to the basic tone oscillator, where it modifies the frequency of the output oscillator/driver.

TOUCH ALARM

A touch alarm circuit is shown in Fig. 1-12. The alarm will sound if the metal object being protected (at left in the schematic diagram) is touched. The alarm will continue to sound until the reset-arm switch is pressed which turns off the alarm and rearms the unit. The sensitivity of the touch alarm can be adjusted with potentiometer R_1. While an inexpensive piezo buzzer B_1 is used as the alarm in this circuit, any warning device may be switched on with the relay.

The touch alarm circuit in Fig. 1-12 can be divided into three main parts: the touch detector, the trigger circuit, and the alarm. The touch detector section includes transistors Q_1, Q_2, and Q_3. When the protected object is touched, it momentarily turns on the junction field-effect transistor (FET) Q_1. The change in

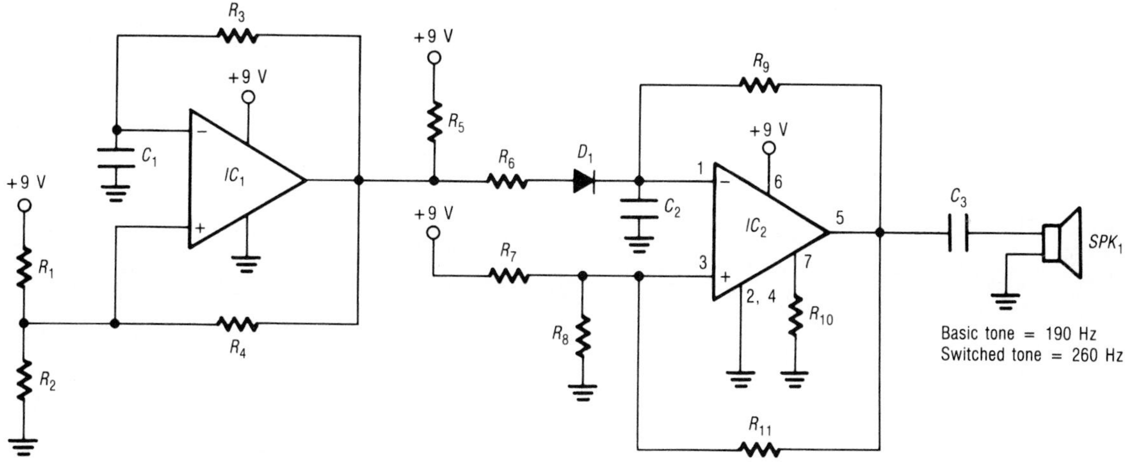

PARTS LIST

C_1	5-μF, 25-V electrolytic capacitor
C_2	0.05-μF, 50-V disk capacitor
C_3	200-μF, 25-V electrolytic capacitor
D_1	1N914 signal diode
IC_1	LM393 dual-voltage comparator IC
IC_2	LM13080 programmable power op-amp IC
R_1,R_2,R_7,R_8	10-kΩ, ½-W resistor
R_3,R_9	75-kΩ, ½-W resistor
R_4,R_{11}	15-kΩ, ½-W resistor
R_5	5.1-kΩ, ½-W resistor
R_6	200-kΩ, ½-W resistor
R_{10}	680-kΩ, ½-W resistor
SPK_1	8-Ω speaker

FIG. 1-11 Two-tone siren. (Used by permission of National Semiconductor Corporation.)

voltage at the drain of the FET activates Q_2 and Q_3. In the trigger circuit, the voltage at the gate of the silicon controlled rectifier (SCR) goes positive, turning on the SCR and snapping the relay closed. The alarm is activated, and B_1 sounds. The SCR latches in the conduction state and the alarm continues to sound even when the finger touching the protected object is removed. If the reset-arm switch is opened, the buzzer is turned off and the touch alarm is once again armed.

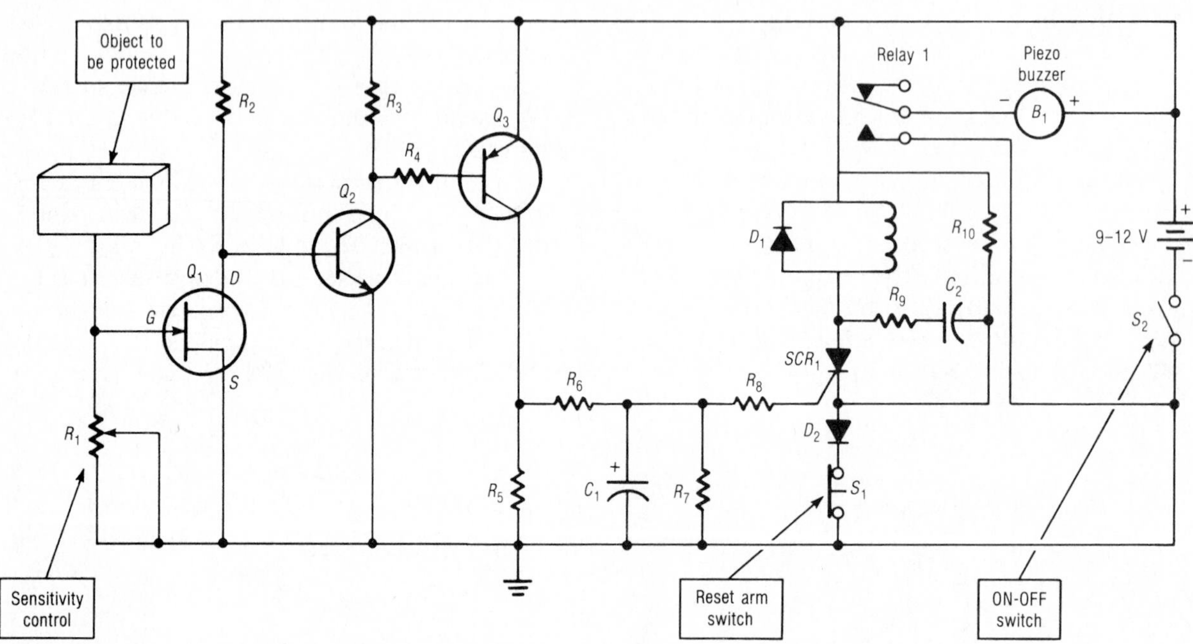

PARTS LIST

B_1	Piezo buzzer (Radio Shack 273-060)
C_1	4.7-μF, 25-V electrolytic capacitor
C_2	0.05-μF, 50-V disk capacitor
D_1,D_2	1N4003 silicon diode, 1 A, 200 PIV
Q_1	2N5458 junction field-effect transistor (similar to Radio Shack 276-2062)
Q_2	2N3904 NPN transistor
Q_3	2N3906 PNP transistor
R_1	2.5-MΩ potentiometer
R_2	47-kΩ, ½-W resistor
R_3	120-kΩ, ½-W resistor
R_4	2.2-kΩ, ½-W resistor
R_5	330-Ω, ½-W resistor
R_6	470-Ω, ½-W resistor
R_7	1-kΩ, ½-W resistor
R_8	330-Ω, ½-W resistor
R_9	10-Ω, ½-W resistor
R_{10}	10-kΩ, ½-W resistor
Relay 1	Relay, 5-V coil (similar to Radio Shack 275-215)
S_1	Normally closed push-button switch
S_2	SPST switch
SCR_1	C106B1 silicon controlled rectifier (or similar SCR)

FIG. 1-12 Touch alarm. (John E. Cunningham, *Building & Installing Electronic Intrusion Alarms,* 3d ed. Used with permission of the publisher. The Howard W. Sams and Co., Inc., Indianapolis, 1982.)

chapter 2

Apple II/IIe Game Port Interface Circuits

INTRODUCTION

Two choices are available for interfacing with the Apple II/II+/IIe microcomputer. The easiest method involves use of the 16-pin game port. The more difficult method involves use of one of the 50-pin slots across the rear of the computer (inside the case). There are eight peripheral connector slots (seven on the Apple IIe).

The game input-output (I/O) connector (game port) is a 16-pin dual in-line package (DIP) socket located on the main pc board at the right rear inside the case of the computer (when standing facing the keyboard). The location of the game port socket is sketched in Fig. 2-1(*a*). For easy access to the pins of the game port and safety, a 16-pin jumper can be used to connect to an external solderless breadboard as illustrated. Only two of eight (seven on the Apple IIe) peripheral connector slots are shown in Fig. 2-1(*a*).

More detail on the game I/O connector is shown in Fig. 2-1(*b*). A description of the signals available at the game I/O connector is detailed in Table 2-1. Note that +5 V and GND are available at pins 1 and 8. The 5-V supply may be used to power your circuits up to 100 mA. Pins 12 through 15 are called *annunciator outputs* and can be driven high or low with computer instructions. For instance, to drive pin 15 high (on), you could apply the following BASIC statement:

```
POKE 49241, 1
```

However, to drive pin 15 low (off), the following BASIC statement could be used:

```
POKE 49240, 0
```

Pins 2 through 4 are 1-bit inputs such as those used to indicate whether a push-button switch is closed or open. To check the condition of the

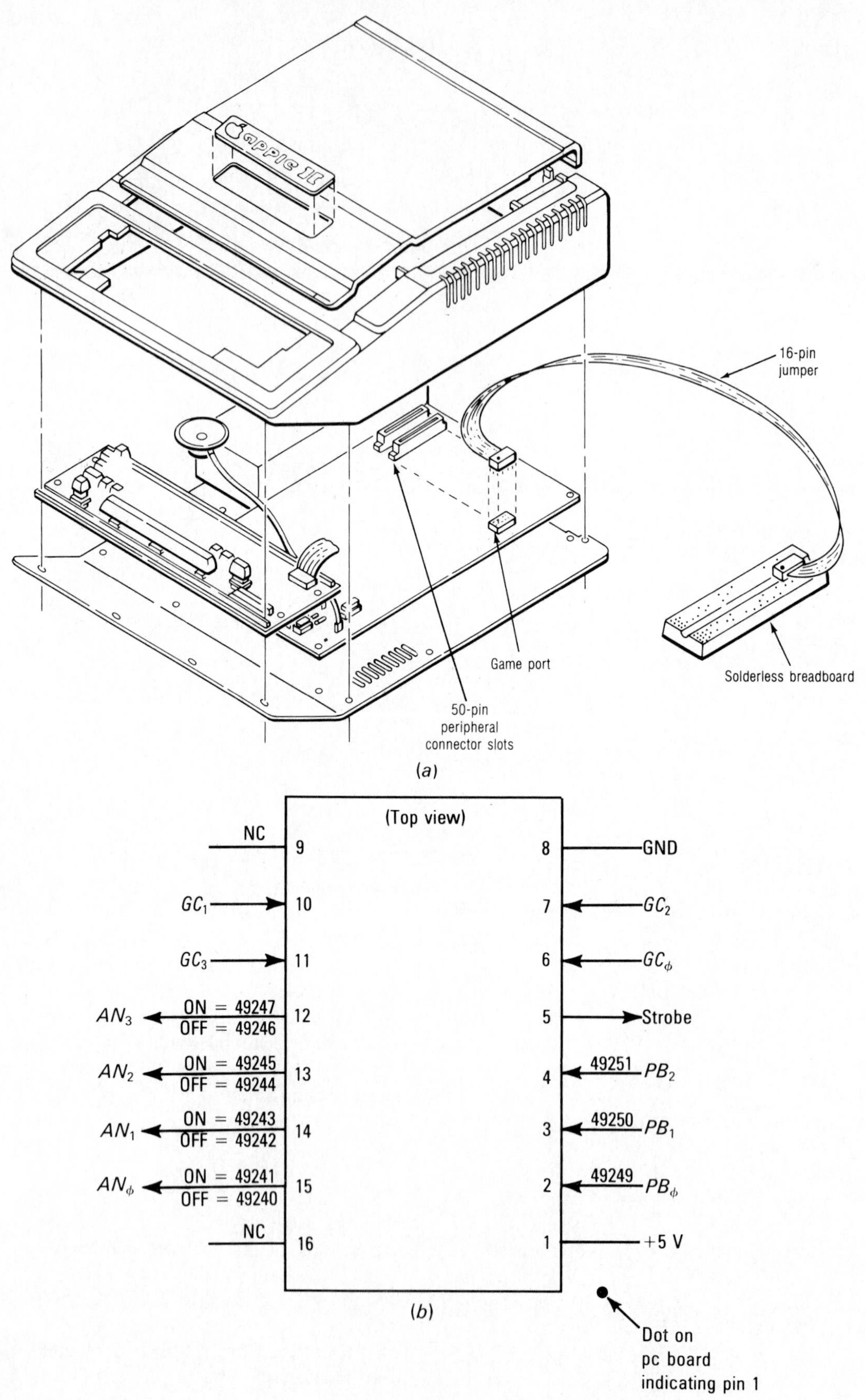

FIG. 2-1 ***(a)*** Connecting to the Apple II game port. ***(b)*** Game I/O connector pinouts.

TABLE 2-1 Game I/O Connector Signal Descriptions

Pin	Name	Description
1	+5 V	+5-V power supply; total current drain on this pin must be less than 100 mA
2–4	PB_0–PB_2	Single-bit (push-button) inputs; these are standard 74LS series TTL* inputs
5	$\overline{C040}$ $\overline{STROBE}$	A general-purpose strobe; this line, normally high, goes low during Φ0 of a read or write cycle to any address from $C040 through $C0F; this is a standard 74LS TTL output
6,7,10,11	GC_0–GC_3	Game controller inputs; these should each be connected through a 150-kΩ variable resistor to +5 V
8	GND	System electrical ground
12–15	AN_0–AN_3	Annunciator outputs; these are standard 74LS series TTL outputs and must be buffered if used to drive other than TTL inputs
9,16	NC	No internal connection

* Transistor-transistor logic.

PB_0 input, the following BASIC statement could be used:

```
X = PEEK (49249)
```

If the number returned from memory location 49249 is > 127, input PB_0 is high; however, if the number is < 128, input PB_0 is low.

Input pins 6, 7, 10, and 11 on the game I/O connector are game controller or paddle inputs. A 150-kΩ variable resistor is connected from one of these inputs and +5 V. Depending on the resistance, a number from 0 to 255 will be returned when the following BASIC statement checks the condition of GC_1 (pin 10):

```
X = PDL(1)
```

On execution of this BASIC statement, the position of the potentiometer will be converted into a number between 0 and 255 and stored in X. More resistance between +5 V and the paddle input yields a greater number such as 255. Less resistance returns a smaller number such as 0.

Great care must be taken in wiring. Use the computer's power supply (+5 V and GND). *Do not use a separate supply.** Harm could be done to the computer if voltages above 5 V or below GND (negative) are applied to the pins of the game port.

* *Exception:* A relay (or other device) can be used to *isolate* an external supply from the computer power supply. An example is shown later in this chapter.

FLASHING LED

The simple LED circuit in Fig. 2-2 can be controlled by use of simple computer programs.

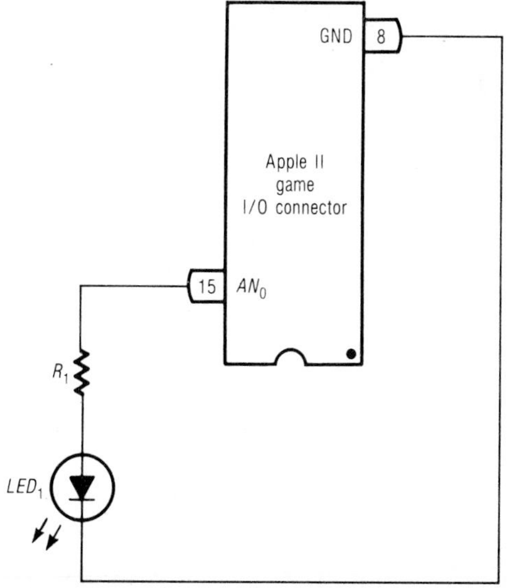

PARTS LIST

Apple II/IIe game I/O connector
LED_1 Light-emitting diode
R_1 150-Ω, ¼-W resistor

FIG. 2-2 LED flasher.

When the annunciator output (AN_0) goes high, the LED lights. However, when pin 15 goes low, the LED does not light.

Program 2-1. This program will ask the user if the light is to be on or off. The circuit will respond by lighting or extinguishing the LED. The POKE 49241, 1 command in line 20 turns on the light. The POKE 49240, 0 in line 30 turns off the LED. Line 40 is a short delay before the question is asked again.

```
10   INPUT "DO YOU WANT LED ON OR
       OFF?";L$
15   REM SET PIN 15 TO HIGH IF "O
       N"
20   IF L$ = "ON" THEN POKE 49241
       ,1
25   REM SET PIN 15 TO LOW IF "OF
       F"
30   IF L$ = "OFF" THEN POKE 4924
       0,0
40   FOR J = 1 TO 500: NEXT J
50   GOTO 10
```

PROGRAM 2-1 Turning a light on or off.

Program 2-2. This program flashes the LED on and off. Lines 20 and 40 are time delays.

```
5    REM SET PIN 15 TO HIGH
10   POKE 49241,1
20   FOR I = 1 TO 100: NEXT I
25   REM SET PIN 15 TO LOW
30   POKE 49240,0
40   FOR I = 1 TO 100: NEXT I
50   GOTO 10
```

PROGRAM 2-2 Flashing LED.

SEQUENCE FLASHER

The LED circuit in Fig. 2-3, along with Program 2-3, will turn on each LED, one at a time from right to left on the schematic. A high at the annunciator outputs (pins 12 through 15) will turn on a light. A low (or GND) on an annunciator output will extinguish that LED.

Program 2-3. This program will first turn all LEDs off in lines 10 through 40. Line 50 turns

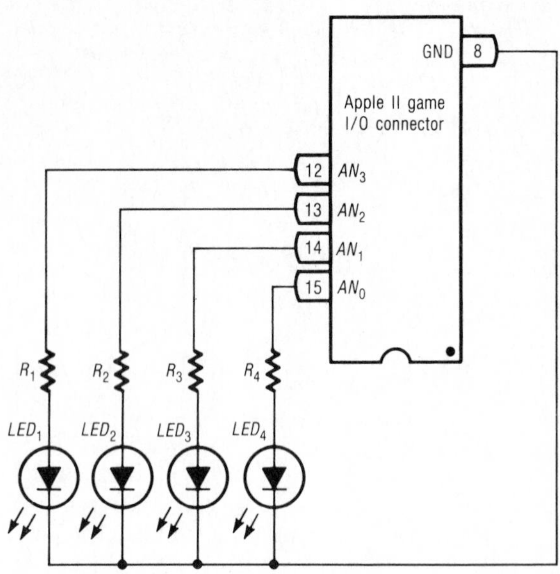

PARTS LIST

Apple game I/O connector	
LED_1 through LED_4	Light-emitting diode
R_1 through R_4	150-Ω, ¼-W resistor

FIG. 2-3 Sequence flasher.

```
5    REM SET OUTPUTS TO LOW
10   POKE 49240,0
20   POKE 49242,0
30   POKE 49244,0
40   POKE 49246,0
45   REM ",0" SYMBOLIZES LOW AND
       ",1" SYMBOLIZES HIGH IN THE
       REST OF THE PROGRAM, BUT ANY
       NUMBER WILL WORK
50   POKE 49241,1
60   Z = 200: REM SPEED ADJUST
70   FOR I = 1 TO Z: NEXT I
80   POKE 49240,0
90   POKE 49243,1
100  FOR I = 1 TO Z: NEXT I
110  POKE 49242,0
120  POKE 49245,1
130  FOR I = 1 TO Z: NEXT I
140  POKE 49244,0
150  POKE 49247,1
160  FOR I = 1 TO Z: NEXT I
170  POKE 49246,0
180  GOTO 50
```

PROGRAM 2-3 Lighting LEDs in sequence.

on LED_4. (Pin 15 goes high.) Line 70 is a time delay whose length can be adjusted by changing the 200 in line 60. Line 80 turns off LED_4. (Pin 15 goes low.) The rest of the program repeats this process for each LED in sequence and then the program repeats.

MOVING A DOT

The circuit in Fig. 2-4 is using the push-button inputs on the game port. When a switch is pressed, the input (pin 2 or 3) is grounded. When the push-button switches are not pressed, pins 2 and 3 of the game port float high.

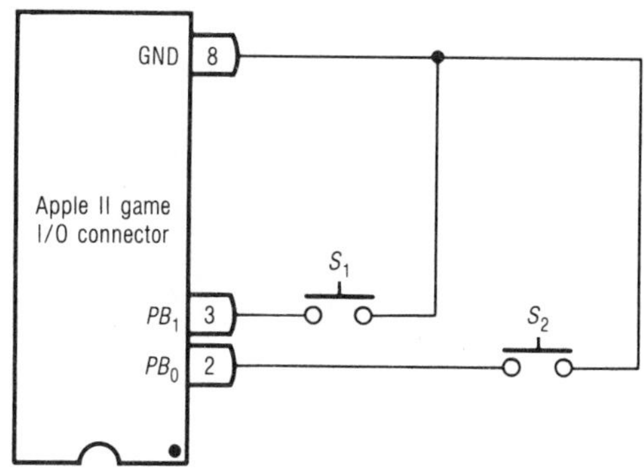

PARTS LIST

Apple game I/O connector
S_1,S_2 Normally open push-button switch

FIG. 2-4 Moving dot circuit.

Program 2-4. When switch S_1 is pressed, a dot on the CRT moves to the left. When S_2 is pressed, the dot moves to the right on the screen. Line 10 looks at pin 2 to determine whether it is high or low. Lines 20 and 30 assign a 1 or a 0 to variable R based on the input. Line 40 looks at pin 3 to determine whether it is high or low. Lines 50 and 60 assign a 1 or a 0 to variable L based on the input. Lines 70 and 80 add or subtract 1 or 0 to variable X, which is the horizontal position of the dot of light on the CRT. Lines 90 and 100 check to see whether the dot will be off the screen when plotted and, if so, start it again at the other end. Lines 110 through 130 place the computer in the graphic mode and plot a white rectangle of light on the screen. Then the process is repeated.

```
10   R = PEEK (49249): REM READ
       PB0 INPUT
20   IF R < 128 THEN R = 1
30   IF R > 127 THEN R = 0
40   L = PEEK (49250): REM READ
       PB1 INPUT
50   IF L < 128 THEN L = 1
60   IF L > 127 THEN L = 0
70   X = X + R: REM MOVE RIGHT
80   X = X - L: REM MOVE LEFT
90   IF X > 39 THEN X = 0
100  IF X < 0 THEN X = 39
110  GR
120  COLOR = 15
130  PLOT X,15
140  GOTO 10
```

PROGRAM 2-4 Moving dot on CRT using push-button inputs.

INPUT AND OUTPUT

The circuit in Fig. 2-5 is used to demonstrate both input and output from the game port. The push-button PB_0 is the input and the LED at

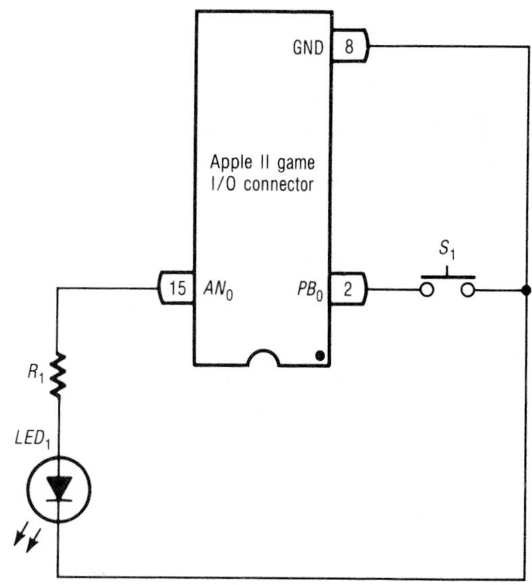

PARTS LIST

Apple II/IIe game I/O connector
LED_1 Light-emitting diode
R_1 150-Ω, 1/4-W resistor
S_1 Normally open push-button switch

FIG. 2-5 Input and output from the same game port.

pin 15 (AN_0) is the output. When the push-button switch is pressed, the PB_0 input will be activated with a low (ground). The computer must set the annunciator output AN_0 high to light the LED.

Program 2-5. The effect of running this program with the circuit in Fig. 2-5 is that when switch S_1 is pressed, the LED will light. When S_1 opens, the LED will go out. Line 20 of the program checks pin 2 to determine whether the input is high or low. Line 30 turns the LED on if the input is low (if the switch was closed). If the input at pin 2 is high (switch not closed), line 40 turns the LED off. Then the process is repeated.

```
10   REM READ INPUT PBO
20   Y = PEEK (49249)
30   IF Y < 128 THEN POKE 49241,1
       : REM IF INPUT IS LOW THEN
       SET OUTPUT TO HIGH
40   IF Y > 127 THEN POKE 49240,0
       : REM IF INPUT IS HIGH THEN
       SET OUTPUT TO LOW
50   GOTO 10
```

PROGRAM 2-5 Input-output demonstration program.

TRAFFIC LIGHT

The circuit in Fig. 2-6 simulates a simple traffic light. Because only four outputs were available on the game port, only the green and red lights for the main and side streets are used. If any of the annunciator outputs (pins 12 through 15) go high, they will turn on their respective LEDs. Push-button switches S_1 and S_2 simulate the sensor switches buried in the streets.

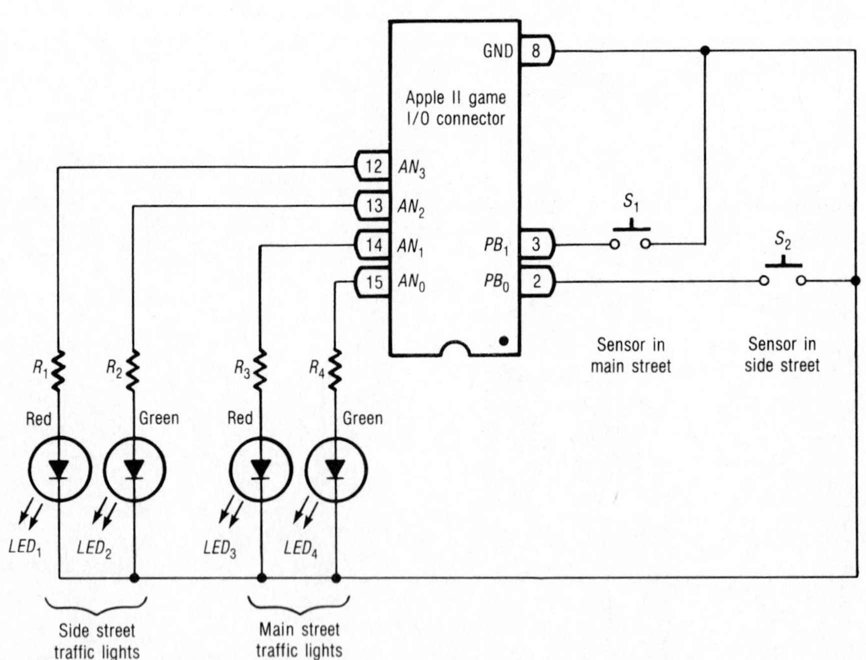

PARTS LIST

Apple II/IIe game I/O connector

LED_1, LED_3	Red light-emitting diode
LED_2, LED_4	Green light-emitting diode
R_1 through R_4	150-Ω, ¼-W resistor
S_1, S_2	Normally open push-button switch

FIG. 2-6 Traffic light circuit.

Program 2-6. The program operates the set of traffic lights in Fig. 2-6. Line 20 turns off all LEDs. Lines 40 and 50 set the main street light green (LED_4 on) and the side street light red (LED_1 on). Line 70 starts a timing loop for the on time for the main street. Line 80 checks the sensor switch in the side street. If S_2 is activated (checked in line 90), the main street light is changed to red. After a short delay (line 130), the side street light turns green. At line 170, another timing loop is entered. In this loop, the sensor in the main street is checked (line 180). If the sensor is activated, the program jumps to line 210, where the side street light is changed back to red and finally the main street light is changed to green. The program is then repeated.

```
10   REM CLEAR OUTPUTS TO LOW
20   POKE 49240,0: POKE 49242,0: POKE
      49244,0: POKE 49246,0
30   REM MAIN PROGRAM
40   POKE 49241,1: REM START WITH
      MAIN ST. GREEN
50   POKE 49247,1: REM SIDE ST.
      RED
60   Y = 1000: REM MAIN ST. TIME
      ADJUST
70   FOR I = 1 TO Y
80   A = PEEK (49249): REM CHECK
      SENSOR IN SIDE ST.
90   IF A < 128 THEN GOTO 110: REM
      IF A IS LOW THEN CHANGE
      SIGN
100  NEXT I
110  POKE 49240,0: REM MAIN ST.
      TO RED
120  POKE 49243,1: REM MAIN ST.
      TO RED
130  FOR J = 1 to 300: NEXT J: REM
      SLIGHT DELAY IN CHANGING
      SIGN
140  POKE 49246,0: REM SIDE ST.
      TO GREEN
150  POKE 49245,1: REM SIDE ST.
      TO GREEN
160  U = 500: REM SIDE ST. TIME
      ADJUST
170  FOR L = 1 TO U
180  B = PEEK (49250): REM CHECK
      SENSOR IN MAIN ST.
190  IF B < 128 THEN GOTO 210: REM
      IF B IS LOW THEN CHANGE
      SIGN
200  NEXT L
210  POKE 49244,0: REM SIDE ST.
      TO RED
220  POKE 49247,1
230  FOR S = 1 TO 300: NEXT S: REM
      SLIGHT DELAY IN CHANGING
      SIGN
240  POKE 49242,0: REM MAIN ST.
      TO GREEN
250  GOTO 40
```

PROGRAM 2-6 Traffic light.

DOWNHILL RUN

The circuit in Fig. 2-7 shows how a game paddle is connected to the game controller input of the computer. Note that the 150-kΩ variable resistor is connected to +5 V and the game controller input GC_0.

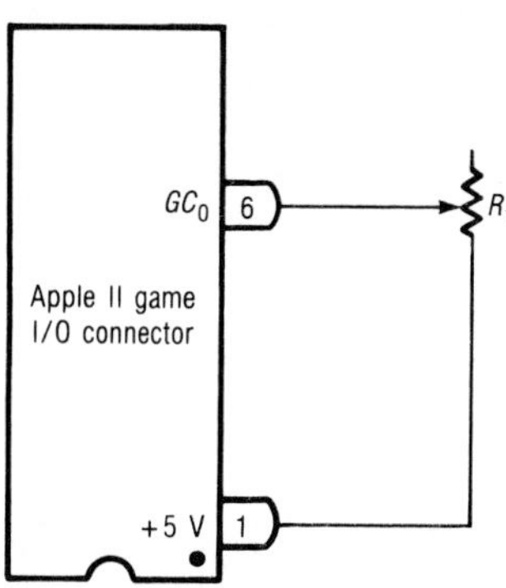

PARTS LIST

Apple game I/O connector
R_1 150-kΩ linear potentiometer (or game paddles for Apple II/IIe)

FIG. 2-7 Downhill (paddle) circuit.

Program 2-7. The program reads the GC_0 input using the BASIC command PDL(0). Vari-

```
10   N = PDL (0): REM SET UP A/D
20   IF N < 7 THEN GOTO 50: REM
      SET LIMITS
30   PRINT TAB( INT (N / 6.8));N
      : REM SET LIMITS
40   GOTO 10
50   PRINT N
60   GOTO 10
```

PROGRAM 2-7 Paddle demonstration.

able N is made equal to a number between 0 and 255. Variable N is used to horizontally tab to the position where the value of N is printed. The program then loops and rereads the value of paddle zero.

Program 2-8. This program is similar to Program 2-7. This program prints a character resembling a sled or skier instead of the number. Programmers can use their imaginations and make changes in this program to make a downhill race game.

```
10   N = PDL (0): REM SET UP A/D
20   IF N < 7 THEN GOTO 50: REM
30   IF N < 7 THEN GOTO 80
40   C$ = "'T'"
50   PRINT TAB( INT (N / 6.8));C
      $
60   N = N1
70   GOTO 10
80   PRINT "'T'"
90   FOR I = 1 TO 20: NEXT I
100  GOTO 10
110  HTAB 35
120  PRINT "'T'"
130  FOR J = 1 TO 20: NEXT J
140  GOTO 10
```

PROGRAM 2-8 Downhill run.

ADJUSTABLE BEEPER

The circuit in Fig. 2-8 will emit beeps from the piezo buzzer. The pause between the beeps can be adjusted with the variable resistance R_1. Pin 11 is acting as a paddle input where the resistance setting of R_1 determines the value returned to the computer. Pin 15 is an output that turns the buzzer on when it is high or off when it is low.

Program 2-9. This program reads the position of R_1 in line 10. The buzzer is activated by line 30 with a variable time delay in line 40. The time delay is dependent on the number returned from the GC_3 input. This was printed on the CRT by program line 20. Line 50 turns the buzzer off. Line 60 is another variable time delay. The process is then repeated.

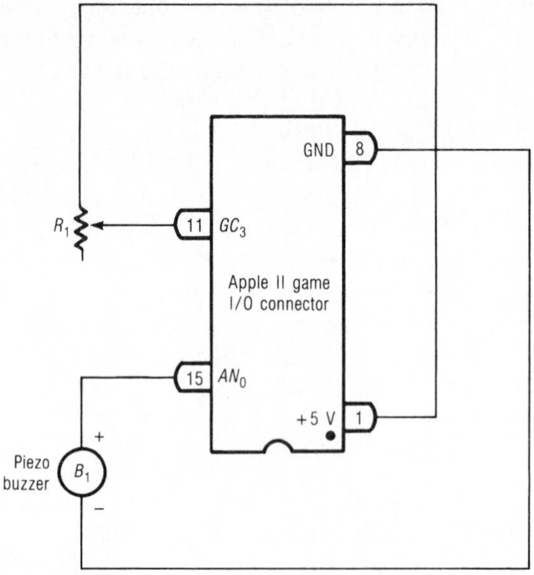

PARTS LIST

Apple II/IIe game I/O connector
B_1 Piezo buzzer (similar to Radio Shack 273-060)
R_1 150-kΩ linear potentiometer (or game paddle for Apple II/IIe)

FIG. 2-8 Adjustable beeper.

```
10  Z = PDL (3): REM SET UP A/D
20  PRINT Z
30  POKE 49241,1: REM SET OUTPUT
      TO HIGH
40  FOR I = 1 TO Z: NEXT I
50  POKE 49240,0: REM SET OUTPUT
      TO LOW
60  FOR T = 1 TO Z: NEXT T
70  GOTO 10
```

PROGRAM 2-9 Adjustable beeper.

ELECTRIC EYE BREAK-BEAM ALARM

The electric eye circuit in Fig. 2-9 reacts to a beam of light. The cadmium sulfide photocell R_1 has low resistance when light is striking its surface. If the light is interrupted, its resistance increases. The changing resistance of the photocell is changed into digital form by the computer as it enters the "paddle" input GC_1.

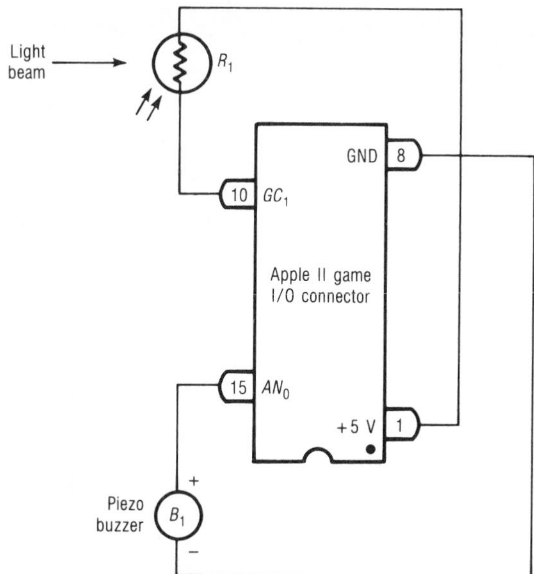

PARTS LIST

Apple II/IIe game I/O connector
B_1 Piezo buzzer (similar to Radio Shack 273-060)
R_1 Cadmium sulfide photocell (Radio Shack 276-116)

FIG. 2-9 Electric eye break-beam alarm.

The piezo buzzer will sound when the computer drives annunciator output AN_0 high.

Program 2-10. The program will activate the buzzer in Fig. 2-9 each time the light beam falling on photocell R_1 is interrupted. Line 10 reads input GC_1 and converts it to a number between 0 and 255 which is printed on the CRT for reference. Lines 30 and 40 turn the buzzer on or off based on the input from the photocell. If a low number is returned from input GC_1, the buzzer is turned off by line 40. The resistance of the photocell is low as a result of the light beam striking its surface. If a high number is returned from input GC_1, the buzzer is turned on by line 30. The resistance of the photocell is high because the light beam is broken. The process is then repeated. The constant 3 can be changed in lines 30 and 40 to adjust the sensitivity of the circuit.

```
10  X = PDL (1)
20  PRINT X
30  IF X > 3 THEN POKE 49241,1: REM
     SET OUTPUT HIGH IF LIGHT IS
     < BEAMED LIGHT
40  IF X < 3 THEN POKE 49240,0: REM
     SET OUTPUT TO LOW IN HIGH I
     NTENSITY LIGHT
50  GOTO 10
```

PROGRAM 2-10 Electric eye break-beam alarm.

LIGHT INTENSITY ALARM

The circuit in Fig. 2-10 is similar to that in Fig. 2-9. The circuit in Fig. 2-10 has two output devices instead of one. Several warning devices can be activated based on the relative intensity of light striking the photocell R_2. More light striking the photocell causes it to have less resistance, causing the computer to return a

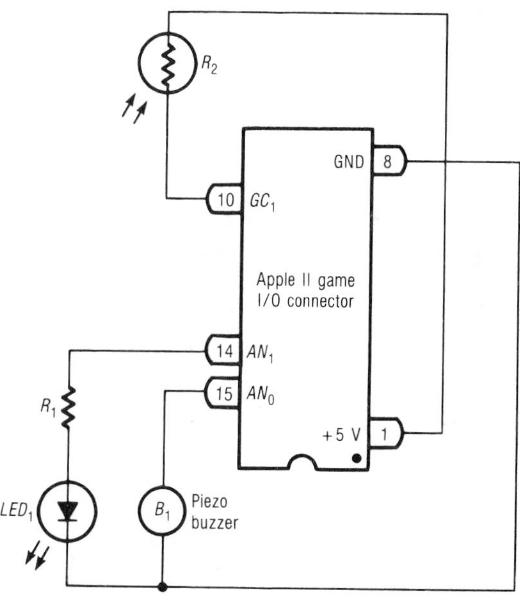

PARTS LIST

Apple II/IIe game I/O connector
B_1 Piezo buzzer (similar to Radio Shack 273-060)
LED_1 Light-emitting diode
R_1 150-Ω, 1/4-W resistor
R_2 Cadmium sulfide photocell (Radio Shack 276-116)

FIG. 2-10 Light intensity alarm.

smaller number to the computer from the paddle input GC_1.

Program 2-11. The program is a two-step alarm that first activates LED_1 in Fig. 2-10 if it is moderately dark. It then activates the buzzer as it becomes even darker. Lines 10 and 20 read paddle input GC_1 and convert the relative resistance of the photocell to a number. The higher the light intensity, the lower the number. Lines 30 and 40 turn the LED on when the light is dim and off when it is bright. Lines 50 and 60 turn the buzzer on if the light becomes very dim. Then the process is repeated.

```
10  X = PDL (1)
20  PRINT X
30  IF X > 10 THEN POKE 49243,1:
     REM SET OUTPUT 2 TO HIGH
     IF MODERATELY LIGHT
40  IF X < 10 THEN POKE 49242,0:
     REM SET OUTPUT 2 TO LOW IN
     BRIGHT LIGHT
50  IF X > 30 THEN POKE 49241,1:
     REM SET OUTPUT 1 TO HIGH
     IF RELATIVELY DARK
60  IF X < 30 THEN POKE 49240,0:
     REM SET OUTPUT 1 TO LOW IF
     FAIR OR BETTER LIGHTING
70  GOTO 10
```

PROGRAM 2-11 Light intensity alarm.

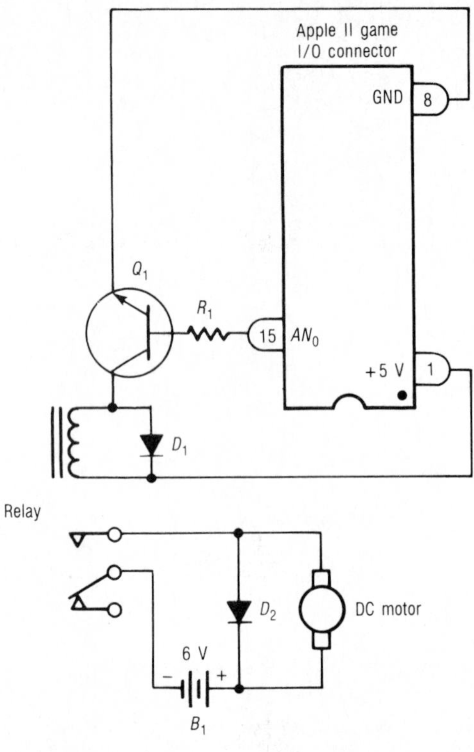

PARTS LIST

Apple II/IIe game I/O connector

B_1	6-V battery (four 1.5-V cells)
D_1,D_2	1N4002 silicon diode, 1 A, 100 PIV
Motor	1.5- to 6-V dc hobby motor
Q_1	2N3904 NPN transistor (or similar)
R_1	1-kΩ, ½-W resistor
Relay	5-V coil relay (similar to Radio Shack 275-215)

FIG. 2-11 Controlling a motor with the computer.

CONTROLLING A MOTOR

The circuit in Fig. 2-11 shows a method of controlling an electric motor using the Apple II/IIe game port. Great caution must be taken to make sure there is *no connection between the motor circuit and the computer circuit.* The computer can be damaged if external voltages enter its circuitry. A high output (AN_0) turns Q_1 on, thus activating the relay and turning on the dc electric motor circuit. Diodes D_1 and D_2 are essential to eliminate or reduce voltage spikes in the system, which can affect the computer circuitry.

Program 2-12. This program turns the dc motor in Fig. 2-11 either on or off. Line 20 clicks the relay and turns on the motor. Line 30 turns the motor off.

```
10  INPUT "DO YOU WANT THE MOTOR
     ON OR OFF? ";R$
20  IF R$ = "ON" THEN POKE 49241
     ,1
30  IF R$ = "OFF" THEN POKE 4924
     0,0
40  GOTO 10
```

PROGRAM 2-12 Turning a motor on and off.

Program 2-13. This program asks the operator how long the motor should be left on and off. It turns the motor on in line 20 and introduces a variable time delay in lines 40 and 50 (motor on time). The program turns the motor off in line 60. The time delay in lines 80 and 90 is the off time of the motor. The timed on-off sequence continues.

```
10    INPUT "HOW MANY SECONDS OF TI
       ME DELAY BETWEEN TURNING MOT
       OR ON AND OFF? ";Z
20    POKE 49241,1: REM TURN ON MO
       TOR
30    REM *TIME DELAY*
40    FOR I = 1 TO (Z * 800)
50    NEXT I
60    POKE 49240,0: REM TURN OFF M
       OTOR
70    REM *TIME DELAY*
80    FOR I = 1 TO (Z * 800)
90    NEXT I
100   GOTO 20
```

PROGRAM 2-13 Controlling on-off time of the motor.

CONTROLLING A MOTOR WITH LIGHT

The circuit in Fig. 2-12 is the same as that in Fig. 2-11 except a photocell has been added as an input to the computer. The photocell R_2 will have low resistance and return a low number to the computer when light strikes its surface. As it becomes darker, the photocell's resistance will increase, as will the number returned to the computer. Great caution must be taken to make sure there is *no connection between the motor circuit and the computer circuit.* The computer can be damaged if external voltages enter its circuitry. A high output (AN_0) turns Q_1 on, thus activating the relay and turning on the dc motor. Diodes D_1 and D_2 are essential to eliminate or reduce voltage spikes in the system which can affect the computer circuitry.

```
10    Z = PDL (3): REM READ GC3
20    PRINT Z
30    POKE 49241,1: REM TURN ON MO
       TOR
40    REM *TIME DELAY*
50    FOR I = 1 TO (Z * 20)
60    NEXT I
70    POKE 49240,0: REM TURN OFF M
       OTOR
80    REM *TIME DELAY*
90    FOR I = 1 TO (Z * 20)
100   NEXT I
110   GOTO 10
```

PROGRAM 2-14 Controlling motor on-off time with light.

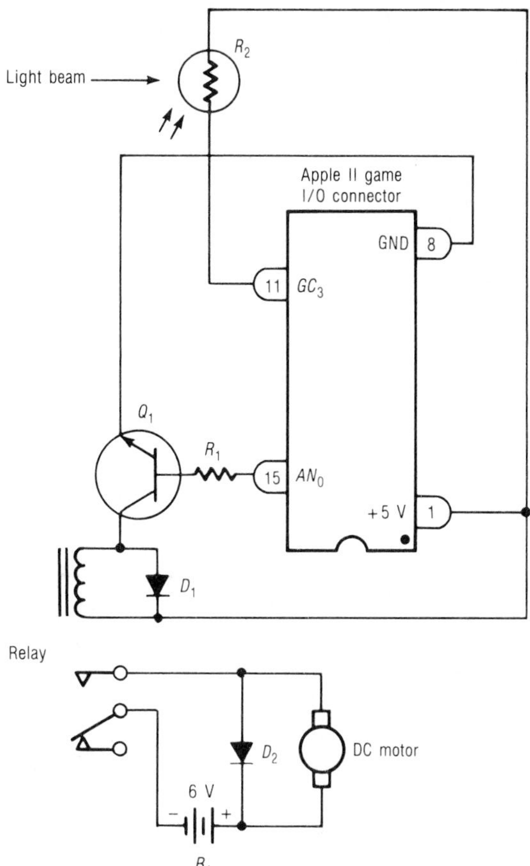

PARTS LIST

Apple II/IIe game I/O connector

B_1	6-V battery (four series 1.5-V cells)
D_1,D_2	1N4002 silicon diode, 1 A, 100 PIV
Motor	6-V dc hobby motor
Q_1	2N3904 NPN transistor (or similar)
R_1	1-kΩ, ½-W resistor
R_2	Cadmium sulfide photocell (Radio Shack 276-116)
Relay	5-V coil relay (similar to Radio Shack 275-215)

FIG. 2-12 Controlling a motor with a photocell.

Program 2-14. When used with the circuit in Fig. 2-12, the program will turn the motor on and off based on the amount of light striking the photocell. The motor will cycle faster when more light is striking the photocell. Its cycle time will be slower with less light striking the photocell. Line 10 reads the photocell input at the paddle input GC_3. This number is then used to calculate the two time delays (time motor is on and time motor is off). The process is then repeated.

chapter 3

Apple II/IIe Peripheral Slot Interface Circuits

INTRODUCTION

Besides the game port, the other interfacing choice on the Apple II/IIe is to use the slots across the back (inside the computer). This is much more difficult and costly than using the game port, but gives more flexibility. The location of slot 7 (used in the following programs) is shown in Fig. 3-1. Also shown is a typical method of extending the slot outside the computer to a socket end connector taped to the solderless breadboard. Detailed instructions for such a connection are given in an excellent book by John E. Uffenback in *Hardware Interfacing with Apple II Plus,* Prentice-Hall, Englewood Cliffs, NJ, 1983.

When using the Apple II/IIe peripheral slots, *great care must be taken in wiring.* Keep wiring neat on the solderless breadboard. Be sure to double-check all connections before connecting to the peripheral connector. For safety, be sure to use the computer's power supply connections available at the slot connector. Application of external voltages above and below the computer power supply voltages can damage the computer.

The signals available at each peripheral connector are shown in Fig. 3-2. Table 3-1 describes each of these signals. Most interfacing projects will not require every signal.

USING THE PPI FOR OUTPUT

The circuit in Fig. 3-3 shows the use of the programmable peripheral interface (PPI) chip IC_2 from Intel. The open arrow symbols on the left represent direct connections to the pins of the peripheral connector in the Apple II/IIe computer. Note that all eight data lines D_0 through D_7 are used while only address lines A_0 and A_1 are connected to the PPI. Only a

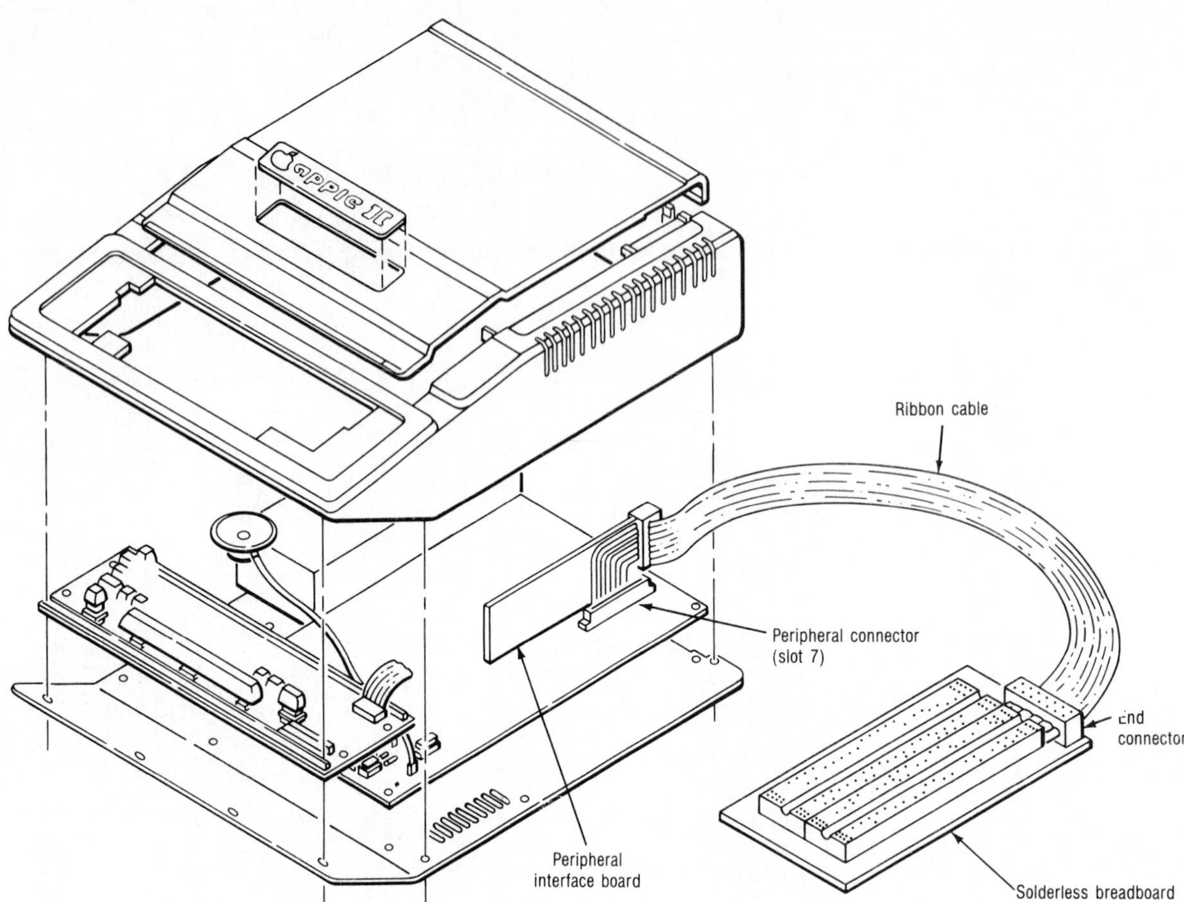

FIG. 3-1 Connecting to peripheral slot connector for breadboarding of interface circuit. (Copyright, 1981, Apple Computer, Inc. Used by permission of Apple Computer, Inc.)

single output bit (pin 4 of IC_2) will be used in this circuit. The 8255 PPI has 24 I/O lines.

Program 3-1. This program asks if the LED is to be turned on or off. Based on the answer, the computer either turns LED_1 on or off. Line 20 initializes or programs the 8255 PPI to be an output device. The PPI must always be initialized before it is used as an input, output, or both. Line 40 turns on the LED using the POKE 49392,1 statement. Line 50 turns the LED off. The program is then repeated.

```
10  REM PPI INITIALIZE
20  POKE 49395,128
30  INPUT "DO YOU WANT THE LED ON
     OR OFF?";A$
40  IF A$ = "ON" THEN POKE 49392
     ,1
50  IF A$ = "OFF" THEN POKE 4939
     2,0
60  GOTO 30
```

PROGRAM 3-1 Turning LED on and off.

Program 3-2. This program flashes LED_1 in Fig. 3-3 on and off. Line 20 initializes the PPI so that it is an output. Line 30 turns on the LED; while lines 40 and 50 cause a short time delay. Line 60 turns off the LED and lines 70 and 80 are a time delay. The process is then repeated.

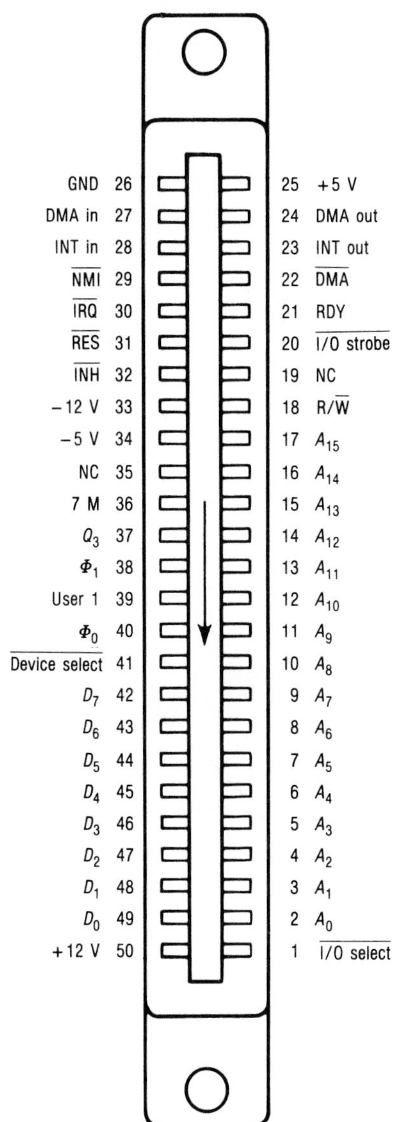

FIG. 3-2 Peripheral connector pinout. (Copyright, 1981, Apple Computer, Inc. Used by permission of Apple Computer, Inc.

```
10  REM PPI INITIALIZE
20  POKE 49395,128
30  POKE 49392,1
40  FOR I = 1 TO 100
50  NEXT I
60  POKE 49392,0
70  FOR J = 1 TO 100
80  NEXT J
90  GOTO 30
```

PROGRAM 3-2 Flashing LED.

USING SEVERAL PPI PORTS

The circuit in Fig. 3-4 demonstrates the use of all of the I/O ports on the 8255 PPI chip. The LEDs will indicate the logic state of each of these outputs. The computer user will try to light different LEDs at the output of the PPI. Notice that the outputs from the 8255 IC are grouped by eights. Each group of eight is called a *port.* The top group would be port *A,* the center group port *C,* and the bottom group port *B.* It is customary to output parallel data from these ports. With the proper initialization, the PPI ports can serve as either outputs or inputs.

Program 3-3. This program causes one or more than one LED in Fig. 3-4 to be lit. Line 20 initializes the PPI so that all the ports are outputs. Line 30 asks which port is to be used. Line 40 asks which of the 8 bits should be lit. If bit 1 at port *A* is selected, for instance, LED_1 would be lit by line 50. If port *C* is selected and 12 is entered, then light-emitting diodes LED_{11} and LED_{12} would be lit by line 70.

```
10  REM INITIALIZE PPI (MAKE ALL
     OUTPUTS)
20  POKE 49395,128
30  INPUT "WHICH PORT WOULD YOU L
     IKE TO CONTROL [A,B,OR C] ?"
     ;X$
40  INPUT "WHICH BIT(S) WOULD YOU
     LIKE TO CONTROL (1,2,4,8,16
     ,32,64,128)? ";Y: REM SINGL
     E BITS CONTROLLED BY TYPING
     PLACE VALUE- MULTIBITS CONTR
     OLLED BY ADDING PLACE VALUES
50  IF X$ = "A" THEN POKE 49392,
     Y
60  IF X$ = "B" THEN POKE 49393,
     Y
70  IF X$ = "C" THEN POKE 49394,
     Y
80  GOTO 30
```

PROGRAM 3-3 LED select switch.

TABLE 3-1 Peripheral Connector Signal Descriptions

Pin	Name	Description
1	$\overline{\text{I/O SELECT}}$	This line, normally high, will become low when the microprocessor references page \$$C_n$, where *n* is the individual slot number. This signal becomes active during Φ_0 and will drive 10 LSTTL loads*—this signal is not present on peripheral connector 0
2–17	A_0–A_{15}	The buffered address bus; the address on these lines becomes valid during Φ_1 and remains valid through Φ_0—these lines will each drive 5 LSTTL loads*
18	R/$\overline{\text{W}}$	Buffered Read/$\overline{\text{Write}}$ signal. This becomes valid at the same time the address bus does and goes high during a read cycle and low during a write; this line can drive up to 2 LSTTL loads*
19	SYNC	On peripheral connector 7 *only,* this pin is connected to the video timing generator's SYNC signal
20	$\overline{\text{I/O STROBE}}$	This line goes low during Φ_0 when the address bus contains an address between \$C800 and \$CFFF; this line will drive 4 LSTTL loads*
21	RDY	The 6502's RDY input; pulling this line low during Φ_1 will halt the microprocessor, with the address bus holding the address of the current location being fetched
22	$\overline{\text{DMA}}$	Pulling this line low disables the 6502's address bus and halts the microprocessor; this line is held high by a 3-kΩ resistor to +5 V
23	INT OUT	Daisy-chained interrupt output to lower-priority devices; this pin is usually connected to pin 28 (INT IN)
24	DMA OUT	Daisy-chained DMA output to lower-priority devices; this pin is usually connected to pin 27 (DMA IN)
25	+5 V	+5-V power supply; 500-mA current is available for *all* peripheral cards
26	GND	System electrical ground
27	DMA IN	Daisy-chained DMA input from higher-priority devices; usually connected to pin 24 (DMA OUT)
26	INT IN	Daisy-chained interrupt input from higher-priority devices; usually connected to pin 23 (INT OUT)
29	$\overline{\text{NMI}}$	Nonmaskable interrupt: when this line is pulled low, the Apple begins an interrupt cycle and jumps to the interrupt handling routine at location \$3FB
30	$\overline{\text{IRQ}}$	Interrupt request: when this line is pulled low, the Apple begins an interrupt cycle only if the 6502's I (interrupt disable) flag is not set; if so, the 6502 will jump to the interrupt handling subroutine whose address is stored in locations \$3FE and \$3FF
31	$\overline{\text{RES}}$	When this line is pulled low the microprocessor begins a RESET cycle.
32	$\overline{\text{INH}}$	When this line is pulled low, all ROMs on the Apple board are disabled; this line is held high by a 3-kΩ resistor to +5 V
33	−12 V	−12-V power supply; maximum current is 200 mA for all peripheral boards
34	−5 V	−5-V power supply; maximum current is 200 mA for all peripheral boards
35	COLOR REF	On peripheral connector 7 *only,* this pin is connected to the 3.5-MHz COLOR REFerence signal of the video generator
36	7 M	7-MHz clock: this line will drive 2 LSTTL loads*
37	Q_3	2-MHz asymmetrical clock; this line will drive 2 LSTTL loads*
38	Φ_1	Microprocessor's phase one clock: this line will drive 2 LSTTL loads*
39	USER 1	This line, when pulled low, disables *all* internal I/O address decoding

TABLE 3-1 (Continued)

Pin	Name	Description
40	Φ_0	Microprocessor's phase zero clock; this line will drive 2 LSTTL loads*
41	$\overline{\text{DEVICE SELECT}}$	This line becomes active (low) on each peripheral connector when the address bus is holding an address between \$C0*n*0 and \$C0*n*F, where *n* is the slot number plus \$8; this line will drive 10 LSTTL loads*
42–49	D_0–D_7	Buffered bidirectional data bus; the data on this line becomes valid 300*n*S into Φ_0 on a write cycle and should be stable for no less than 100 nanoseconds before the end of Φ_0 on a read cycle; each data line can drive one LSTTL load
50	+12 V	+12-V power supply; this can supply up to 250 mA total for all peripheral cards

* Loading limits are for each peripheral card.

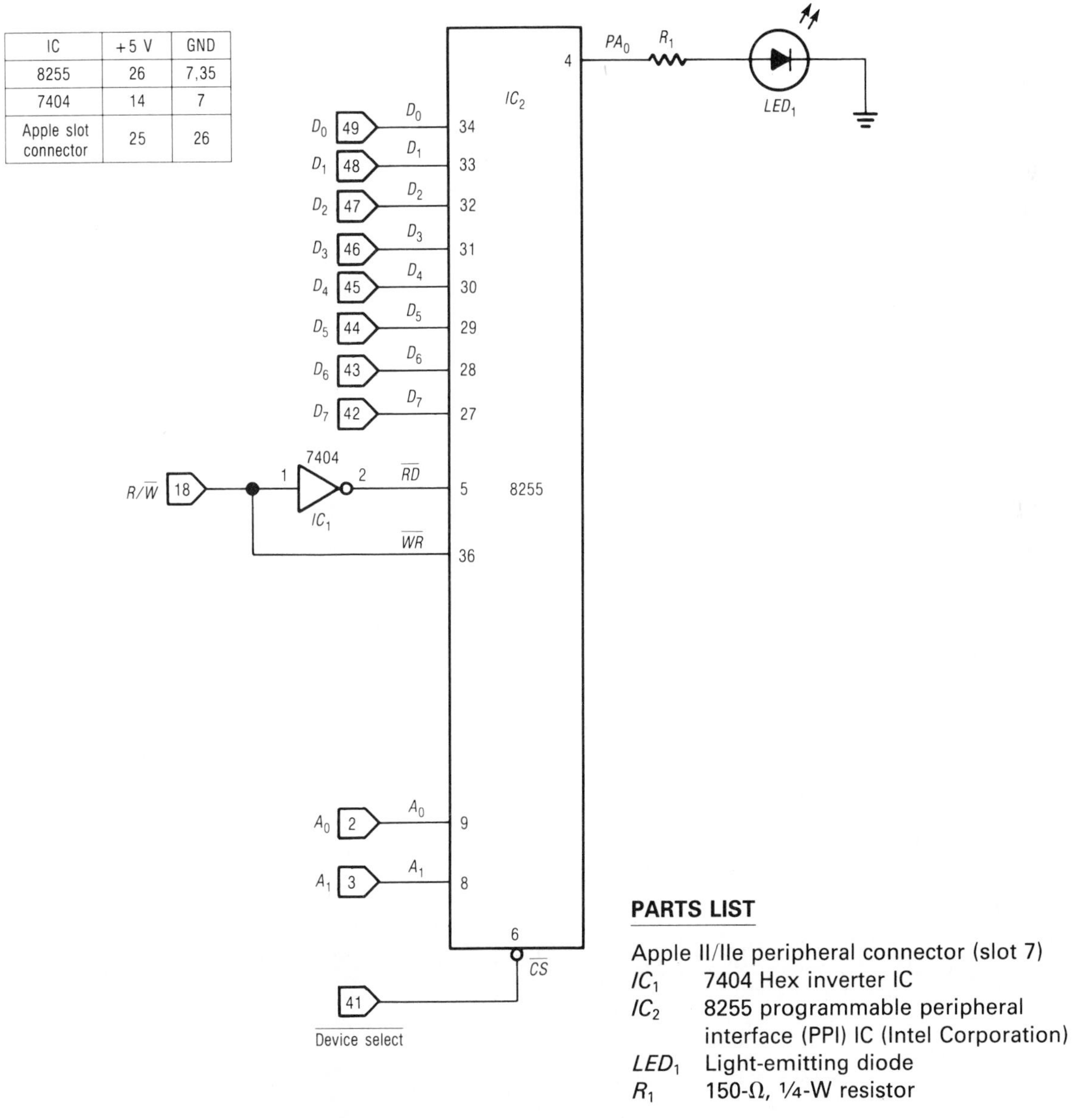

PARTS LIST

Apple II/IIe peripheral connector (slot 7)
IC_1 7404 Hex inverter IC
IC_2 8255 programmable peripheral interface (PPI) IC (Intel Corporation)
LED_1 Light-emitting diode
R_1 150-Ω, ¼-W resistor

FIG. 3-3 Using the PPI chip as a simple output.

IC	+5 V	GND
8255	26	7,35
7404	14	7
Apple slot connector	25	26

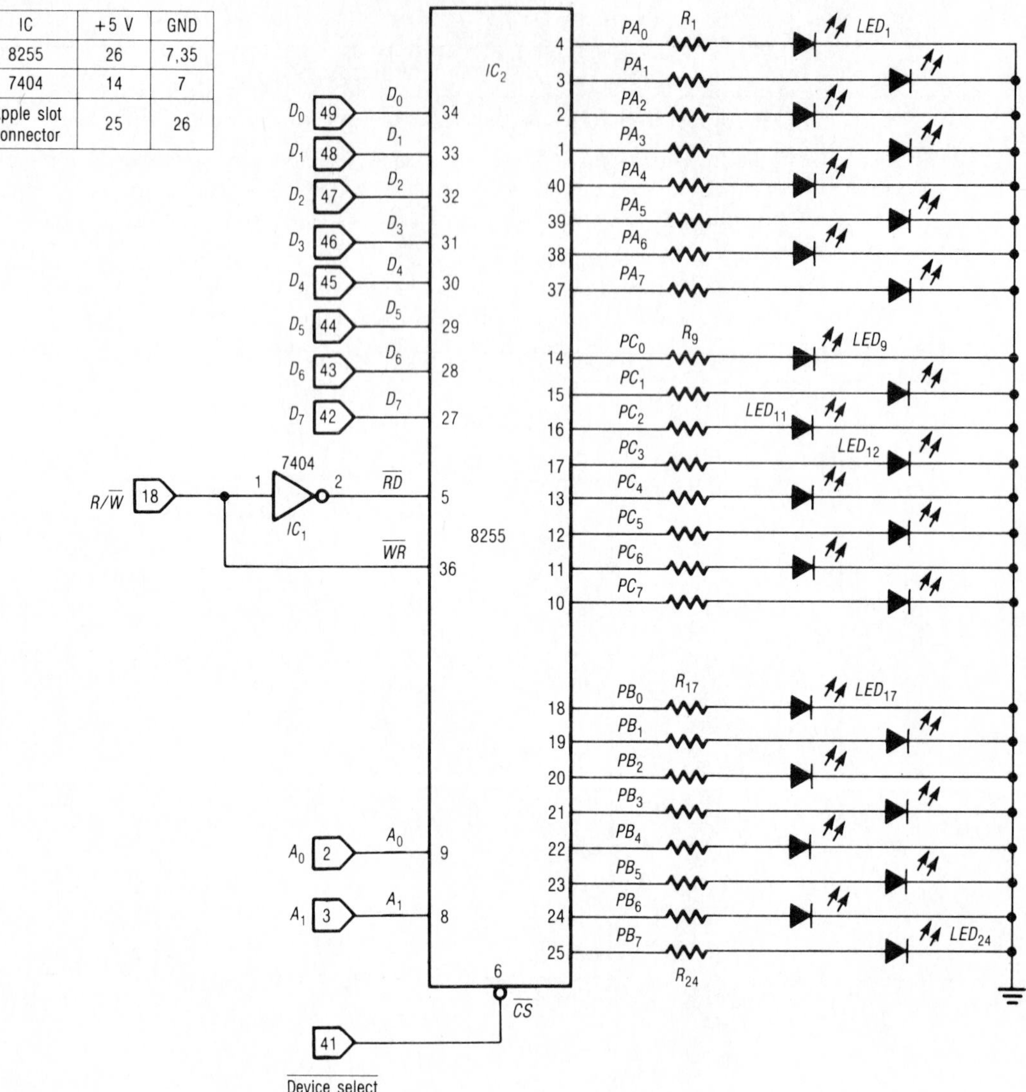

PARTS LIST

Apple II/IIe peripheral connector	
IC_1	7404 Hex inverter IC
IC_2	8255 programmable peripheral interface IC (Intel Corporation)
LED_1 through LED_{24}	Light-emitting diode
R_1 through R_{24}	150-Ω, ¼-W resistor

FIG. 3-4 Using all the PPI ports as outputs.

REAL-TIME CLOCK

The circuit in Fig. 3-5 gives the computer a real-time clock capability. The clock may be used for purposes such as displaying the time of the day, sounding an alarm, or turning machines on or off at a specific time. All three ports of the 8255 PPI are used to send data to and gather time information from the clock. The MSM5832 real-time clock chip IC_3 is a CMOS unit by OKI Semiconductor made to be interfaced with microprocessor-based systems such as the Apple II/IIe computer. An external crystal (XTAL) controls the time base of the clock chip. Capacitor C_1 can be adjusted if the clock is running slightly fast or slow. The backup battery B_1 may be omitted, but the clock chip must be reset each time the computer is turned off. If the backup battery B_1 is omitted, D_1, D_2, and R_{13} can also be eliminated.

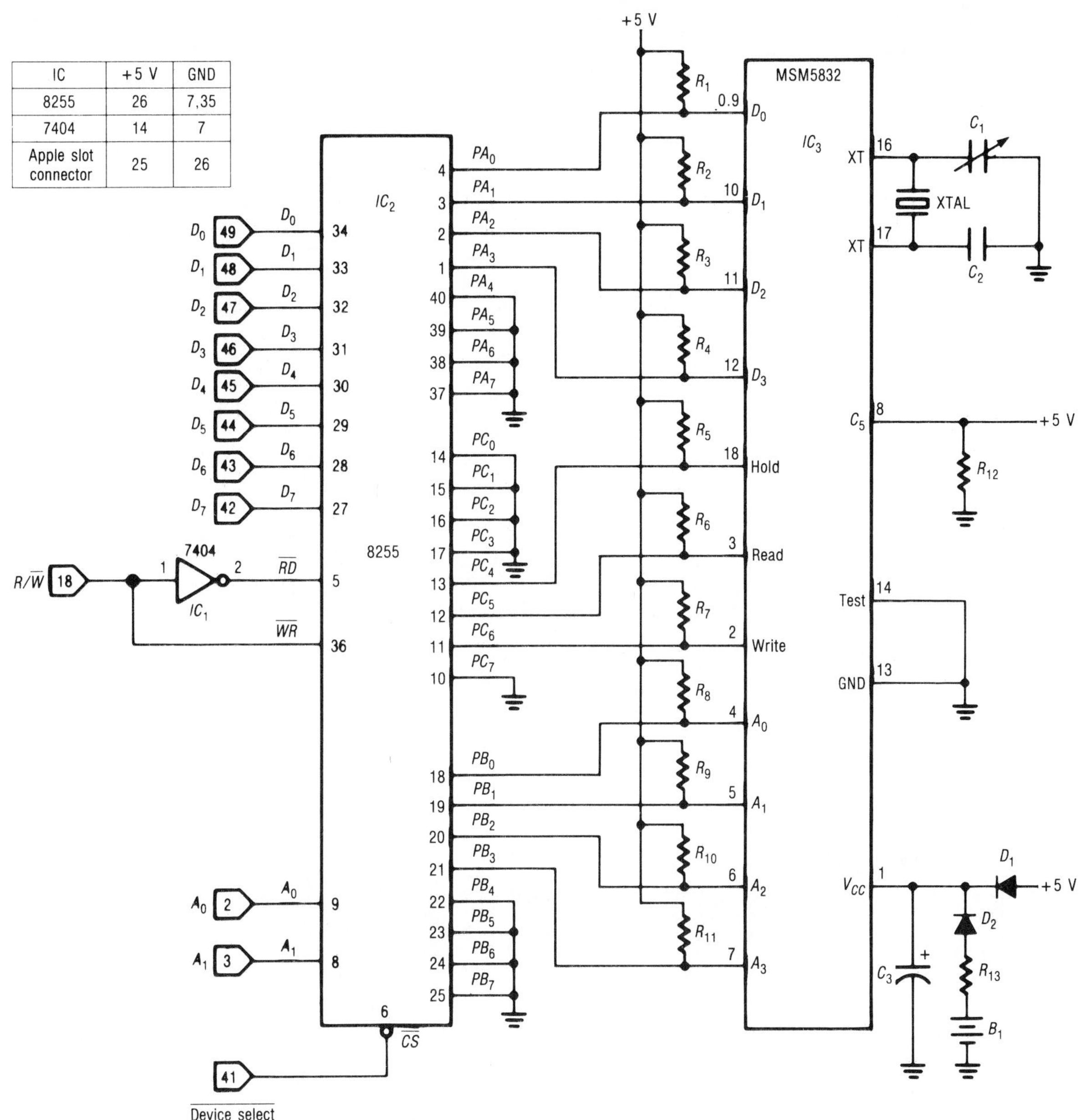

PARTS LIST

Apple II/IIe peripheral connector

B_1	3.6-V rechargeable battery (three ni-cad cells)
C_1	5- to 35-pF variable capacitor
C_2	15-pF, 100-V disk capacitor
C_3	4.7-μF, 25-V electrolytic capacitor
D_1, D_2	1N34A germanium diode
IC_1	7404 Hex inverter IC
IC_2	8255 programmable peripheral interface IC (Intel Corporation)
IC_3	MSM5832 real-time clock calendar IC (OKI Semiconductor)
R_1 through R_{12}	10-kΩ, ¼-W resistor
R_{13}	1.6-kΩ, ¼-W resistor
XTAL	32.768-kHz crystal (Digi-Key #KF38G)

FIG. 3-5 Interfacing a real-time clock using the PPI chip. (Used by permission of OKI Semiconductor.)

Program 3-4. This program reads the contents of the MSM5832 clock and displays hours, minutes, and seconds on the screen. The display is updated every second.

```
10   REM INITIALIZE PPI
20   POKE 49395,145
30   POKE 49394,32: REM READ
40   REM SET ADDR.'S & VARIABLES
50   POKE 49393,5: REM ADDR. H10
60   A = PEEK (49392): REM READ
      H10
70   POKE 49393,4: REM ADDR. H1
80   B = PEEK (49392): REM READ
      H1
90   POKE 49393,3: REM ADDR. M10
100  C = PEEK (49392): REM READ
      M10
110  POKE 49393,2: REM ADDR. M1
120  D = PEEK (49392): REM READ
      M1
130  POKE 49393,1: REM ADDR. S10
140  E = PEEK (49392): REM READ
      S10
150  POKE 49393,0: REM ADDR. S1
160  F = PEEK (49392): REM READ
      S1
170  REM MAIN PROGRAM
180  HOME
190  VTAB 12
200  HTAB 18
210  PRINT "TIME"
220  VTAB 14
230  HTAB 16
240  PRINT A;B":"C;D":"E;F
250  REM "TIMED CLICK"
260  FOR J = 1 TO 2
270  SPK = PEEK ( - 16336)
280  NEXT J
290  FOR W = 1 TO 600: NEXT W
300  GOTO 30
```

PROGRAM 3-4 Reading the clock.

Program 3-5. This program is used for writing data into the MSM5832 clock chip to set the current time. The user enters the tens of hours, ones of hours, tens of minutes, and ones of minutes. The seconds are not set in this program. Use Program 3-4 to check the results of setting the clock.

```
400  REM INITIALIZE PPI
410  POKE 49395,128
420  REM SET UP WRITE & HOLD
430  POKE 49394,80
440  REM MAIN PROGRAM
450  POKE 49393,5: REM H10 ADDR.
460  INPUT "INPUT TENS OF HOURS(0
      -1) ";SA
470  POKE 49392,SA: REM DATA
480  POKE 49393,4: REM H1 ADDR.
490  INPUT "INPUT ONES OF HOURS(0
      -9) ";SB
500  POKE 49392,SB: REM DATA
510  POKE 49393,3: REM M10 ADDR.
520  INPUT "INPUT TENS OF MINUTES
      (1-5) ";SC
530  POKE 49392,SC: REM DATA
540  POKE 49393,2: REM M1 ADDR.
550  INPUT "INPUT ONES OF MINUTES
      (0-9) ";SD
560  POKE 49392,SD: REM DATA
570  END
```

PROGRAM 3-5 Setting the clock.

Program 3-6. This program will sound an alarm at the time set by the user. It will also display the current time and the alarm time on the CRT.

```
10   HOME
20   INPUT "ALARM SET- ENTER THE T
      ENS OF HOURS- ";H
30   INPUT "ALARM SET- ENTER THE O
      NES OF HOURS- ";HO
40   INPUT "ALARM SET- ENTER THE T
      ENS OF MINUTES- ";M
50   INPUT "ALARM SET- ENTER THE O
      NES OF MINUTES- ";MO
60   REM INITIALIZE PPI
70   POKE 49395,145
80   POKE 49394,32: REM READ
90   REM SET ADDR.'S & VARIABLES
100  POKE 49393,5: REM ADDR. H10
110  A = PEEK (49392): REM READ
      H10
120  POKE 49393,4: REM ADDR. H1
130  B = PEEK (49392): REM READ
      H1
140  POKE 49393,3: REM ADDR. M10
150  C = PEEK (49392): REM READ
      M10
160  POKE 49393,2: REM ADDR. M1
170  D = PEEK (49392): REM READ
      M1
180  POKE 49393,1: REM ADDR. S10
190  E = PEEK (49392): REM READ
      S10
200  POKE 49393,0: REM ADDR. S1
210  F = PEEK (49392): REM READ
      S1
```

PROGRAM 3-6 Using the clock as an alarm.

```
220  REM MAIN PROGRAM
230  HOME
240  VTAB 12
250  HTAB 18
260  PRINT "TIME","ALARM"
270  VTAB 14
280  HTAB 16
290  PRINT A;B":"C;D":"E;F,H;HO":
      "M;MO":00"
300  REM "TIMED CLICK"
310  FOR J = 1 TO 2
320  SPK = PEEK ( - 16336)
330  NEXT J
340  FOR W = 1 TO 600: NEXT W
350  IF A = H AND B = HO and C =
      M AND D = MO THEN GOTO 370
360  GOTO 80
370  FOR I = 1 TO 30: REM TIMING
      LOOP FOR ALARM BUZZER
380  SPK = PEEK ( - 16336)
390  NEXT I
400  GOTO 80
```

PROGRAM 3-6 (***Continued***)

MELODY GENERATOR

The circuit in Fig. 3-6 will generate one of 25 short tunes and several chimes. The heart of the circuit is the AY-3-1350 tunes generator IC_4 by General Instrument Corporation. The tunes that can be generated by the AY-3-1350 are listed with their codes in Table 3-2. The AY-3-1350 is a microcomputer-based synthesizer with preprogrammed tunes. The interfacing to the computer is done through the general-purpose peripheral interface adapter (PIA) chip IC_1. Ports A and B are being used as outputs on the 6820 PIA in Fig. 3-6. The 4016 bilateral switch IC_3 is being used something like a rotary switch in tune selection. Variable resistor R_1 and associated circuitry set the tempo of the tune. The pitch is set by using R_9 (adjusts the oscillator frequency). The output of the tune generator drives Q_5 and the speaker. The envelope output (pin 13 of IC_4) modifies the output for a better sound.

Program 3-7. This program causes the melody generator circuit in Fig. 3-6 to sequence through each of the tunes listed in Table 3-2. This program tests the circuit for proper operation and lets the operator hear all the tunes available on the AY-3-1350 IC.

```
10  POKE 49405,0: REM PIA INITIAL
     IZED
20  POKE 49404,64: REM CALL ON P
     ORT A
30  GOTO 20
```

PROGRAM 3-7 Play all tunes.

TABLE 3-2 Codes for Tunes on the AY-3-1350 Melody Generator

Tunes		Tunes	
A0	"Toreador"	A3	"O Sole Mio"
B0	"William Tell"	B3	"Santa Lucia"
C0	"Hallelujah Chorus"	C3	"The End"
D0	"Star-Spangled Banner"	D3	"Blue Danube"
E0	"Yankee Doodle"	E3	"Brahms' Lullaby"
A1	"John Brown's Body"	A4	"Hell's Bells"
B1	"Clementine"	B4	"Jingle Bells"
C1	"God Save the Queen"	C4	"La Vie en Rose"
D1	"Colonel Bogey"	D4	"Star Wars"
E1	"Marseillaise"	E4	"Beethoven's 9th"
A2	"America, America"	Chime X	Westminster Chime
B2	"Deutschland Lied"	Chime Y	Simple Chime
C2	"Wedding March"	Chime Z	Descending Octave Chime
D2	"Beethoven's 5th"		
E2	"Augustine"		

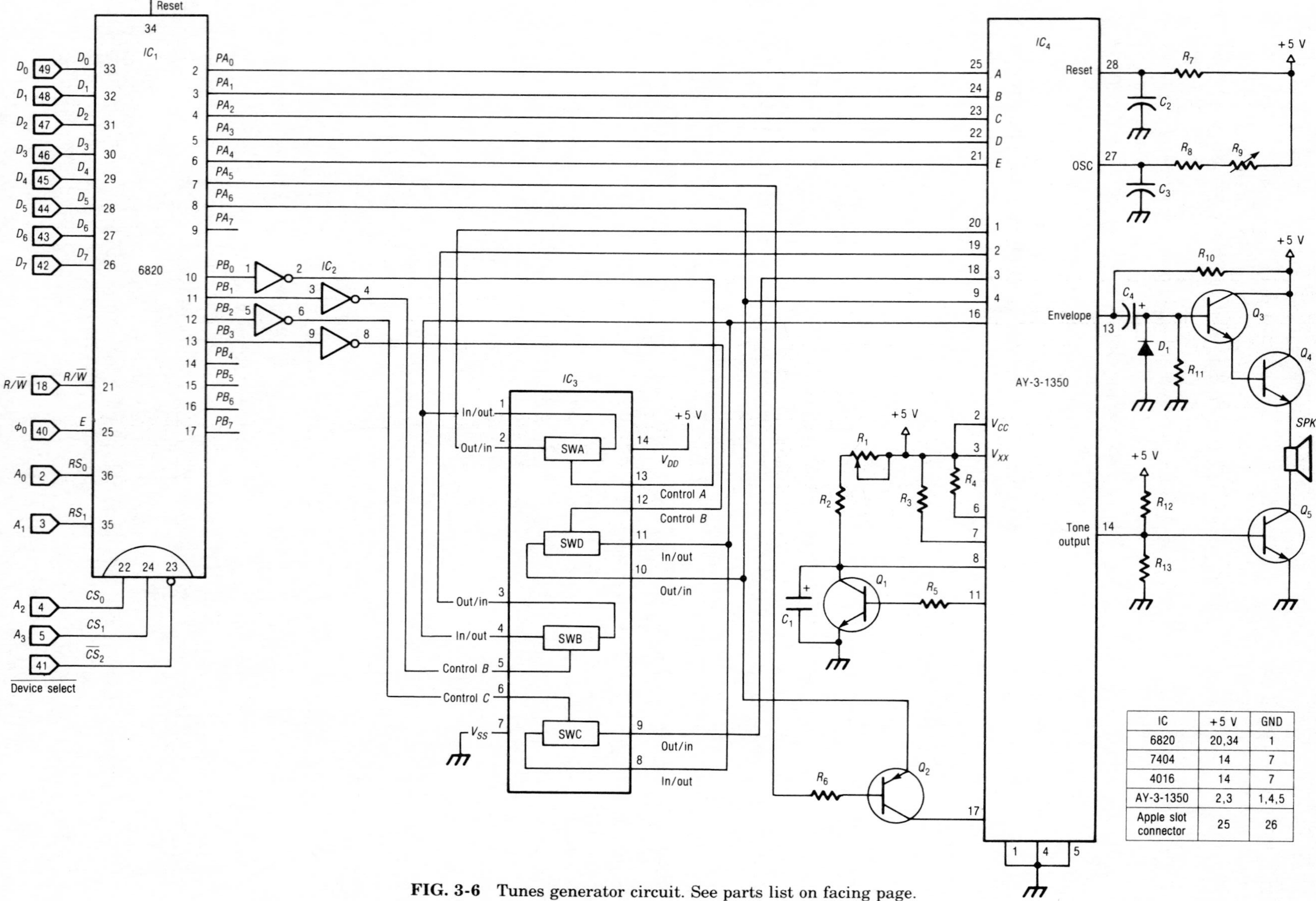

IC	+5 V	GND
6820	20,34	1
7404	14	7
4016	14	7
AY-3-1350	2,3	1,4,5
Apple slot connector	25	26

FIG. 3-6 Tunes generator circuit. See parts list on facing page.

PARTS LIST FOR FIG. 3-6

Apple II/IIe peripheral connector

C_1	1-μF, 25-V tantalum capacitor
C_2	0.1-μF, 50-V disk capacitor
C_3	47-pF, 100-V disk capacitor
C_4	10-μF, 25-V electrolytic capacitor
D_1	1N4148 signal diode
IC_1	6820 peripheral interface adapter (PIA) IC (Motorola)
IC_2	7404 Hex inverter IC
IC_3	CD4016 CMOS quad bilateral switch IC
IC_4	AY-3-1350 tunes generator IC (General Instrument)
Q_1	2N3904 NPN transistor
Q_2	2N3906 PNP transistor
Q_3, Q_4, Q_5	2N2222 NPN transistor
R_1	500-kΩ linear potentiometer
R_2, R_7	100-kΩ, ¼-W resistor
R_3, R_4, R_{13}	33-kΩ, ¼-W resistor
R_5	10-kΩ, ¼-W resistor
R_6	27-kΩ, ¼-W resistor
R_8	3.9-kΩ, ¼-W resistor
R_9	20-kΩ trimmer potentiometer
R_{10}	3.3-kΩ, ¼-W resistor
R_{11}	47-kΩ, ¼-W resistor
R_{12}	2.7-kΩ, ¼-W resistor
SPK_1	8-Ω speaker

Program 3-8 This program lets the computer operator select which tune the melody generator will play. The program asks the operator to identify the selected tune by its code letter and number as given in Table 3-2. The tune will play over and over again until a new tune is selected.

```
10    POKE 49405,0: POKE 49407,0
20    POKE 49404,X + 32
30    POKE 49406,Y
40    INPUT "ENTER LETTERS A-E,THEN
       NUMBERS 1-4";X$,Y
50    IF X$ = "A" THEN X = 1
60    IF X$ = "B" THEN X = 2
70    IF X$ = "C" THEN X = 4
80    IF X$ = "D" THEN X = 8
90    IF X$ = "E" THEN X = 16
100   FOR N = 1 TO 100: NEXT N
110   IF Y = 4 THEN Y = 8
120   IF Y = 3 THEN Y = 4
130   GOTO 20
```

PROGRAM 3-8 Select tune to be played.

chapter 4

Automotive Circuits

AUTOMATIC HEADLIGHT REMINDER

The automatic headlight-reminder circuit shown in Fig. 4-1 is simple to build and can save the car owner the frustration of a run-down battery when the car is parked with its lights left on. The headlight-reminder warning buzzer will sound only if the ignition switch is turned off while the headlights are still on.

The automatic headlight-reminder circuit in Fig. 4-1 has three connections to the car's electrical system: the ground connection, the ignition switch, and the headlights. The headlight reminder will buzz only when the ignition terminal is disconnected or opened *and* the headlights' terminal is at +12 V (the lights are on). Current can now flow up through R_1 and the primary of the transformer, alternately turning transistor Q_1 on and off at 400 to 800 Hz. The oscillator drives the speaker through transformer T_1, creating the sound of a buzzer.

A parts kit and a pc board for the automatic headlight reminder are currently available from Mode Electronics.

AUTOMOBILE BACKUP SOUNDER

The backup sounder produces a loud beeping noise when the automobile is placed in reverse. As on many trucks and heavy equipment, it serves as a warning to those around the vehicle that it is about to back up. The schematic diagram and parts list for the automobile backup sounder are shown in Fig. 4-2. The backup sounder is divided into two sections in Fig. 4-2. The power supply/decoupler section furnishes 5 V of regulated dc to the sounder circuit. The power supply also decouples the sounder circuit from transient voltage spikes in the automotive electrical system.

The sounder circuit is a low-frequency oscillator. The input to the sounder section is the connection to the backup light. When the backup light is activated, the positive voltage at the base of Q_1 turns on the transistor. This applies nearly the full 5 V across the LM3909 IC circuit. The LM3909 oscillator IC alternately makes pin 4 positive and near ground, turning on and off transistor Q_2. This, in turn, makes the piezo buzzer beep. The tone

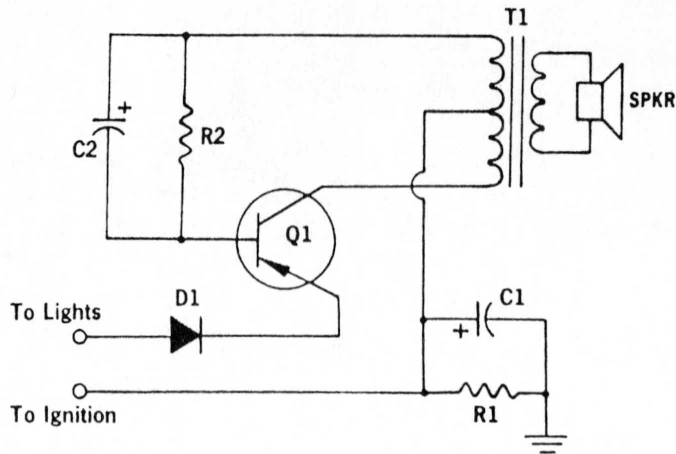

PARTS LIST

C_1	33-μF, 25-V electrolytic capacitor
C_2	0.22-μF, 200-V capacitor
D_1	1N4005 silicon diode, 1 A, 600 PIV
Q_1	2N4126 PNP transistor (similar to Radio Shack 276-2023)
R_1	680-Ω, ½-W resistor
R_2	15-kΩ, ½-W resistor
SPKR	8-Ω speaker
T_1	1-kΩ (center-tapped) to 8-Ω secondary audio output transformer (similar to Radio Shack 273-1380)

FIG. 4-1 Automatic headlight reminder. (Used by permission of Mode Electronics.)

adjust control R_5 changes the frequency of the beeping.

DIGITAL AUTOMOTIVE TUNEUP METER

The digital tuneup meter in Fig. 4-3 can measure engine speed from 0 to 9999 rpm, dwell angles from 0 to 99.9 degrees (0 to 99.9°), and dc voltage levels from −9.9 to +99.9 V. This tuneup meter can be used with most four-, six-, or eight-cylinder factory ignition systems.

The RCA chip set IC_1 and IC_2 is basically a digital voltmeter with full-scale sensitivity of 999 mV. IC_2 and the display-enable transistors Q_1, Q_2, and Q_3 drive the LED seven-segment displays DIS_1, DIS_2, and DIS_3.

For measuring VOLTS, the input potential is connected to J_3 and J_4. The 100:1 voltage divider R_{14} and R_{15} sends the divided voltage through switch S_{1b} to the input of the A/D converter IC_1. The voltage is converted to digital form and displayed on DIS_3, DIS_2, and DIS_1. The decimal point (such as 12.9) is placed on DIS_1 by switch S_{1c}.

The TACH and DWELL modes share the wave-shaping section C_8, R_{10}, D_2, R_8, and Q_4 shown in Fig. 4-3. In the DWELL mode, the signal is routed through switch S_{1a} to the R_{16} through the R_{20} and C_9 group, which averages the value of the distributor duty cycle (changes the pulses to an average dc voltage). The voltage passes through switches S_{2b} and S_{1b} to input pin 11 of the A/D converter (IC_1).

In the TACH mode, the one-shot multivibrator IC_3 is triggered each time the points close. The signal is routed through R_9 to switch S_{1a} to the averaging section. The average dc voltage is applied to the input of the A/D converter. The digital tuneup meter in Fig. 4-3 is powered by a 9-V internal battery (alkaline or ni-cad) or may be attached to a 12-V battery by means of jack J_1.

PARTS LIST

B_1	Piezo buzzer (Radio Shack 273-060)
C_1	2000-μF, 25-V electrolytic capacitor
C_2	10-μF, 25-V electrolytic capacitor
C_3	0.1-μF, 50-V disk capacitor
C_4	100-μF, 25-V electrolytic capacitor
D_1	1N4004 silicon diode, 1 A, 400 PIV
D_2,D_3	1N4001 silicon diode, 1 A, 50 PIV
F_1	¼-A slow-blow fuse
IC_1	7805 or LM340-5 5-V regulator IC
IC_2	LM3909 oscillator IC
Q_1,Q_2	2N2222 NPN transistor (or similar NPN)
R_1	50-Ω, 1-W resistor
R_2	27-kΩ, ¼-W resistor
R_3	470-Ω, ¼-W resistor
R_4	33-Ω, ¼-W resistor
R_5	5-kΩ, linear-taper potentiometer
R_6	10-kΩ, ¼-W resistor
S_1	SPST switch
Z_1,Z_2	12-V, 1-W zener diode

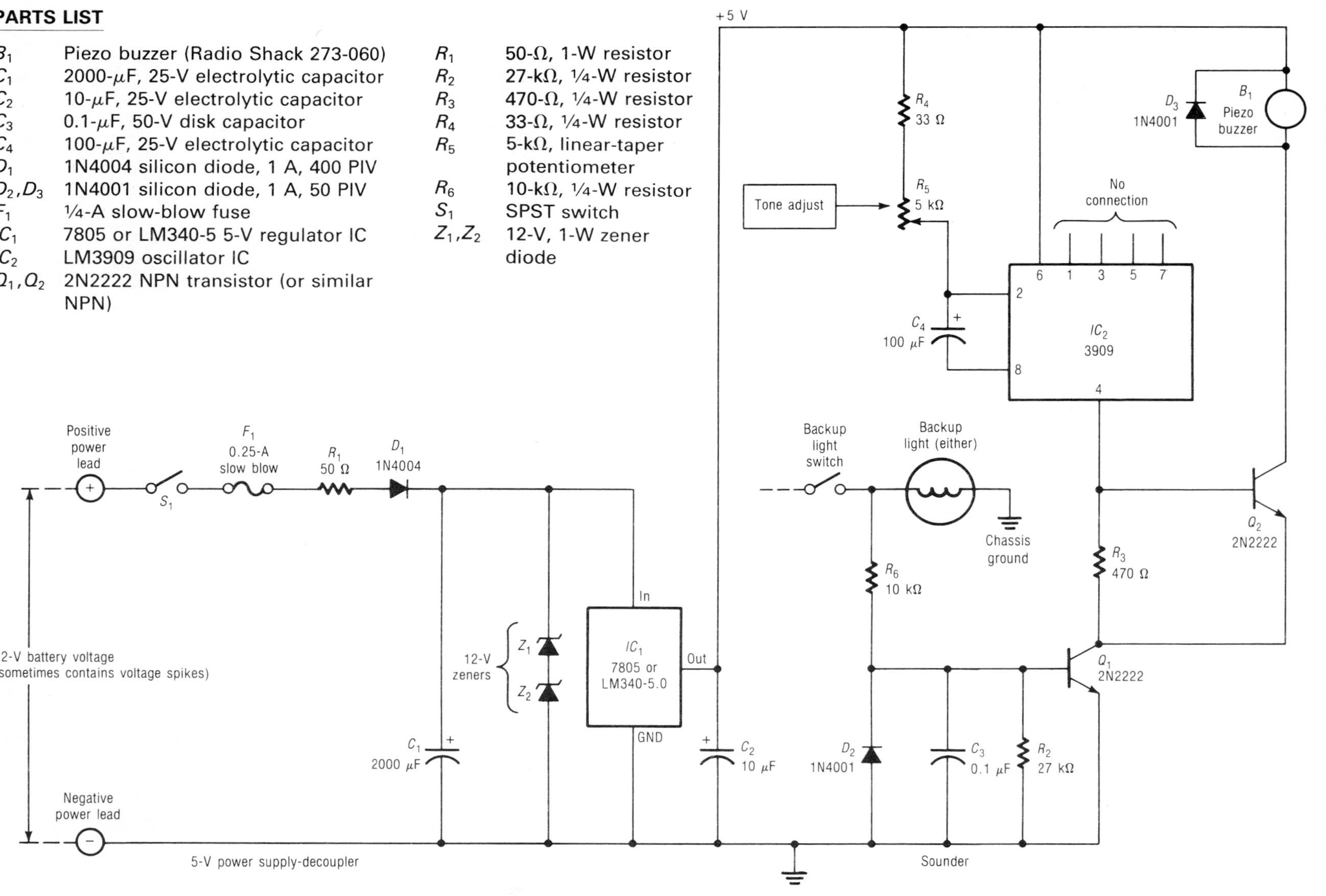

FIG. 4-2 Automobile backup sounder. (Michael Gannon, *Workbench Guide to Semiconductor Circuits and Projects,* Prentice-Hall, New Jersey, 1982, pp. 136–137. Used by permission of Prentice-Hall, Inc.)

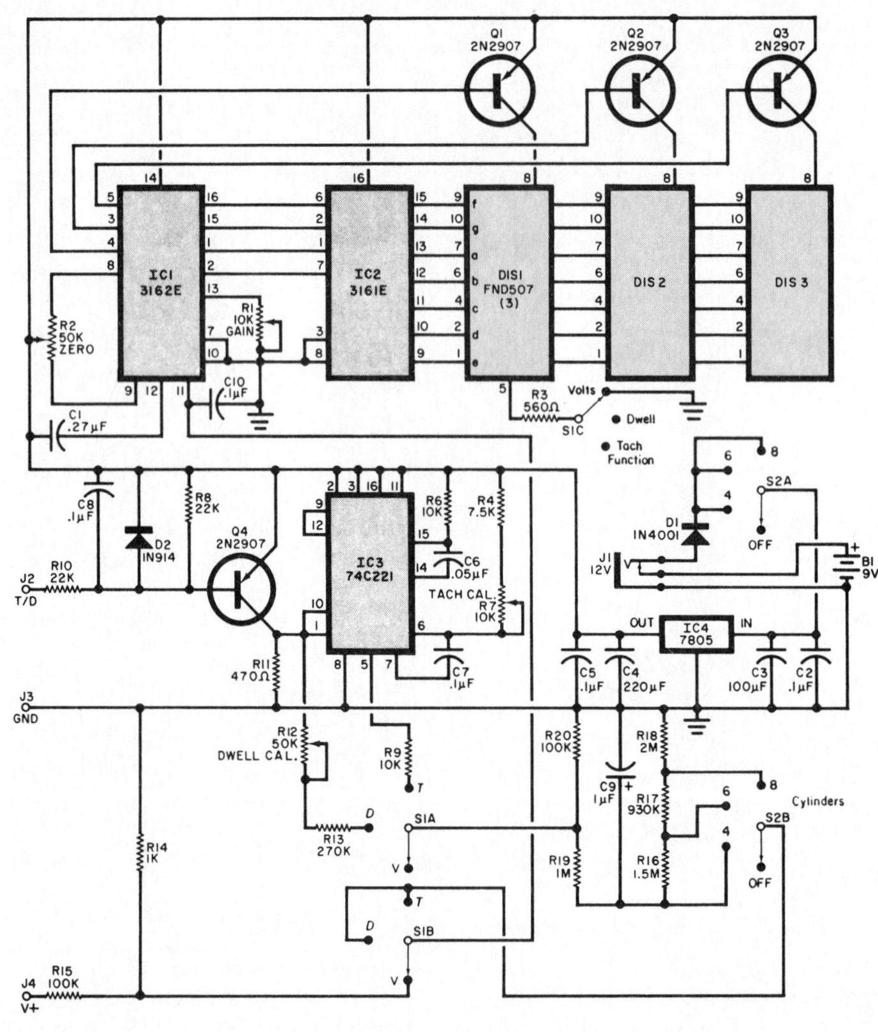

PARTS LIST

B_1	9-V battery (six AA alkaline or ni-cad cells)
C_1	0.27-μF, 100-V Mylar capacitor
C_2,C_5,C_8,C_{10}	0.1-μF, 25-V disk capacitor
C_3	100-μF, 16-V electrolytic capacitor
C_4	220-μF, 16-V electrolytic capacitor
C_6	0.05-μF, 100-V Mylar capacitor
C_7	0.1-μF, 100-V Mylar capacitor
C_9	1.0-μF, 100-V Mylar capacitor
D_1	1N4001 silicon diode, 1 A, 50 PIV
D_2	1N914 signal diode
DIS_1,DIS_2,DIS_3	Common-anode seven-segment LED display (Fairchild FND507 or similar)
IC_1	CA3162E A/D converter IC (RCA)
IC_2	CA3161E seven-segment decoder-driver IC (RCA)
IC_3	74C221 dual monostable multivibrator IC
IC_4	7805, 5-V regulator IC
J_1	Miniature closed-circuit jack
J_2,J_3,J_4	Banana jack
Q_1 through Q_4	2N2907A PNP transistor (or similar PNP)
R_1,R_7	10-kΩ trimmer potentiometer
R_2,R_{12}	50-kΩ trimmer potentiometer
R_3	560-Ω, 1/4-W resistor
R_4	7.5-kΩ, 1/4-W resistor
R_6,R_9	10-kΩ, 1/4-W resistor
R_8,R_{10}	22-kΩ, 1/4-W resistor
R_{11}	470-Ω, 1/4-W resistor
R_{13}	270-kΩ, 1/4-W resistor
R_{14}	1-kΩ, 1 percent resistor
R_{15},R_{20}	100-kΩ, 1 percent resistor
R_{16}	1.5-MΩ, 1 percent resistor
R_{17}	930-kΩ, 1 percent resistor
R_{18}	2-MΩ, 1 percent resistor
R_{19}	1-MΩ, 1 percent resistor
S_1	Three-pole, four-position, rotary switch
S_2	Two-pole, four-position, rotary switch

FIG. 4-3 Digital tuneup meter. (Reprinted from *Popular Electronics*. Copyright © May 1982, Ziff-Davis Publishing Company.)

IDIOT LIGHT AUDIO BACKUP

The circuit shown in Fig. 4-4 is an audio backup for the oil pressure and temperature "idiot" lights found on many automobiles. The circuit will sound an alarm and light the LED D_3 if either the oil or the temperature lamps are activated by their sensors (switches in Fig. 4-4). The idiot light audio backup circuit does have a convenient time delay of about 10 to 20 seconds, so the alarm will not sound when the car is routinely started and turned off.

A regulated power supply/decoupler circuit is shown at the left in Fig. 4-4. It is based on a +5-V voltage regulator IC_1.

The 4011 CMOS NAND gate IC_2 senses whether either the oil or temperature switches are closed (either pins 1 or 2 are grounded). If either one or both inputs (pins 1 and 2) are grounded, the NAND gate in the 4011 IC raises output pin 3 to near +5 V. Capacitor C_3 and resistor R_4 form the time delay circuit. If the top of C_3 charges to about +2.7 V, the voltage at pins 10 and 11 of IC_2 goes high (+5 V). This turns on transistor Q_1, thus activating the buzzer and LED.

The wiring to the oil lamp/switch and temperature lamp/switch are diagrammed near the top in Fig. 4-4. The zener diodes Z_3 and Z_4 make the input voltages from the auto electrical system compatible with that of the 4011 IC. The sound portion of the alarm may be turned off by using switch S_1.

MUSICAL HORN

The musical horn can be used in a boat or automobile or at home to generate a preprogrammed tune. The circuit is powerful enough to drive a 10-watt (10-W) speaker or horn.

A block diagram of the musical horn is shown in Fig. 4-5. The song is stored in the song PROM IC while the tones are produced by the tone generator section. When the push-button switch is pressed, voltage is applied to all the circuits (+5 V to ICs and +12 V to the amplifier). The one-shot reset clears the 8-bit binary counter to zero and the slow clock sequences the song PROM memory locations. The song PROM specifies the notes or rests to the comparator. The tone generator (clock-counter-tone PROM) along with the comparator and delay flip-flop produce the proper pitch specified by the song PROM's output. This tone is sent to the audio amplifier and speaker. The duration of the tone is also controlled by the song PROM. The specific ICs that perform the functions in the musical horn are also shown on the block diagram.

The schematic diagram and parts list for the musical horn are detailed in Fig. 4-6. The physical placement of the ICs on the schematic diagram in Fig. 4-6 correspond to the functional blocks in Fig. 4-5. The speed at which the song generator sequences through the notes is controlled by potentiometer R_1. Potentiometer R_5 controls the pitch of the tones generated by the tone generator section of the musical horn. The parts list for Fig. 4-6 specifies the values and wattage of resistor R_{10} based on the power ratings of the speaker or horn.

As an aid in gathering parts for the musical horn, Electronic Kits International, Inc., currently has the following available:

1. No. 882 musical horn parts kit
2. No. 782 musical horn pc board
3. No. EKI-0 IC (tone PROM)
4. No. EKI-1 IC ("La Cucaracha" song PROM)
5. No. EKI-2 IC ("Dixie" song PROM)

VEHICLE ALARM

The vehicle alarm in Fig. 4-7 may be used in automobiles, trucks, vans, or motor homes. This alarm is triggered by either a switch or by a change in the voltage in the vehicle's electrical system. This voltage spike in the electrical system may be caused by something as simple as the courtesy light turning on as a door is opened. The unit also features automatic alarm shutoff and automatic reset. The circuit in Fig. 4-7(a) is divided into the vehicle alarm and the siren sections.

If the auxiliary (AUX) input were grounded, the saturated transistor Q_1 would stop conducting. The collector of Q_1 would go positive, triggering the silicon controlled rectifier SCR_1 into conduction. Pin 2 of IC_1 would be grounded, turning on the 555 timer (wired as a one-shot multivibrator). The output of IC_1 (pin 3) would go high for about 3 minutes, turning on transistor Q_2. Relay RL_1 would be activated, causing the siren to sound. If the alarm is turned off midway through its 3-minute cycle, IC_1 must be reset when power is restored. This is accomplished by R_7 and C_4. When power is first turned on, capacitor C_4 is not charged, so the

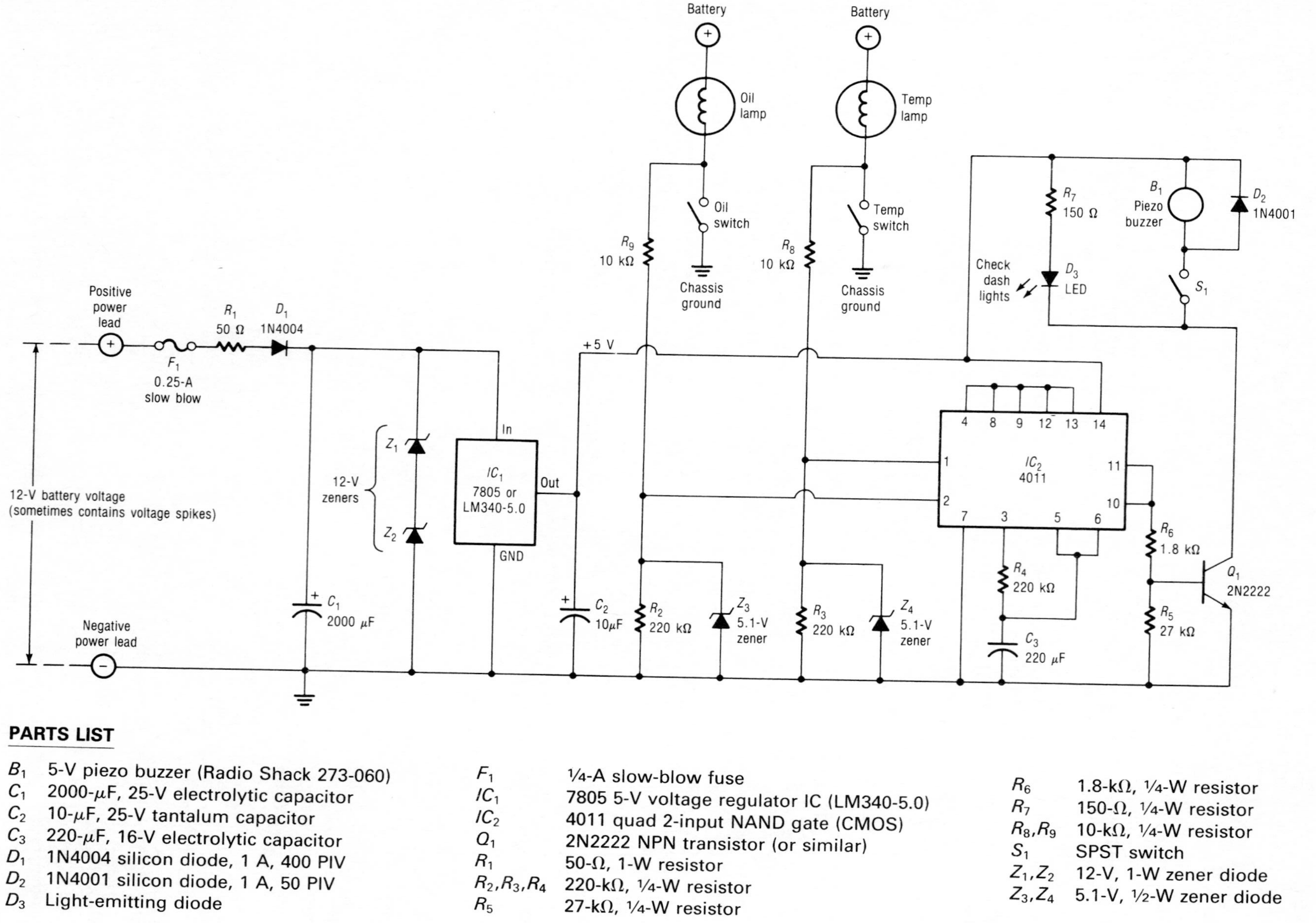

PARTS LIST

B_1	5-V piezo buzzer (Radio Shack 273-060)
C_1	2000-μF, 25-V electrolytic capacitor
C_2	10-μF, 25-V tantalum capacitor
C_3	220-μF, 16-V electrolytic capacitor
D_1	1N4004 silicon diode, 1 A, 400 PIV
D_2	1N4001 silicon diode, 1 A, 50 PIV
D_3	Light-emitting diode
F_1	¼-A slow-blow fuse
IC_1	7805 5-V voltage regulator IC (LM340-5.0)
IC_2	4011 quad 2-input NAND gate (CMOS)
Q_1	2N2222 NPN transistor (or similar)
R_1	50-Ω, 1-W resistor
R_2,R_3,R_4	220-kΩ, ¼-W resistor
R_5	27-kΩ, ¼-W resistor
R_6	1.8-kΩ, ¼-W resistor
R_7	150-Ω, ¼-W resistor
R_8,R_9	10-kΩ, ¼-W resistor
S_1	SPST switch
Z_1,Z_2	12-V, 1-W zener diode
Z_3,Z_4	5.1-V, ½-W zener diode

FIG. 4-4 Idiot light audio backup. (Michael Gannon, *Workbench Guide to Semiconductor Circuits and Projects*, Prentice-Hall, New Jersey, 1982, pp. 141–142, 229–230. Used by permission of Prentice-Hall, Inc.)

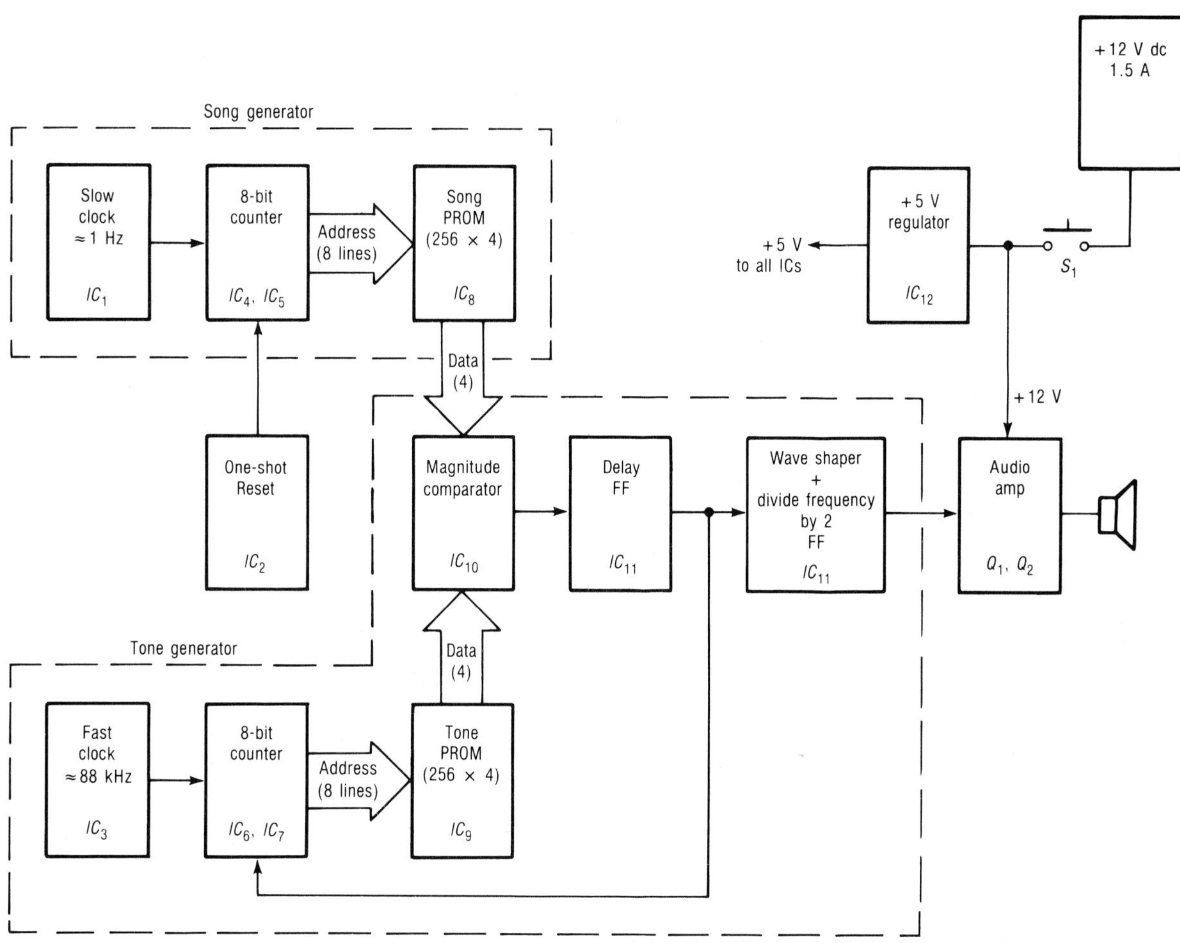

FIG. 4-5 Block diagram of the musical horn.

reset (pin 4) of IC_1 is low or grounded. As capacitor C_4 becomes charged, pin 4 of IC_1 becomes high and the vehicle alarm is ready to operate. A voltage spike in the vehicle electrical system will be coupled through C_1 to the base of Q_1. If this voltage spike is longer than 5 milliseconds, Q_1 will turn off, gating SCR_1 and turning on the one-shot (555 timer). The one-shot will saturate Q_2 for several minutes, activating the relay and siren.

The basic tone of the siren shown in Fig. 4-7(*a*) is produced by the 555 timer (IC_2) wired as a free-running multivibrator. The unijunction transistor Q_3 is connected as a ramp generator with its output (from emitter) being fed into pin 5 of IC_2. This low-frequency signal modulates the basic tone of the timer IC. The modulated tone signal coming from pin 3 of IC_2 is sent to driver transistor Q_4. The power transistor operates as an amplifier, switching up to 1 A through the 5- to 10-W speaker. This unit produces a very loud siren sound.

Installation details of a typical system in a vehicle are shown in Fig. 4-7(*b*). The numbers on the panel correspond to numbers on the schematic diagram. The ON/OFF switch may be a key switch mounted on the outside of the vehicle or a hidden switch for deactivating the alarm before entering. The auxiliary switches are optional and are activated by opening such components as the hood, doors, or hatch or trunk lid.

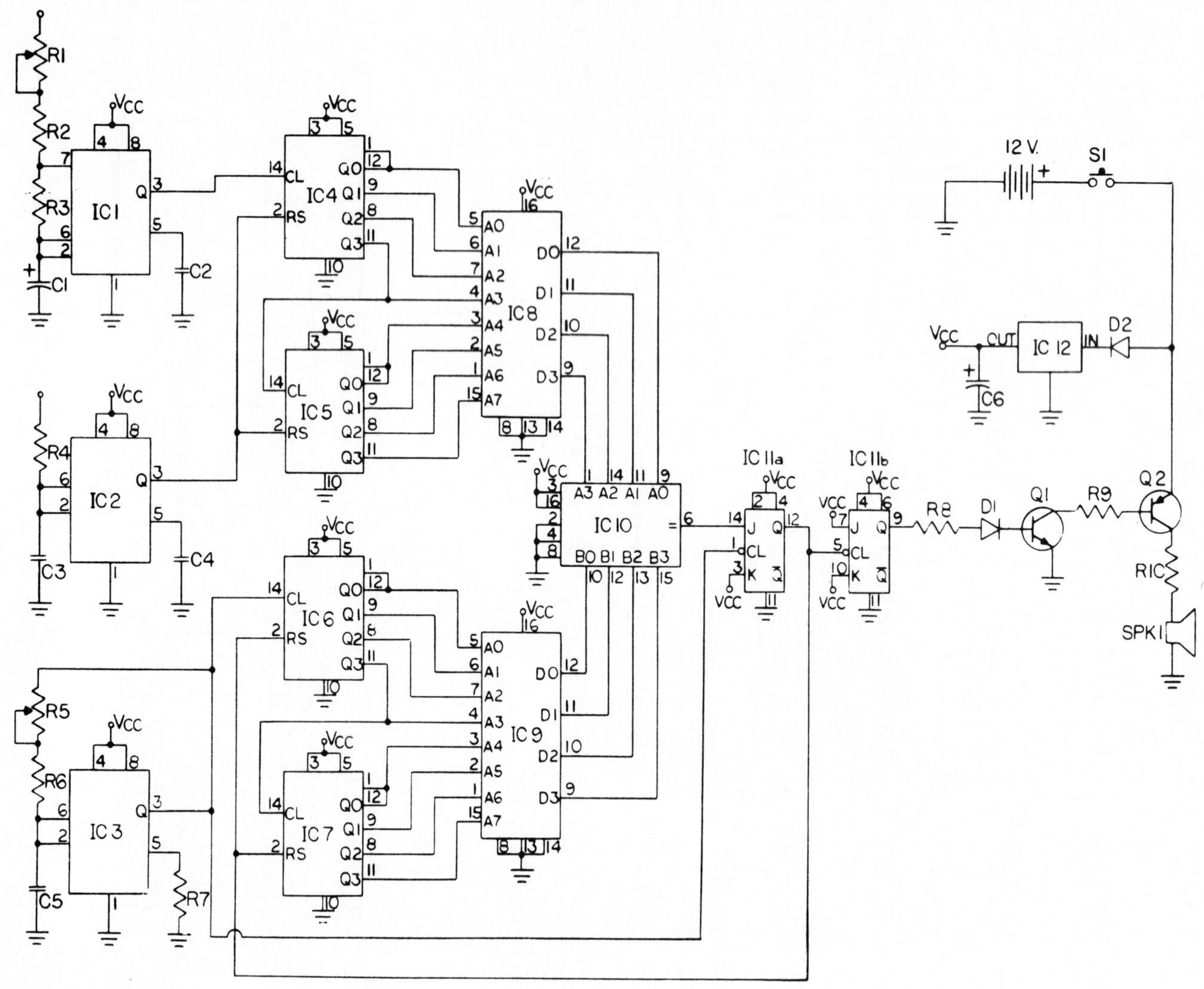

PARTS LIST

B_1	12-V automotive battery (or 12-V dc supply that will furnish 1.5 A)
C_1	1-μF, 25-V electrolytic capacitor
C_2,C_4,C_5	0.01-μF, 50-V disk capacitor
C_3	0.1-μF, 50-V disk capacitor
C_6	10-μF, 25-V electrolytic capacitor
D_1,D_2	1N4003 silicon diode, 1 A, 200 PIV
IC_1,IC_2,IC_3	555 timer IC
IC_4,IC_5,IC_6,IC_7	7493 binary counter IC
IC_8	Song PROM IC (two songs are currently available from Electronic Kits International, Inc.)
IC_9	Tone PROM IC (currently available as EKI-0 IC from Electronic Kits International, Inc.)
IC_{10}	7485 4-bit magnitude comparator IC
IC_{11}	7473 *J-K* flip-flop IC
IC_{12}	LM309 +5-V voltage regulator IC
Q_1	2N3904 NPN transistor
Q_2	2N301 PNP power transistor
R_1	100-kΩ potentiometer
R_2,R_7	10-kΩ, ½-W resistor
R_3	1-kΩ, ½-W resistor
R_4	100-kΩ, ½-W resistor
R_5	500-Ω potentiometer
R_6,R_9	100-Ω, ½-W resistor
R_8	330-Ω, ½-W resistor
R_{10}	Optional limiting resistor*
S_1	Normally open push-button switch
SPK_1	8-Ω speaker or horn (up to 10 W)

* If a 10-W speaker is used, R_{10} is not needed. If a 5-W speaker is used, $R_{10} = 3\ \Omega$, 2 W. If a 1-W speaker is used, $R_{10} = 16\ \Omega$, 2 W. If a ½-W speaker is used, $R_{10} = 27\ \Omega$, 2 W.

FIG. 4-6 Musical horn. (Used by permission of Electronic Kits International, Inc.)

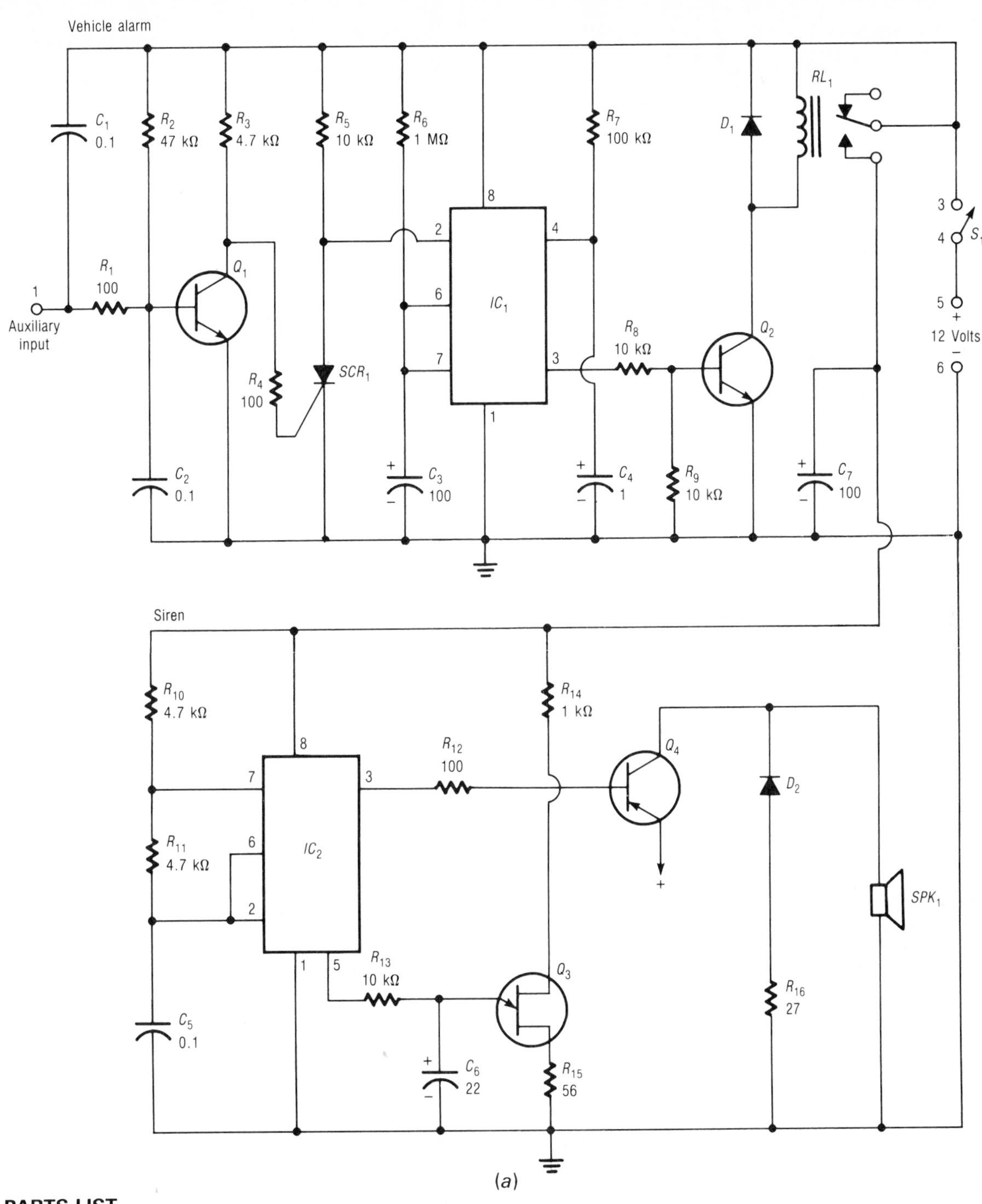

(a)

PARTS LIST

C_1,C_2,C_5	0.1-μF, 50-V disk capacitor
C_3,C_7	100-μF, 16-V electrolytic capacitor
C_4	1-μF, 16-V electrolytic capacitor
C_6	22-μF, 16-V electrolytic capacitor
D_1,D_2	1N4001 silicon diode, 1 A, 50 PIV
IC_1,IC_2	555 timer IC
Q_1	2N3904 NPN transistor
Q_2	2N2222 NPN transistor
Q_3	2N4870 or 2N4871 unijunction transistor (ECG 6410)
Q_4	2N6109 or TIP32 PNP power transistor (Radio Shack 276-2027)
R_1,R_4,R_{12}	100-Ω, ½-W resistor
R_2	47-kΩ, ½-W resistor
R_3,R_{10},R_{11}	4.7-kΩ, ½-W resistor
R_5,R_8,R_9,R_{13}	10-kΩ, ½-W resistor
R_6	1-MΩ, ½-W resistor
R_7	100-kΩ, ½-W resistor
R_{14}	1-kΩ, ½-W resistor
R_{15}	56-Ω, ½-W resistor
R_{16}	27-Ω, ½-W resistor
RL_1	12-V coil, relay
S_1	SPST switch (hidden on outside of vehicle or key operated)
SCR_1	C103Y silicon controlled rectifier (or similar SCR)
SPK_1	8-Ω, 10-W speaker or horn

FIG. 4-7 Vehicle alarm installation: a typical system. (Used by permission of Robert Delp Electronics.)

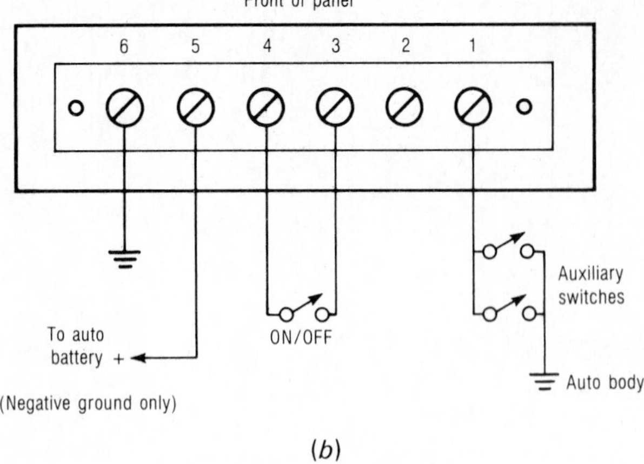

(*b*)

FIG. 4-7 ***(Continued)***

INTERIOR LIGHT DELAY

The delay circuit shown in Fig. 4-8 will keep the interior dome light on for 10 to 20 seconds after the vehicle's door is closed. This is enough time to find your ignition key and start the engine.

The 555 timer IC is wired as a one-shot multivibrator in this delay circuit. As the door opens, switch S_1 is closed lighting the interior lights and applying a negative-going pulse to the trigger input (pin 2) of the IC starting the one-shot multivibrator. The output of the 555 timer (pin 3) goes high for about 15 seconds, turning on transistor Q_1, causing the dome unit to light even after the door is closed opening switch S_1. After the 15-second delay, output pin 3 of the IC goes back low, turning off Q_1. The time delay can be adjusted by changing the values of R_1 and C_1.

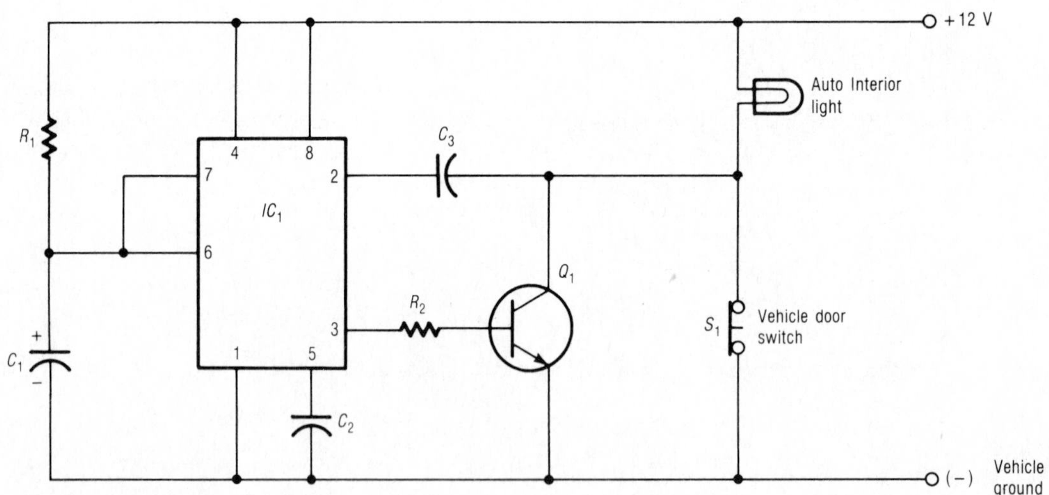

PARTS LIST

C_1	3.3-μF, 25-V electrolytic capacitor
C_2,C_3	0.01-μF, 50-V disk capacitor
IC_1	555 timer IC
Q_1	NPN transistor (Radio Shack 276-2009)
R_1	3.9-MΩ, ½-W resistor
R_2	680-Ω, ½-W resistor

FIG. 4-8 Interior light delay.

chapter 5

Battery Charging Circuits

AUTOMOTIVE BATTERY CHARGER

A simple 12-V automotive battery charger is shown in Fig. 5-1. This unit will have a charge rate of about 2 to 6 A depending whether the battery is charged. The red output clip is placed on the positive (+), while the black clip is attached to the negative (−) terminal of the automotive battery. This charger is designed to bring the normal discharged battery up to full charge in 8 to 16 hours. This unit is not for use with small lawn equipment and motorcycle batteries.

In the battery charger circuit in Fig. 5-1, household 117 V ac is stepped down to about 15 V ac by transformer T_1. The neon lamp L_1 and series resistor R_1 act as a pilot lamp. Rectifier $Rect_1$ is a full-wave bridge rectifier. Ammeter M_1 monitors the battery charging rate, and breaker CB_1 protects the circuit if the output leads are short-circuited. It is common to use color-coded clips at the output. If the available transformer has a secondary voltage of 13 to 16 V ac, it will work but the charging rate will be either lower or higher than expected. The rectifier module must be attached to a suitable heat sink.

The high-voltage wiring must be done carefully to make sure that it is completely isolated from the enclosure. The green ground wire on the three-pronged plug must be attached to the metal chassis. Use proper strain relief techniques where cords enter and exit the case.

WARNINGS: when charging lead-acid batteries:

1. To avoid an explosion, do not allow flames or sparks near a charging battery.

2. To prevent personal injury or damage to clothing, do not allow battery fluid to contact eyes, skin, or fabrics.

CURRENT-LIMITED 6-V BATTERY CHARGER

The battery charger shown in Fig. 5-2 can be used on most 6-V lead-acid batteries. The red lead is connected to the positive terminal of the battery and the black lead to the negative.

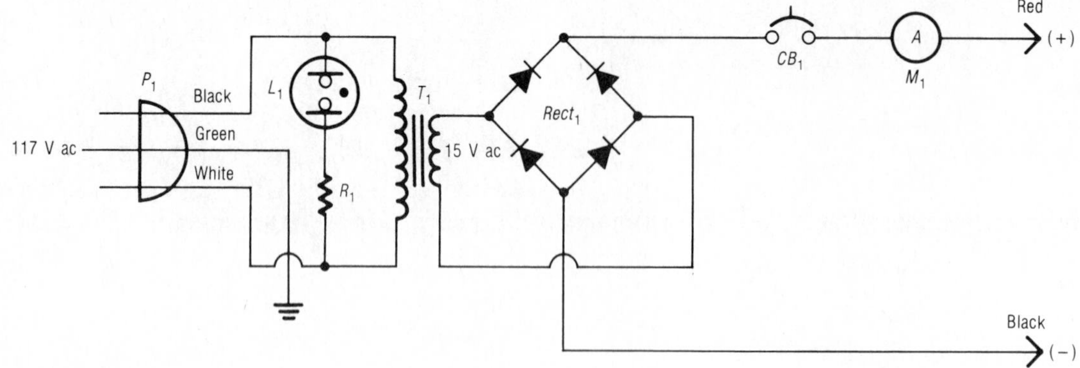

PARTS LIST

CB_1	10-A circuit breaker
L_1	NE-2 neon lamp
M_1	0- to 10-A ammeter
P_1	Three-pronged ac plug
R_1	47-kΩ, ½-W resistor
$Rect_1$	25-A, 200-V rectifier module
T_1	15-V, 5-A secondary, power transformer

FIG. 5-1 Automotive battery charger.

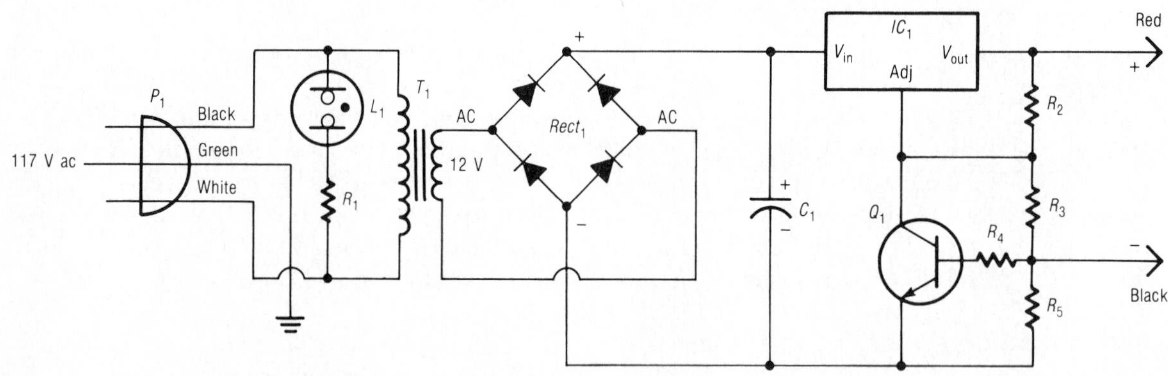

PARTS LIST

C_1	1000-μF, 25-V electrolytic capacitor
IC_1	LM338 5-A power voltage regulator IC
L_1	*NE*-2 neon lamp
P_1	Three-pronged ac plug
Q_1	2N2222 NPN transistor
R_1	47-kΩ, ½-W resistor
R_2	120-Ω, ½-W resistor
R_3	1-kΩ, ½-W resistor
R_4	100-Ω, ½-W resistor
R_5	0.2-Ω, 5-W resistor (0.2 Ω sets maximum charge current to 3 A)
$Rect_1$	10-A, 50-PIV rectifier module
T_1	12-V, 3-A power transformer

FIG. 5-2 Current-limited 6-V battery charger. (Used by permission of National Semiconductor Corporation.)

In the battery charger circuit in Fig. 5-2, household 117 V ac is stepped down to about 12 V ac by transformer T_1. The neon lamp L_1 and series resistor R_1 act as a pilot lamp. Rectifier $Rect_1$ is a full-wave bridge rectifier. Regulator IC_1 and associated circuitry is used to limit the output current to about 3 A. The value of resistor R_5 can be adjusted slightly to change the maximum current output.

The high-voltage wiring must be done carefully to make sure that it is completely isolated from the enclosure. The green ground wire on the three-pronged plug must be attached to the metal chassis. Remember to use proper strain relief techniques where cords enter and exit the case.

WARNINGS: when charging lead-acid batteries:

1. To avoid an explosion, do not allow flames or sparks near a charging battery.
2. To prevent personal injury or damage to clothing, do not allow battery fluid to contact eyes, skin, or fabrics.

NI-CAD BATTERY REJUVENATOR-CHARGER

Many defective nickel-cadmium (ni-cad) batteries can be rejuvenated and then recharged by using the circuit shown in Fig. 5-3. Some nicad batteries will refuse to take a charge after

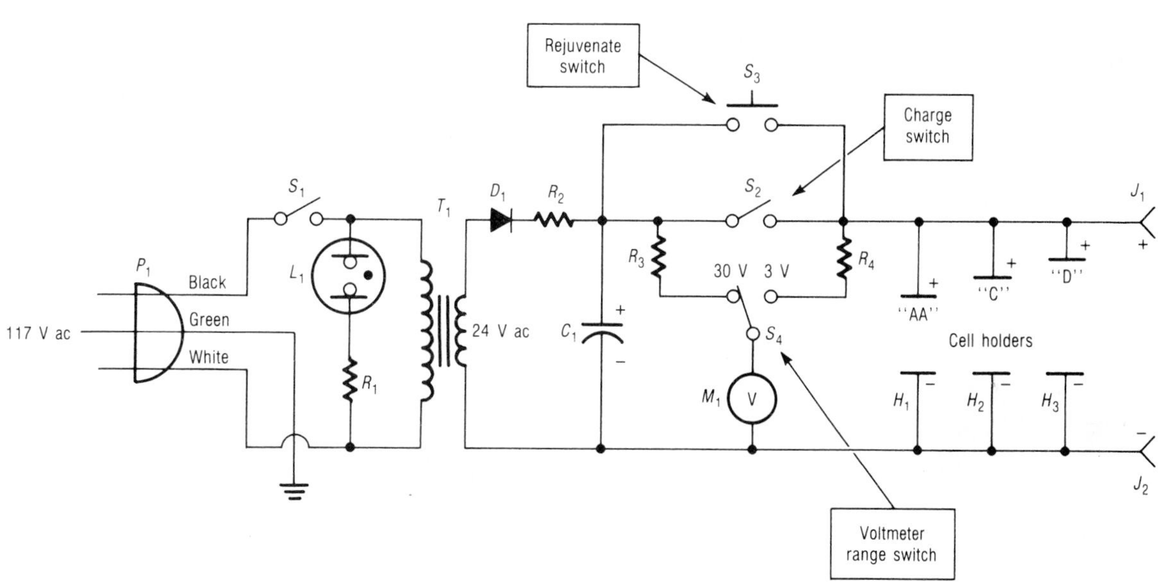

PARTS LIST

C_1	5000-μF, 50-V electrolytic capacitor
D_1	1N5402 silicon diode, 3 A, 200 PIV
H_1	AA (penlight) battery holder
H_2	C-size battery holder
H_3	D-size battery holder
J_1	Banana jack (red)
J_2	Banana jack (black)
L_1	NE-2 neon lamp
M_1	0 to 1-mA dc panel meter (similar to Radio Shack 270-1752)
P_1	Three-pronged ac plug
R_1	33-kΩ, ½-W resistor
R_2	300-Ω, 5-W resistor
R_3	30-kΩ, 1 percent, ½-W resistor
R_4	3-kΩ, 1 percent, ½-W resistor
S_1,S_2	SPST switch
S_3	Normally open push-button switch
S_4	SPDT switch (spring return)
T_1	24-V ac, 1-A power transformer

FIG. 5-3 Ni-cad battery rejuvenator-charger. (Carl Grolle and Michael Girosky, *Workbench Guide to Electronic Projects You Can Build in Your Spare Time,* Parker, New York, 1981, pp. 169–173. Used by permission of Parker Publishing Company, Inc.)

a time. Some of these can be restored by discharging a heavy momentary current through them. This clears any microscopic short circuits that may have developed inside the battery.

The ni-cad battery rejuvenator-charger develops about 25 to 30 dc volts across the large capacitor C_1. This voltage would be monitored by voltmeter M_1 and R_3. To use the unit to "zap" or rejuvenate a faulty battery, place the battery in one of the cell holders with switch S_2 open. When the voltmeter shows its highest voltage (about 30 V), press the rejuvenate switch S_3. This will discharge C_1 through the switch and the cell. Try zapping the cell repeatedly. Now close the charge switch S_2 and hold the voltmeter range switch in the 3-V position. If the short in the cell has been cleared, the voltmeter will gradually increase to about 1.25 V as the cell becomes charged. If the voltage does not increase, the cell is not taking a charge.

To charge a ni-cad battery, place it in its appropriate cell holder, power the circuit, and close the charge switch. Use the 3-V voltmeter range switch to check whether the voltage increases. A normal cell's voltage will gradually increase to 1.25 V as it becomes charged. Several cells can be charged at one time in this unit. Most ni-cad batteries can be recharged in 8 to 16 hours with this unit.

The high-voltage wiring must be done carefully to make sure that it is completely isolated from the enclosure. The green ground wire on the three-pronged plug must be attached to the metal chassis. Use proper strain relief techniques where cords enter and exit the case.

NI-CAD BATTERY CHARGER

A versatile ni-cad battery charger is shown in Fig. 5-4. With this charger, 1 to 12 ni-cad batteries (AA, C, or D sizes) can be charged at a

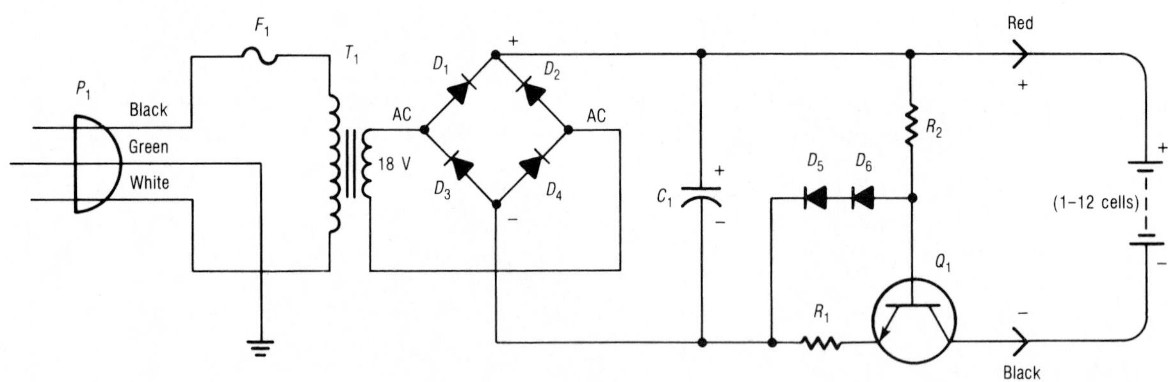

PARTS LIST

C_1	470-μF, 35-V electrolytic capacitor
D_1 through D_6	1N4002 silicon diode, 1 A, 100 PIV
F_1	1-A fuse
Q_1	2N3055 NPN power transistor
R_1	24-Ω, ½-W resistor
R_2	2700-Ω, ½-W resistor
T_1	18-V, 2-A power transformer (similar to Radio Shack 273-1515)

FIG. 5-4 Ni-cad battery charger. (John Edwards, *Exploring Electricity & Electronics with Projects,* TAB Books, Pennsylvania, 1983, pp. 53, 55. Used by permission of TAB Books, Inc.)

single time. Other ni-cad batteries with voltages of up to about 15 V (such as 9-V batteries) can also be charged. The circuit produces a near constant charging current of about 30 mA regardless of how many cells are connected. Most ni-cad batteries can be fully charged in 8 to 20 hours.

The 117-V ac input to the ni-cad battery charger must be safely isolated from the enclosure, and the grounded prong on the plug must be connected to the metal case. Suitable strain relief must be provided where cords enter and exit the enclosure.

TRICKLE CHARGER

The 12-V trickle charger shown in Fig. 5-5 is useful on lawn equipment and in motorcycle, and even automotive, lead-acid batteries. This unit will charge at a very low rate (0.1 to 0.3 A) and therefore may be left connected to most batteries continuously without harm.

The trickle charger is simple and very inexpensive to build, but some safety precautions must be taken. The 117 V ac must be safely isolated from the enclosure, and the grounded prong on the plug must be connected to the metal case. Suitable strain relief must be provided where cords enter and exit the enclosure. It is customary to use color-coded clips at the end of the cables on a battery charger (red for positive and black for negative).

WARNINGS: When charging lead-acid batteries:

1. To avoid an explosion, do not allow flames or sparks near a charging battery.

2. To prevent personal injury or damage to clothing, do not allow battery fluid to contact eyes, skin, or fabrics.

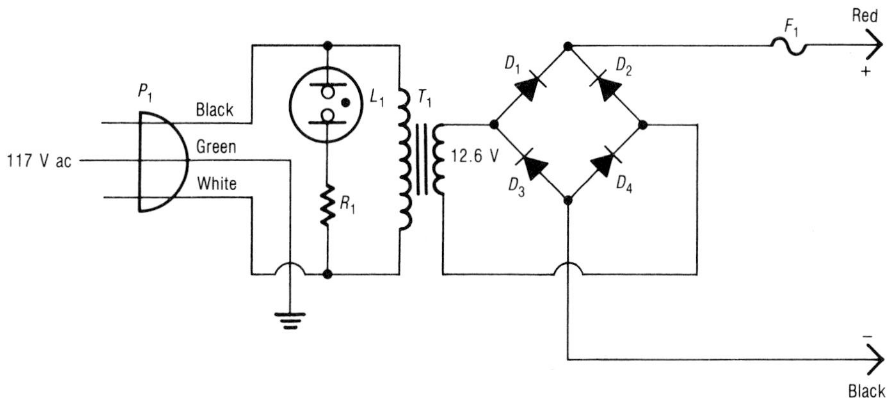

PARTS LIST

D_1, D_2, D_3, D_4	1N4003 silicon diode, 1 A, 200 PIV
F_1	½-A fuse
L_1	NE-2 neon lamp
P_1	Three-pronged ac plug
R_1	47-kΩ, ½-W resistor
T_1	12.6-V, 300-mA power transformer (similar to Radio Shack 273-1385)

FIG. 5-5 Trickle charger for 12-V batteries.

chapter 6

Camping and Outdoor Circuits

BUG SHOO

The simple "bug-shoo" circuit in Fig. 6-1 is an electronic insect repeller for the camper who wants to try a new gadget. The unijunction transistor Q_1 functions as an oscillator to drive the 8-Ω earphone. The oscillator frequency can be adjusted from about 5 to 25 kHz by using R_3. The frequency increases as the value of R_3 decreases. To check operation, listen to the earphone as the potentiometer R_3 is adjusted at the lower frequencies. The circuit is pocket size and uses very little power. A parts kit and a pc board are currently available for the bug-shoo circuit from Mode Electronics.

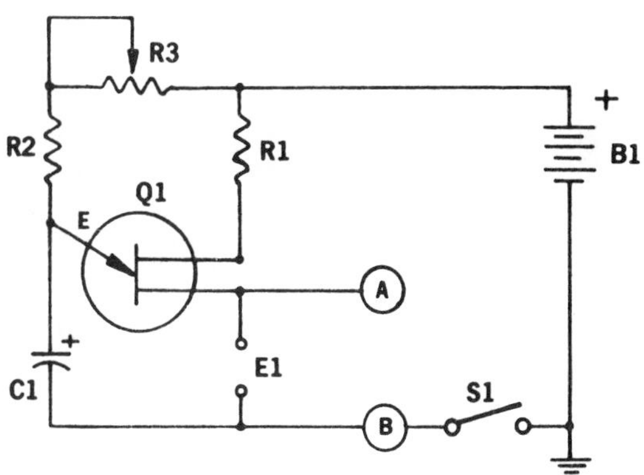

PARTS LIST

- B_1 9-V battery
- C_1 0.005-μF, 50-V disk capacitor
- E_1 8-Ω earphone (similar to Radio Shack 33-174)
- Q_1 2N2646 unijunction transistor (similar to ECG-6401)
- R_1 220-Ω, ¼-W resistor
- R_2 2.7-kΩ, ¼-W resistor
- R_3 50-kΩ, linear-taper trimmer potentiometer
- S_1 SPST switch

FIG. 6-1 Bug shoo. (Used by permission of Mode Electronics.)

ELECTRONIC FLYPAPER

The "bug zapper" or electronic flypaper circuit is shown in Fig. 6-2. The unit is placed in a case along with a black light (not shown on the schematic diagram). The contact grid may be bare tinned-copper wire wrapped tightly around a hollow enclosure. The grid wires

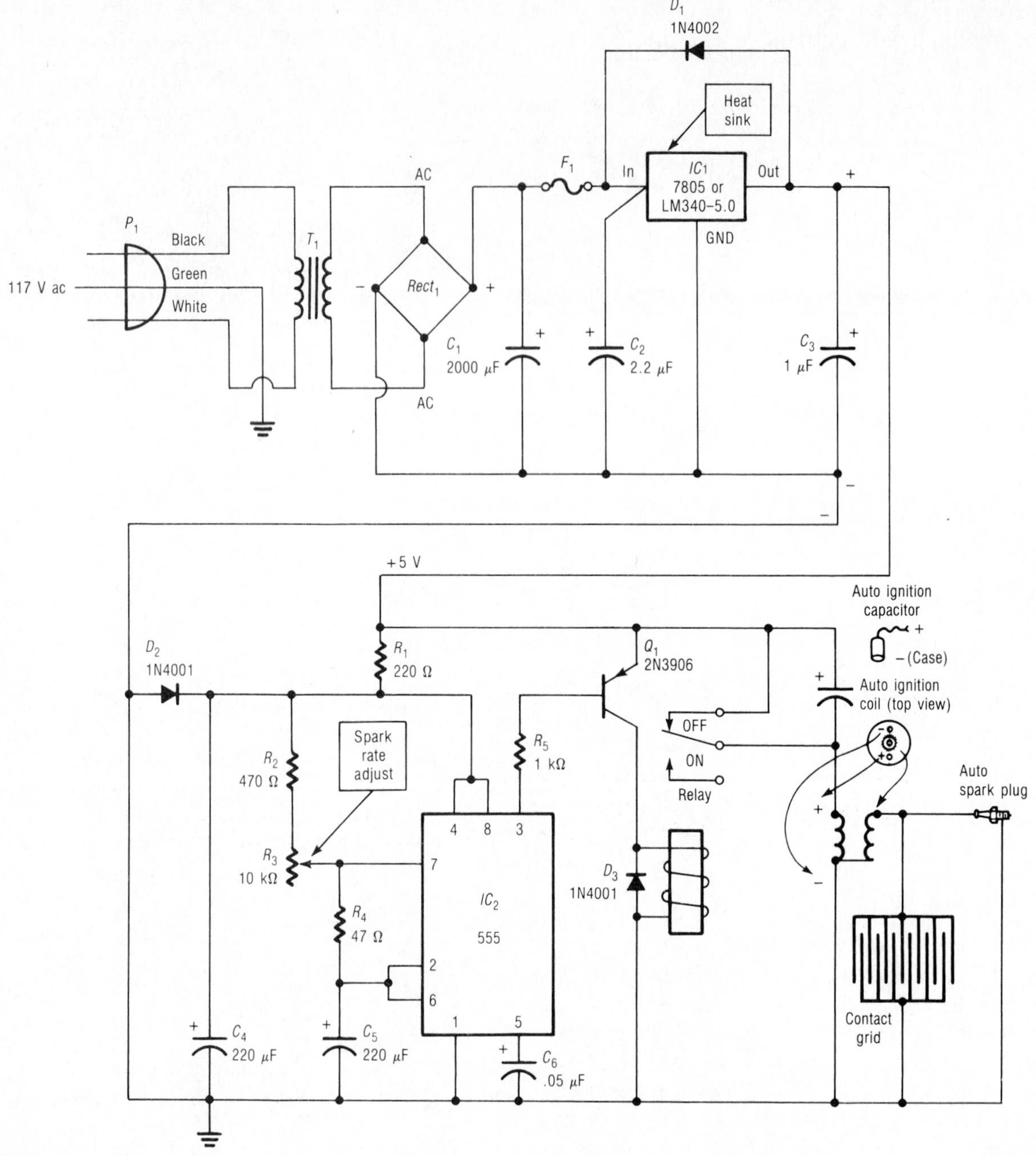

PARTS LIST

Auto ignition capacitor Any 12-V auto condenser
Auto ignition coil Any 12-V coil
Auto spark plug

C_1	2000-μF, 25-V electrolytic capacitor
C_2	2.2-μF, 25-V tantalum capacitor
C_3	1-μF, 25-V tantalum capacitor
C_4,C_5	220-μF, 25-V electrolytic capacitor
C_6	0.05-μF, 50-V disk capacitor
D_1	1N4002 silicon diode, 1 A, 100 PIV
D_2,D_3	1N4001 silicon diode, 1 A, 50 PIV
F_1	½-A fuse
IC_1	7805 or LM340—5.0 voltage regulator IC (similar to Radio Shack 276-1770)
IC_2	555 Timer IC
P_1	Three-pronged ac plug
Q_1	2N3906 PNP transistor (or similar)
R_1	220-Ω, ¼-W resistor
R_2	470-Ω, ¼-W resistor
R_3	10-kΩ, linear-taper potentiometer
R_4	47-Ω, ¼-W resistor
R_5	1-kΩ, ¼-W resistor
$Rect_1$	1-A, 50-PIV bridge rectifier module (similar to Radio Shack 276-1161)
Relay	SPDT 5-V relay (similar to Radio Shack 275-215)
T_1	12- or 12.6-V, 500-mA power transformer (similar to Radio Shack 273-1385)

FIG. 6-2 Electronic flypaper. (Michael Gannon, *Workbench Guide to Semiconductor Circuits and Projects*, Prentice-Hall, New Jersey, 1982, pp. 76–77, 200–201. Used by permission of Prentice-Hall, Inc.)

should be spaced 3 to 5 millimeters (mm) apart [about 1/8 to 3/16 inch (in)]. When the unit is turned on at night, insects attracted to the black-light bulb will contact the grid and be killed.

The electronic flypaper circuit in Fig. 6-2 can be divided into three parts: the power supply, the timer, and the high-voltage section. Transformer T_1 steps down 117 V ac to 12 V ac while the rectifier module changes it to direct current. Circuit IC_1 regulates the voltage to +5 V dc. Timer IC_2 is wired as a free-running multivibrator whose output frequency (0.5 to 20 Hz) is controlled by potentiometer R_3. When output pin 3 of the 555 timer pulses low for a short time, it momentarily turns on the transistor, thus activating the relay. When the relay snaps to the ON position, a pulse of current flows through the auto ignition coil primary winding. As the relay snaps open, a high-voltage pulse is produced in the coil, which is stepped up to several thousand volts in the secondary of the auto ignition coil. The high-voltage pulses will zap any small insects that might be near or on the grid. If nothing is near the grid, the high voltage is dissipated as it jumps the gap of the auto spark plug.

The high-voltage wiring must be done carefully to make sure that it is completely isolated from the enclosure. The green ground wire on the three-pronged plug must be attached to the metal chassis. Use proper strain relief techniques where cords enter and exit the case. Use test probe wire or other suitable wire on the high-voltage wiring.

Even though the shock experienced from touching the grid will be unpleasant, it should not be dangerous. However, it is well to install the bug zapper high enough to be out of the way of pets and children.

FISH CALLER

The inexpensive fish caller is a fun circuit for the person who fishes and who has everything. The fish caller shown in Fig. 6-3 emits an audible clicking sound from the small crystal earphone. In use, the unit is sealed in a jar or plastic bag and lowered into the water.

The fish caller circuit in Fig. 6-3 operates as an oscillator. Transistor Q_1 is alternately turned on and off, causing clicks to be emitted from the high-impedance earphone. The fre-

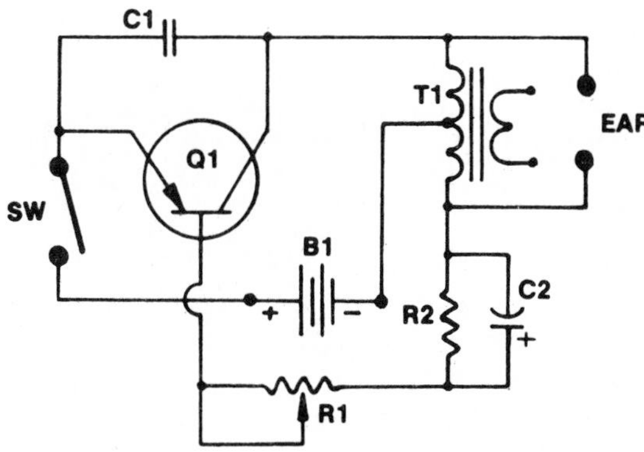

PARTS LIST

B_1	1.5-V battery
C_1	0.01-μF, 50-V disk capacitor
C_2	47-μF, 16-V electrolytic capacitor
EAR	Crystal earphone (high-impedance type)
Q_1	2SB22 PNP germanium transistor (similar to Radio Shack 276-2007)
R_1	4.7-kΩ trimmer potentiometer
R_2	27-kΩ, 1/2-W resistor
SW	SPST switch
T_1	1-kΩ center-tapped coil of audio output transformer

FIG. 6-3 Fish caller. (Used by permission of Mode Electronics.)

quency of the clicks can be adjusted by using potentiometer R_1.

A parts kit and a pc board are currently available for the fish caller circuit from Mode Electronics.

12-V BATTERY-OPERATED FLUORESCENT LIGHT

The fluorescent light circuit in Fig. 6-4 is operated from a 12-V automobile-type battery. The unit may be constructed as a portable unit or built into a commercial fixture in a mobile home or trailer to provide fluorescent lighting from 12 V dc. The fluorescent bulb used with this circuit is limited to 20 W.

When switch SW_1 is closed as shown in Fig. 6-4, transistors Q_1 and Q_2 alternately conduct, creating pulsations in the tapped primary of transformer T_1. The frequency of the pulsations is about 30 kHz at 10 to 15 V peak to peak. The transformer steps up the voltage, firing the fluorescent lamp. Power transistors Q_1 and Q_2 require a very good heat sink, or they will be damaged. Good ventilation is required as the oscillator draws 4 to 6 A from the battery. Resistor R_2 drops the voltage to the bulb somewhat after the fluorescent bulb has fired.

Caution should be used in wiring the battery-powered fluorescent light as a result of the high voltages present. Proper strain relief techniques should be used where cords enter and exit the case.

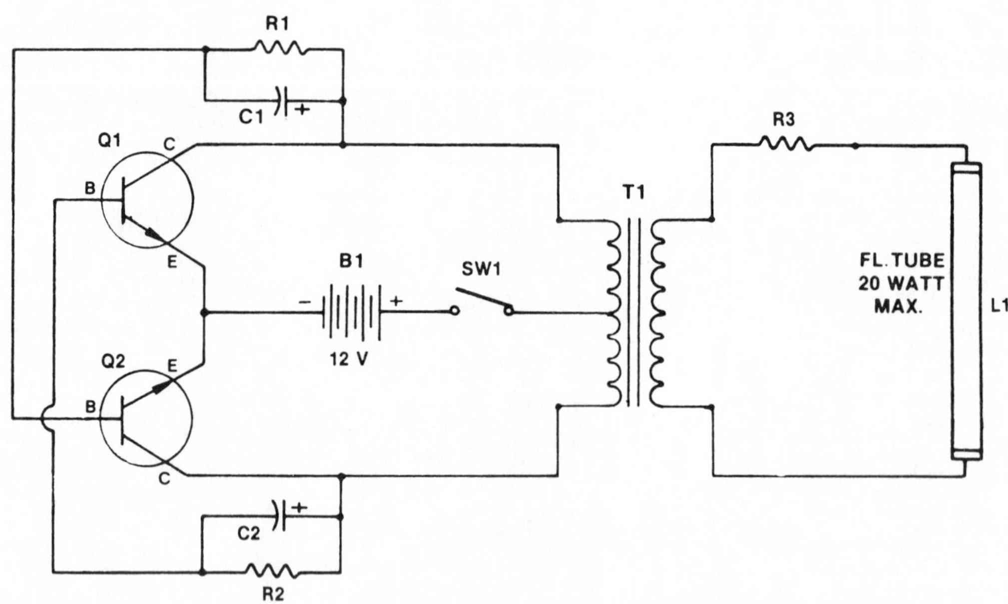

PARTS LIST

B_1	12-V automobile-type battery
C_1,C_2	2.5-μF, 50-V electrolytic capacitor
L_1	10- to 20-W fluorescent bulb
Q_1,Q_2	2N3055 NPN power transistor (similar to Radio Shack 276-2041)
R_1,R_2	120-Ω, 5-W resistor
R_3	100-Ω, 5-W resistor
SW_1	SPST switch, 6 dc A
T_1	6.3-V, 1-A (center-tapped) secondary power transformer (similar to Stancor P-6134)

FIG. 6-4 12-V battery-operated fluorescent light. (Used by permission of Mode Electronics.)

FOUR- TO THREE-WIRE TAILLIGHT CONVERSION

Most foreign automobiles use separate bulbs for brake lights and rear turn signals, unlike most American cars. Therefore, to operate American-made trailers with these foreign cars, four wires must be used. These are for ground, left-turn signal, right-turn signal, and brake. Many American-made trailers use three wires (ground, left-turn signal/brake bulb, and right-turn signal/brake bulb). The four- to three-wire taillight conversion circuit shown in Fig. 6-5 will solve this problem. The schematic diagram in Fig. 6-5 does not show the other bulbs in the trailer taillights, which are for the running lights.

The heart of the four- to three-wire conversion circuit is a CMOS EXCLUSIVE-OR gate IC_1, which drives the power darlington transistors Q_1 and Q_2. Resistors R_4 and R_5 are current-limiting resistors, while the others connected to the input pins of the XOR gates pull the inputs to ground when they might otherwise be floating high.

As an example of how the four- to three-wire conversion circuit in Fig. 6-5 works, assume that the brake input is high (+12 V) and the right signal input alternates between high and low (ground). Pin 5 is high while pin 6 is low because there is no signal at the left signal input. Therefore, the XOR gate will output a high at pin 4, activating the left-side trailer light with a brake signal. As the right signal input alternates to high, the output at pin 3 goes low, turning off Q_1 and the right side trailer light. However, as the right signal alternates low, the output at pin 3 goes high, activating Q_1, which causes the right-side trailer bulb to flash on.

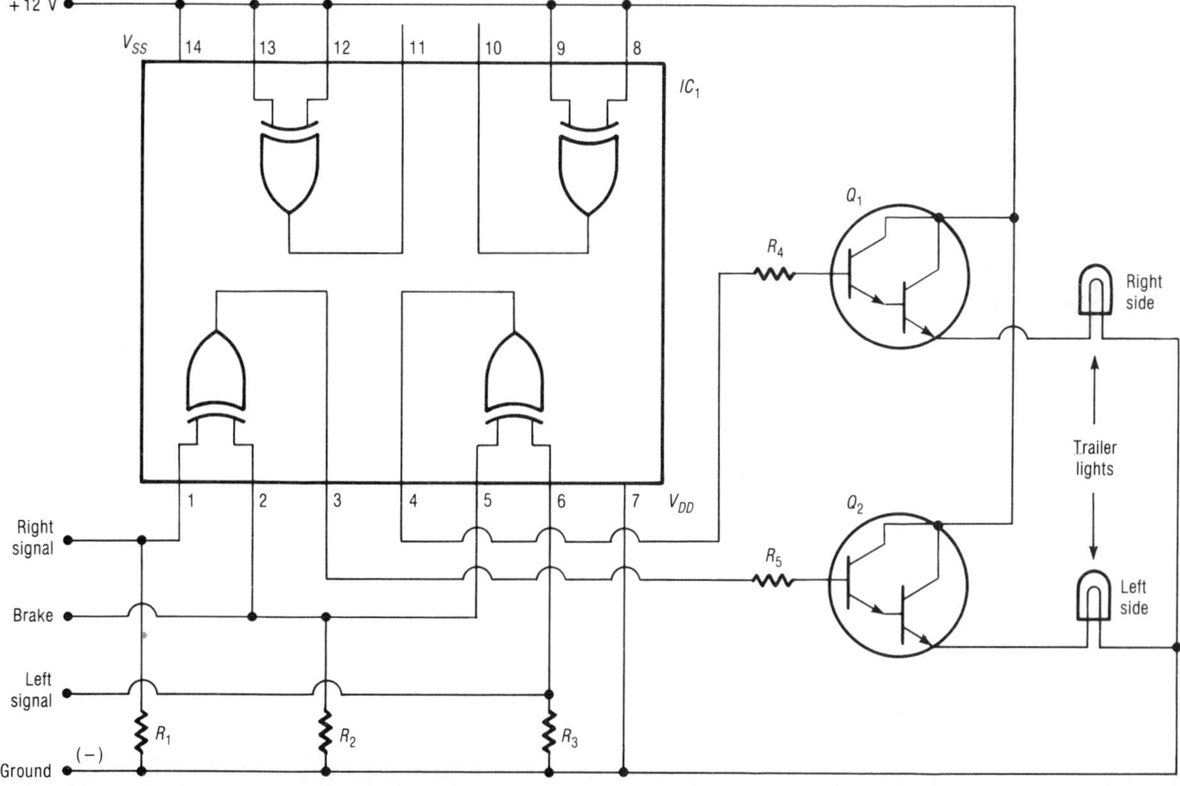

PARTS LIST

IC_1	4030 CMOS quad 2-input EXCLUSIVE-OR gate IC
R_1 through R_5	1-kΩ, ¼-W resistor
Q_1, Q_2	2N6388 power darlington transistors (similar to Radio Shack 276-2068)

FIG. 6-5 Four- to three-wire taillight conversion. (Reprinted from *Popular Electronics*. Copyright © March 1982, Ziff-Davis Publishing Company.)

chapter 7

Clock and Counter Circuits

BINARY CLOCK

A binary clock circuit, based on the National Semiconductor MM5315 metal oxide semiconductor/large-scale integrated circuit (MOS/LSI) clock IC, is shown in Fig. 7-1. The IC multiplexes individual light-emitting diodes (LEDs) instead of the more normal seven-segment displays with the output read in binary-coded decimal (BCD) instead of decimal. The display LEDs L_1 through L_{20} should be physically arranged similarly to those in Fig. 7-1.

Consider the examples in Fig. 7-2 for learning to read the binary display. The binary clock is wired in the 24-hour mode; hence the 21 : 58 : 00 reading in Fig. 7-2(*c*). For a 12-hour clock, pin 13 of the IC must be grounded (to V_{DD}).

The power supply section of the binary clock consists of the plug, the 12-V transformer, bridge rectifier D_1 through D_4, and filter capacitors C_1 and C_2. Resistor R_1 and diode D_5 send a signal to the 60-Hz input (pin 20) of the IC. The combination of R_2 and C_4 set the frequency of the multiplex oscillator in the IC.

The controls on the binary clock in Fig. 7-1 are as follows:

1. *Reset* clears the counters for an output of 00 : 00 : 00.
2. *Hold* stops the clock and holds the current time on the displays.
3. *Fast set* causes the display to advance rapidly when setting the time on the clock.
4. *Slow set* causes the display to advance slowly when setting the time on the clock.

If the reset (pin 16) were grounded, all LEDs in the display would be cleared and the displayed time would read 00 : 00 : 00. After 1 second, the logic in the IC would turn on driver transistor Q_{10}, placing a negative voltage on the cathodes of all the LEDs in the S_1 column (right column in display). Row driver transistor Q_4 would be turned on by the IC, placing a positive voltage at the anodes of the bottom

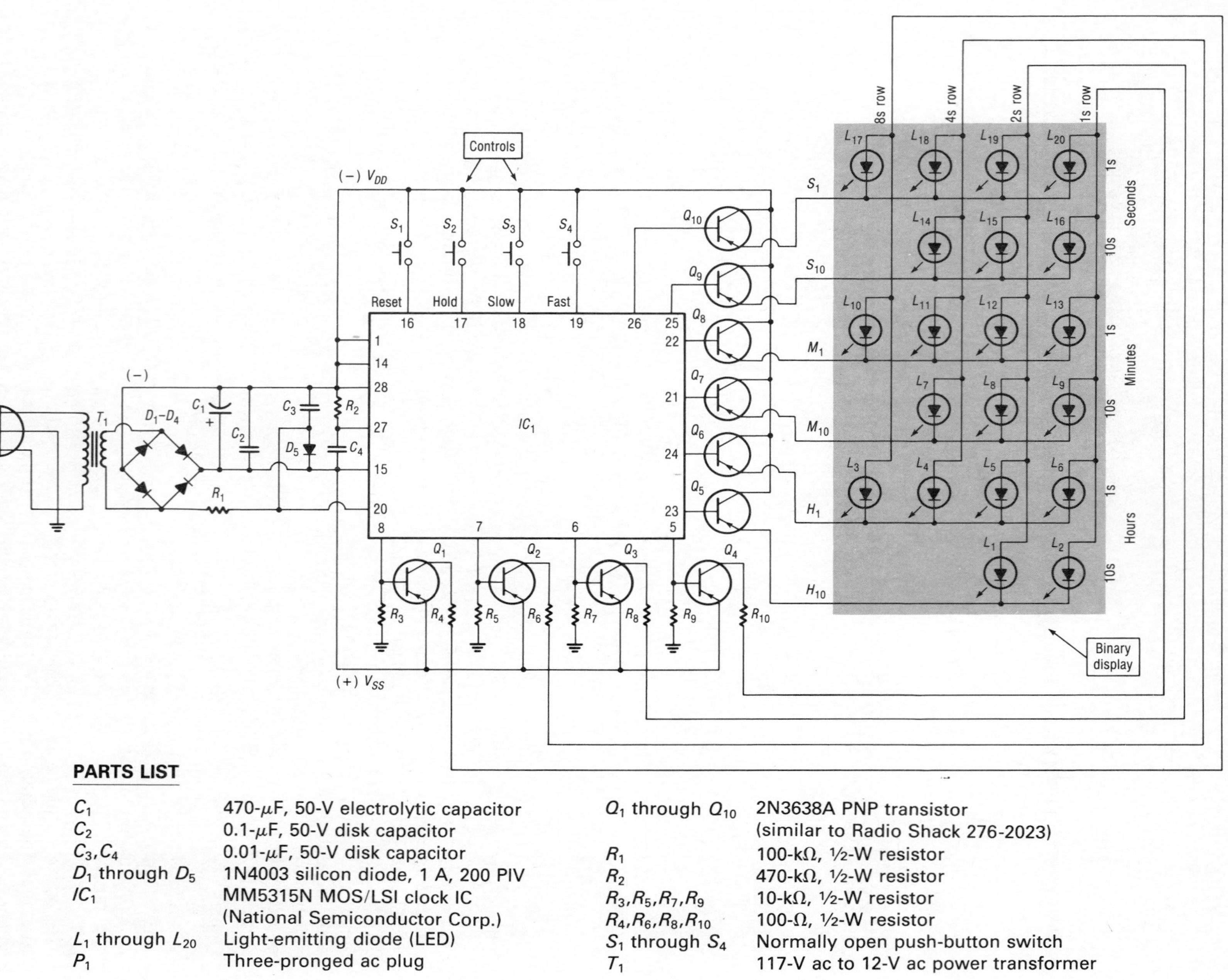

PARTS LIST

C_1	470-μF, 50-V electrolytic capacitor	Q_1 through Q_{10}	2N3638A PNP transistor (similar to Radio Shack 276-2023)
C_2	0.1-μF, 50-V disk capacitor	R_1	100-kΩ, ½-W resistor
C_3,C_4	0.01-μF, 50-V disk capacitor	R_2	470-kΩ, ½-W resistor
D_1 through D_5	1N4003 silicon diode, 1 A, 200 PIV	R_3,R_5,R_7,R_9	10-kΩ, ½-W resistor
IC_1	MM5315N MOS/LSI clock IC (National Semiconductor Corp.)	R_4,R_6,R_8,R_{10}	100-Ω, ½-W resistor
L_1 through L_{20}	Light-emitting diode (LED)	S_1 through S_4	Normally open push-button switch
P_1	Three-pronged ac plug	T_1	117-V ac to 12-V ac power transformer

FIG. 7-1 Binary clock. (Used by permission of Electronic Kits International, Inc.)

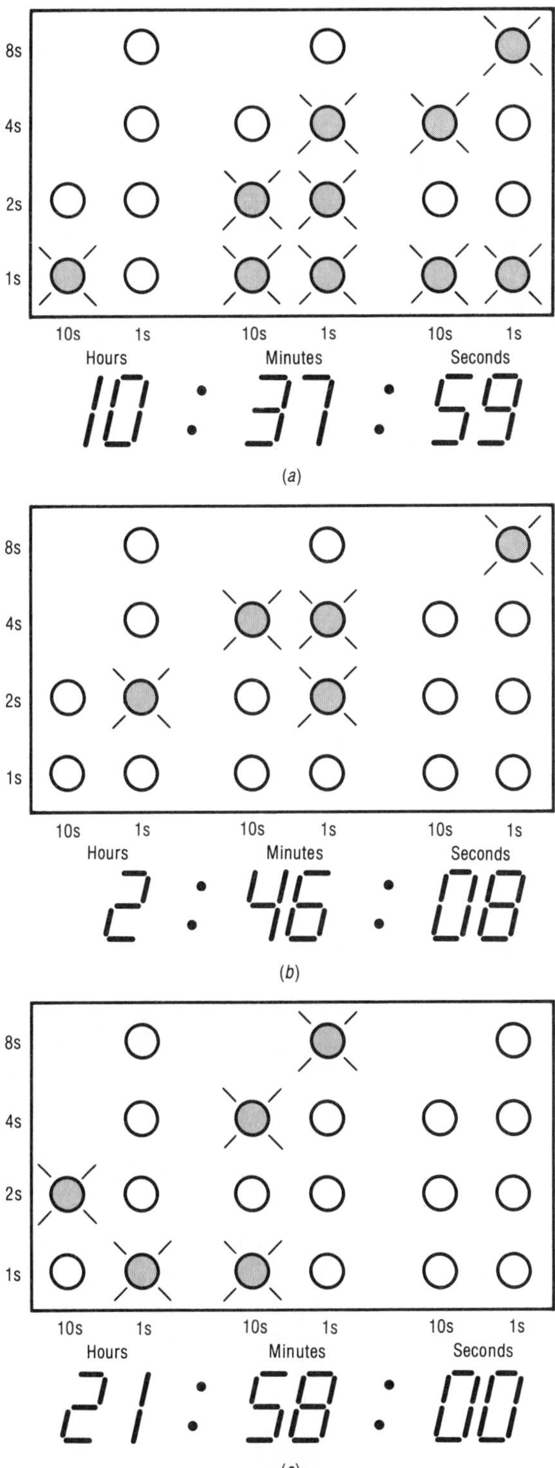

FIG. 7-2 Reading the binary clock's display.

row of LEDs. Light-emitting diode L_{20} is forward-biased and lights. As the row drivers multiplex the rows, only the 1s row is turned on in this example. With only L_{20} lit, the time reads 00:00:01. Any LED in the binary display can be turned on or off by the column and row driver transistors.

The high-voltage wiring must be done carefully to make sure that it is completely isolated from the enclosure. The green ground wire on the three-pronged plug must be attached to the metal chassis. Use proper strain relief techniques where cords enter and exit the case. A parts kit and two-sided pc board are currently available for binary clock from Electronic Kits International, Inc.

SIMPLE BINARY COUNTER

A simple 6-bit binary counter circuit is shown in Fig. 7-3. The output LEDs display the binary count equal to 0 through 63 in decimal. When switch S_1 is closed, it clears the 4024 complementary metal oxide semiconductor (CMOS) binary counter IC and also activates the oscillator circuitry IC_1 and associated parts. The oscillator pulses are fed into the 4024 binary counter IC, and the outputs drive the LED displays. Note that the least-significant bit (LSB) LED is connected to output pin 12, whereas the MSB (32s bit) is connected to pin 4 of the IC. When the counter has counted to binary 111111 (decimal 63), pin 3 of the counter goes high, turning off the oscillator. The frequency of the oscillator can be adjusted by changing the value of capacitor C_2.

LCD DIGITAL CLOCK

A simple-to-build LCD digital clock circuit is shown in Fig. 7-4. The unit can be powered for up to 1 year with a single 1.5-V battery. The circuit is based on the National Semiconductor MM48143 CMOS clock IC, which drives a 0.5-in liquid-crystal display (LCD). A 32.768-kHz quartz crystal serves as the time base. The unit comes as the National Semiconductor MA1032 digital LCD clock module. Other features of the LCD clock include a backlighted display for night viewing, a PM indicator, and an alarm symbol on the display and alarm beeper.

The MA1032 clock module is manufactured by use of a CMOS IC; therefore, care must be

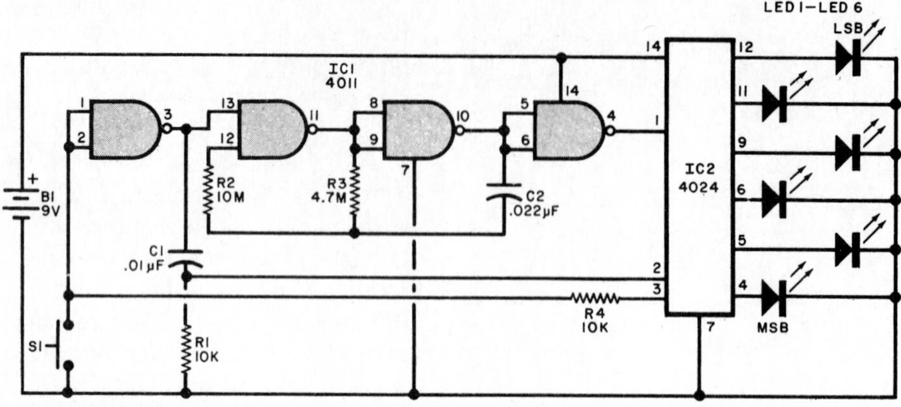

PARTS LIST

B_1	9-V battery
C_1	0.01-μF, 50-V disk capacitor
C_2	0.022-μF, 50-V disk capacitor
IC_1	4011 CMOS quad two-input NAND gate IC
IC_2	4024 CMOS binary counter IC
LED_1 through LED_6	Light-emitting diode
R_1, R_4	10-kΩ, ½-W resistor
R_2	10-MΩ, ½-W resistor
R_3	4.7-MΩ, ½-W resistor
S_1	Normally open push-button switch

FIG. 7-3 Simple 6-bit binary counter. (Reprinted from *Popular Electronics*. Copyright © April 1978, Ziff-Davis Publishing Company.)

taken not to subject the edge connectors to static charges which may harm the integrated circuit. A summary of the LCD digital clock display modes and display indicators are shown in Fig. 7-5.

SIX-DIGIT LED DIGITAL CLOCK

A six-digit clock circuit using seven-segment LED displays is detailed in Fig. 7-6. The digital clock operates on household 117 V ac and continuously displays hours, minutes, and also seconds. The unit is based on the National Semiconductor MOS/LSI MM5314 clock chip.

A block diagram of the six-digit clock is sketched in Fig. 7-7. The power supply is not shown but produces the +12 V dc and the 60-Hz input signal. The MM5314 clock chip then divides the input frequency into a 1-pulse-per-second (1-PPS) signal. The counts are accumulated for all six displays and sent to the multiplexer. The segment and digit drives activate the six LED displays. To reduce wiring complexity, the displays are flashed on in sequence by the IC multiplex circuitry. The multiplexing action occurs at a fast rate; therefore, the eye does not notice a flickering in the displays. The oscillator in the block diagram drives the multiplexer. The decoder and PROM convert from the internal coding to the seven-segment code needed to activate the segment drivers.

The power supply of the digital clock is shown at the left in the schematic in Fig. 7-6. The transformer steps down the household 117 V ac to 12 V while the bridge rectifier D_1 through D_4 changes ac to dc. Capacitor C_1 acts as a power supply filter. The ac is fed into the 60-Hz input (pin 16) through R_3. Parts C_3 and R_4 determine the multiplex oscillator frequency. Light-emitting diodes D_6 and D_7 with limiting resistors R_1 and R_2 form the flashing

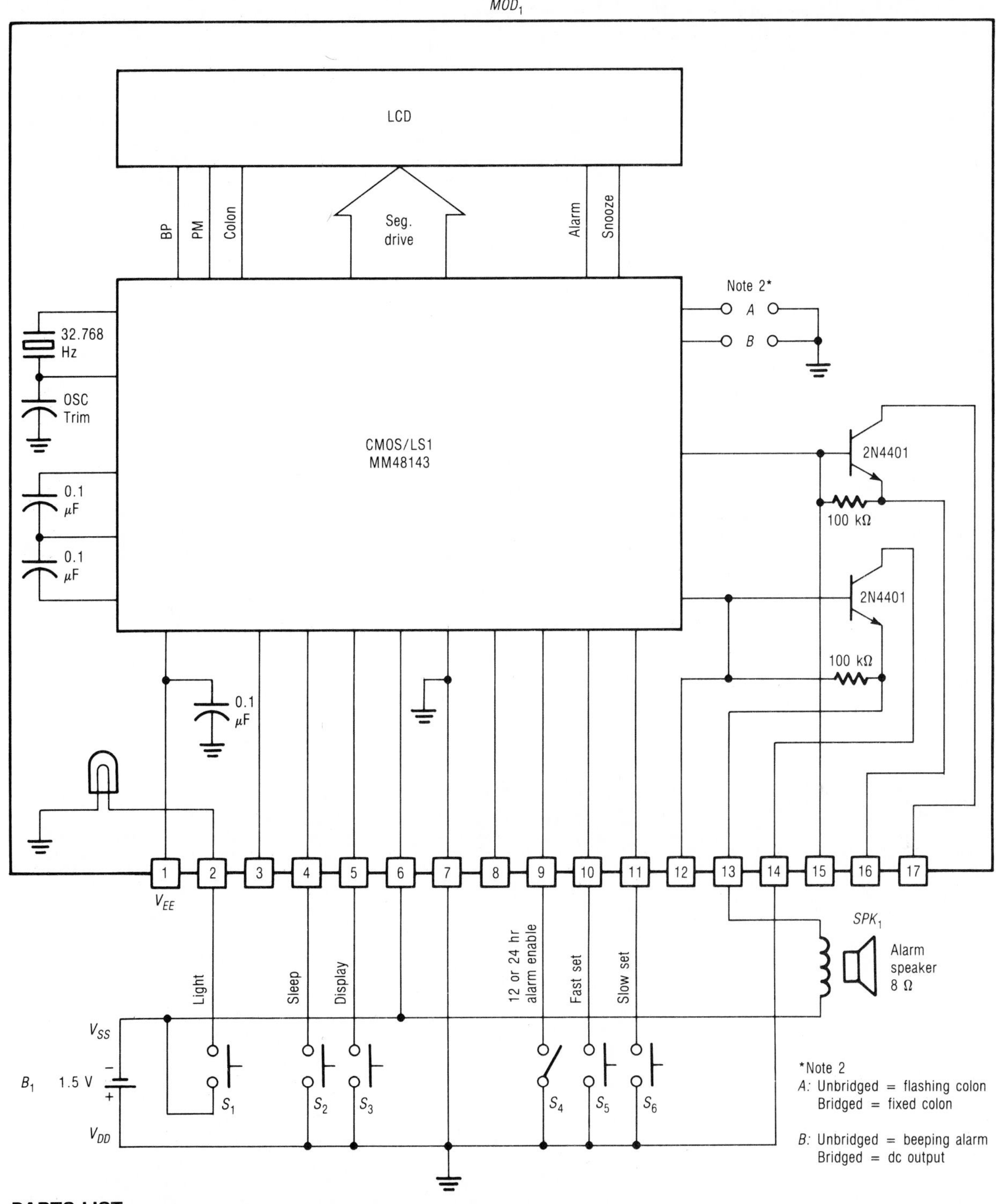

PARTS LIST

B_1	1.5-V alkaline battery (AA, C, or D size)
MOD_1	MA1032 LCD digital clock module (National Semiconductor)
S_1, S_2, S_3, S_5, S_6	Normally open push-button switch
S_4	SPST switch (or push-on push-off switch)
SPK_1	8-Ω speaker

FIG. 7-4 LCD digital clock circuit. (Used by permission of National Semiconductor Corporation.)

Summary of MA1032 Clock IC Display Modes

Mode	Display Identifier
Time display	Flashing colon (unless fixed option) Bell fixed, if alarm enabled Normal four-digit time format
Alarm set	No colon Bell flashing independent of alarm enable Normal four-digit time format
Time set	All four digits, colon, and PM flash (unless fixed option) Bell fixed, if alarm enabled Normal four-digit time format
Time set (hold mode)	Same as time set except colon is fixed
Seconds display	Flashing colon (unless fixed option) Bell fixed, if alarm enabled Hours digits blanked, seconds incrementing
Timer display	Flashing colon (unless fixed option) Bell fixed, if alarm enabled Hours digits blanked, PM blanked, two-digit number represents minutes remaining on timer after 2 seconds of sleep control depression, 10-digit decrements

(*a*)

Display Indicators on LCD of MA1032 Unit

Function	Enunciator	Description
Alarm time displayed	🔔	Flashes
Alarm enabled	🔔	On
Sleep counter displayed	Zzz	Flashes
Sleep mode on	Zzz	On
PM time in 12-hour mode	PM	On

(*b*)

FIG. 7-5 LCD digital clock. (Used by permission of National Semiconductor Corporation.)

colon between the hours and minutes displays. Three controls are shown at the bottom of Fig. 7-6: fast set, slow set, and hold.

The high-voltage wiring must be done carefully to make sure that it is completely isolated from the enclosure. The green ground wire on the three-pronged plug must be attached to the metal chassis. Use proper strain relief techniques where cords enter and exit the case. A parts kit and a two-sided pc board for this six-digit LED clock are currently available from Electronic Kits International, Inc.

0–99 DIGITAL COUNTER

The digital counter sketched in Fig. 7-8 will display a decimal count from 00 to 99 and then

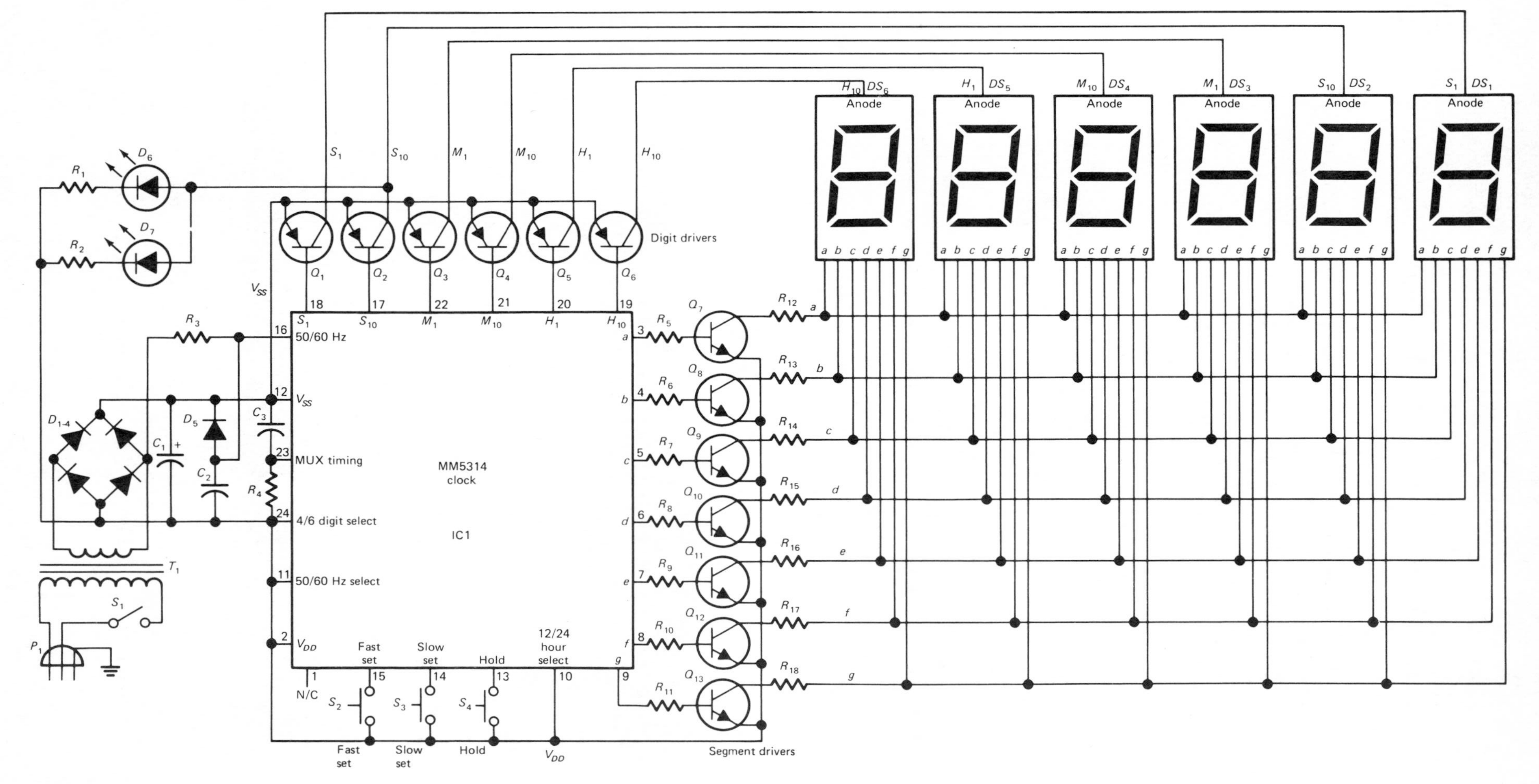

PARTS LIST

C_1	100-μF, 50-V electrolytic capacitor
C_2,C_3	0.01-μF, 50-V disk capacitor
D_1 through D_5	1N4003 silicon diode, 1 A, 200 PIV
D_6,D_7	Light-emitting diode
DS_1 through DS_6	FND 510 or FND 507; common-anode seven-segment LED display
IC_1	MM5314 MOS/LSI digital clock IC (National Semiconductor)
P_1	Three-pronged ac plug
Q_1 through Q_6	2N3638A PNP transistor (similar to Radio Shack 276-2023)
Q_7 through Q_{13}	2N3904 NPN transistor (similar to Radio Shack 276-2009)
R_1,R_2	470-Ω, ¼-W resistor
R_3	100-kΩ, ¼-W resistor
R_4	470-kΩ, ¼-W resistor
R_5 through R_{11}	10-kΩ, ¼-W resistor
R_{12} through R_{18}	220-Ω, ¼-W resistor
S_1	SPST switch
S_2,S_3,S_4	Normally open push-button switch
T_1	12-V, 300-mA power transformer (similar to Radio Shack 273-1385)

FIG. 7-6 Six-digit LED digital clock. (Roger L. Tokheim, *Activities Manual for Digital Electronics*, 2d ed., McGraw-Hill, New York, 1984, pp. 130–133. Used by permission of Electronic Kits International, Inc., and McGraw-Hill Book Company.)

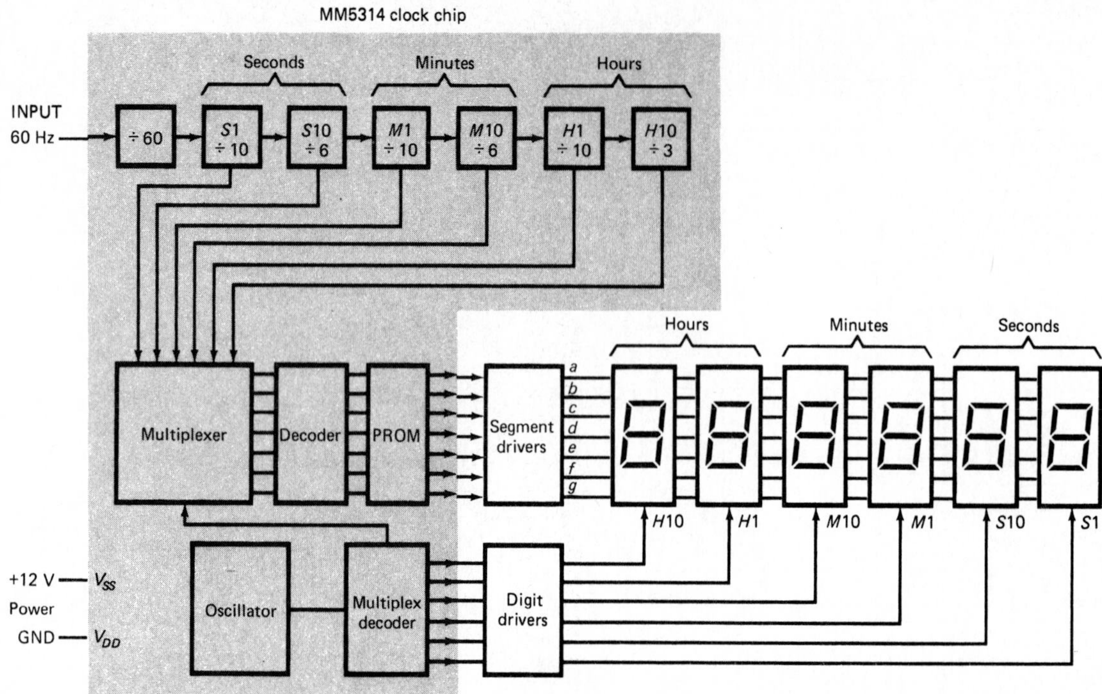

FIG. 7-7 Block diagram of six-digit clock using the MM5314 IC. (Roger Tokheim, *Digital Electronics,* 2d ed., McGraw-Hill, New York, 1984, p. 195. Used by permission of McGraw-Hill Book Company.)

recirculate back to 00. It is an excellent counter demonstration circuit, using easy-to-find ICs. Switch S_2 will clear the displays to 00 when connected to +5 V. The 555 timer IC_1 is wired as an astable multivibrator whose frequency can be adjusted using potentiometer R_2. Decoders IC_3 and IC_5 translate the BCD output of the 1s and 10s counters IC_2 and IC_4 into a decimal output at the LED displays. The 150-Ω resistors R_3 through R_{16} limit the current through the LEDs to a safe level. The power supply on this unit is shown as a 6-V battery (or four 1.5-V cells) with series diode and resistor to produce the +5 V needed to operate the displays and TTL ICs.

DIGITAL AUTOMOBILE CLOCK WITH LED DISPLAY

A digital car clock circuit using an LED display is shown in Fig. 7-9. The National Semiconductor MM5378 auto clock chip is the heart of this 12-hour clock. The clock should be connected to an uninterrupted 12-V supply (bottom two power supply connections in Fig. 7-9). The display enable power supply connection should be connected so that the display turns off with the ignition switch. The controls are the slow-set S_1 and fast-set S_2 switches. Variable capacitor C_2 can be used to make slight adjustments in the accuracy of the clock. The clock consumes very little power with the ignition switch off and the displays not lit. However, the clock chip keeps accurate time even with the ignition switch in the OFF position.

Diodes D_1 and D_2 protect the IC from reverse voltage, while zener diode D_3 protects the IC from voltage spikes from the auto electrical system. The crystal X_1 along with capacitors C_2 and C_3 and resistor R_1 are the external parts associated with the IC's 2.097152-MHz time-base oscillator. Transistors Q_1 through Q_4 and associated resistors are digit drivers to turn on

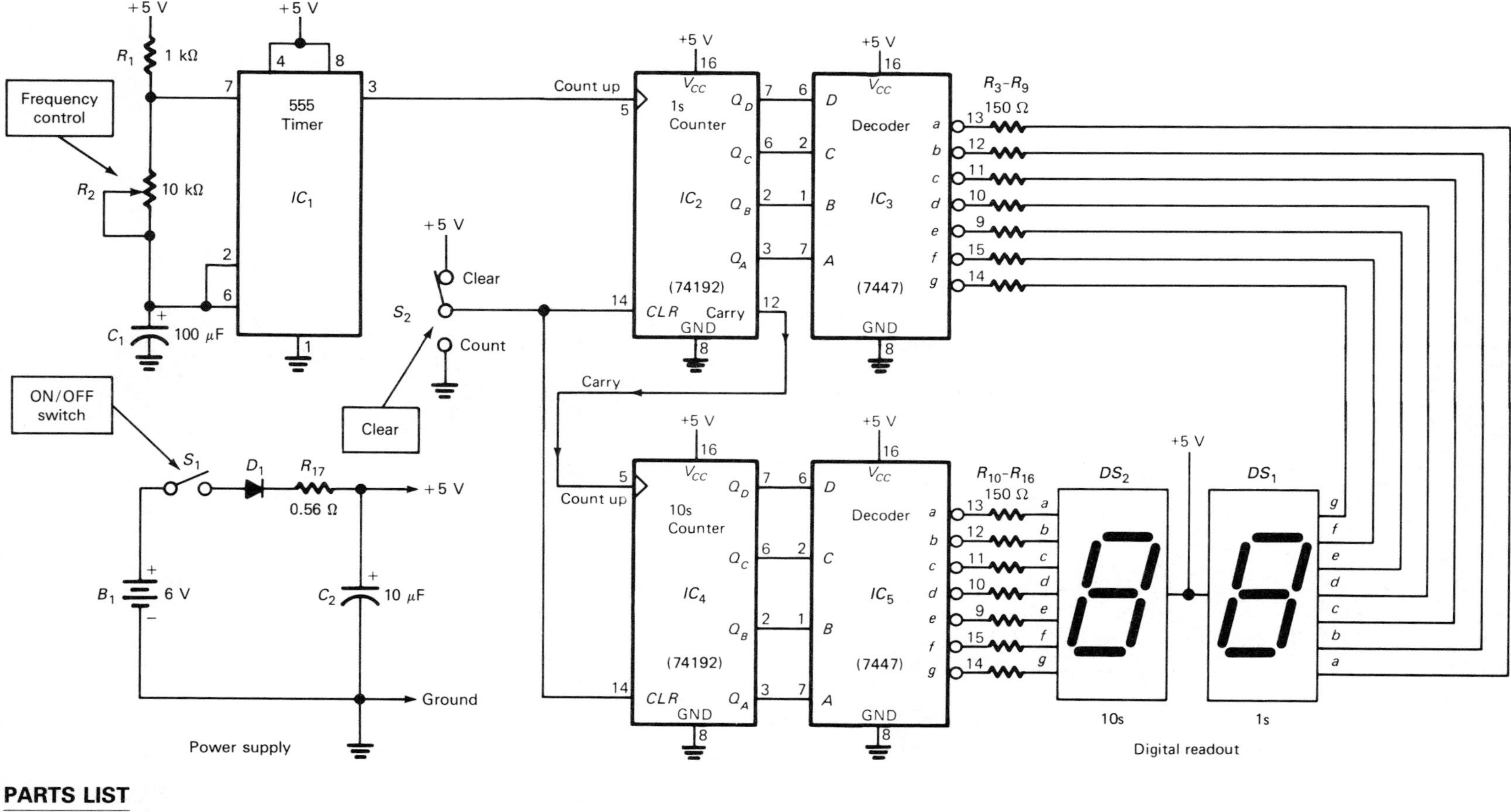

PARTS LIST

B_1	6-V battery (or four 1.5-V batteries)
C_1	100-μF, 25-V electrolytic capacitor
C_2	10-μF, 25-V electrolytic capacitor
D_1	1N4001 silicon diode, 1 A, 50 PIV
DS_1,DS_2	FND507 common-anode seven-segment display (TIL321) (vertical mounting—Radio Shack 276-053)
IC_1	555 Timer IC
IC_2,IC_4	74192 universal counter IC
IC_3,IC_5	7447 decoder-driver IC
R_1	1-kΩ, 1/4-W resistor
R_2	10-kΩ linear-taper potentiometer
R_3 through R_{16}	150-Ω, 1/4-W resistor
R_{17}	0.56-Ω, 1/2-W resistor
S_1	SPST switch
S_2	SPDT switch

FIG. 7-8 0–99 digital counter. (Roger L. Tokheim, *Activities Manual for Digital Electronics*, 2d ed., McGraw-Hill, New York, 1984, pp. 72, 163. Used by permission of McGraw-Hill Book Company.)

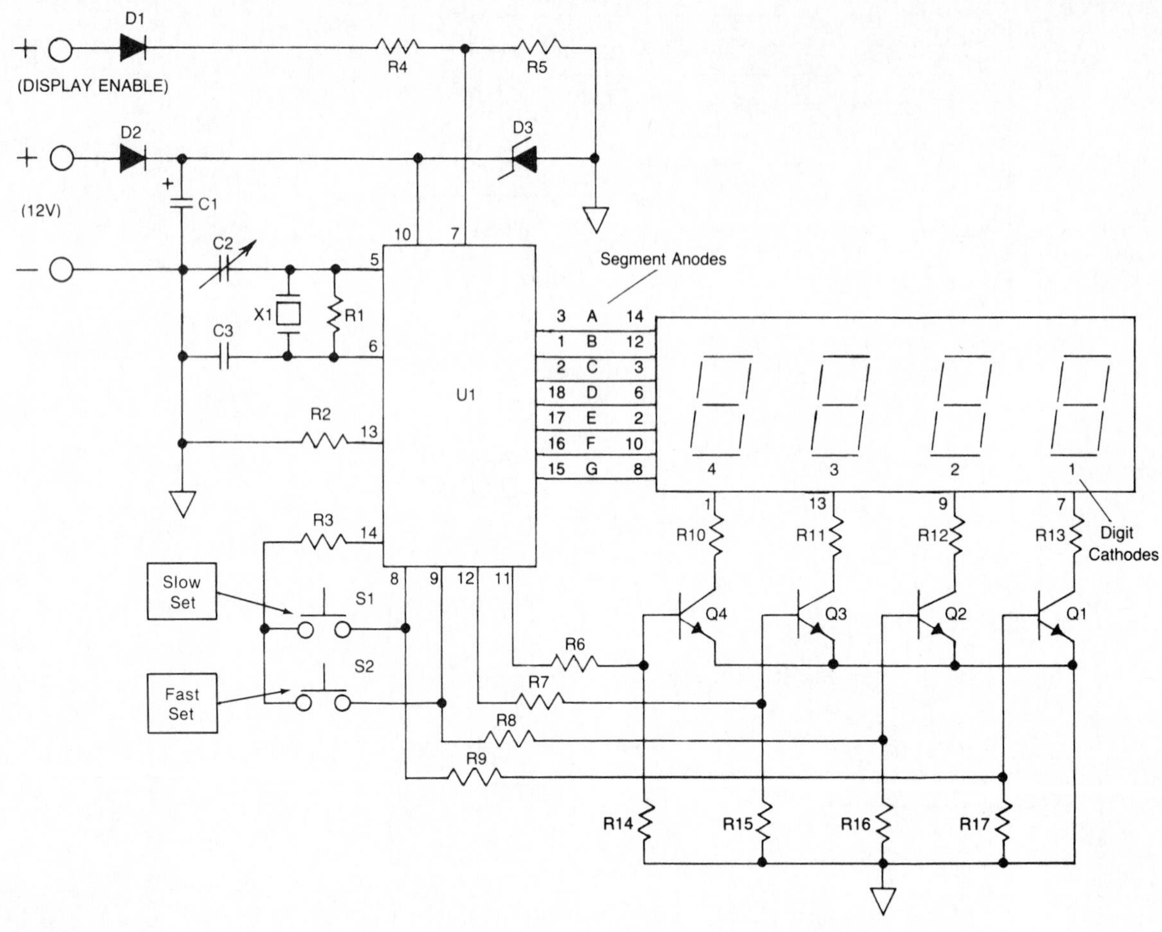

PARTS LIST

C_1	4.7-μF, 50-V electrolytic capacitor
C_2	5- to 45-pF variable capacitor (Part 63051, Graymark International)
C_3	22-pF mica capacitor
D_1,D_2	1N4002 silicon diode, 1 A, 100 PIV
D_3	1N5250 zener diode, 20 V, ½ W
DIS 1	Five-digit LED Display (Part 63048, Graymark International)
Q_1 through Q_4	2N3904 NPN transistor (or similar)
R_1	15-MΩ, ¼-W resistor
R_2	15-Ω, ¼-W resistor
R_3,R_4	100-kΩ, ¼-W resistor
R_5	1-MΩ, ¼-W resistor
R_6 through R_9	10-kΩ, ¼-W resistor
R_{10} through R_{13}	100-Ω, ¼-W resistor
R_{14} through R_{17}	3.9-kΩ, ¼-W resistor
S_1,S_2	Normally open push-button switch
U_1	MM5378 auto clock IC (National Semiconductor)
X_1	2.097152-MHz crystal

FIG. 7-9 Digital car clock with LED display. (Used by permission of Graymark International, Inc.)

the multiplexed LED displays. Resistors R_4 and R_5 form a voltage divider to set the brightness (pin 7 on IC).

A parts kit and a pc board are currently available for the digital car clock from Graymark International, Inc.

chapter 8

Electronic Game Circuits

BINARY HI-LO GAME

A binary hi-lo game using easy-to-find parts is diagrammed in Fig. 8-1. To play the game turn on switch S_3 and press switch S_1 for several seconds. This loads a random number into the 74193 counter IC_2. The player tries to guess the *4-bit binary number* by setting the four switches in S_2. If the number is correct, green LED_2 will light and the game is won. If you guess too high, the red LED_1 lights; if you guess too low, the yellow LED_3 will light. The object of the game is to guess the random binary number in the fewest guesses.

The 555 timer is wired as a free-running multivibrator and serves to clock the 74193 counter when switch S_1 is pressed. When S_1 is released, the counter holds the "random count" at the B input of the 7485 magnitude comparator IC_3. The player selects which 7485 input A pins are to be grounded through switch S_2. If the A and B inputs to IC_3 are of the same binary magnitude, pin 6 goes high, lighting LED_2. If the magnitude of A is greater than B, pin 5 goes high and LED_1 lights, indicating that the guess was too high. If the magnitude of A is less than B, then pin 7 goes high and LED_3 lights, indicating that the guess was too low. Diode D_1 drops the battery voltage to about +5 V for the ICs.

DECISION MAKER

The simple decision maker circuit shown in Fig. 8-2 will indicate true-false or heads-tails when activated. When powered with 117 V ac, neon lamps V_1 and V_2 will alternately flash. If switch S_1 is closed, one of the lamps will be left on. Opening the switch will again cause the neon lamps to flash.

The household 117 V ac is converted into dc by D_1 and filtered by C_1. With switch S_1 open, dc voltage will be applied to both neon lamps. As a result of variations in the lamps, one will

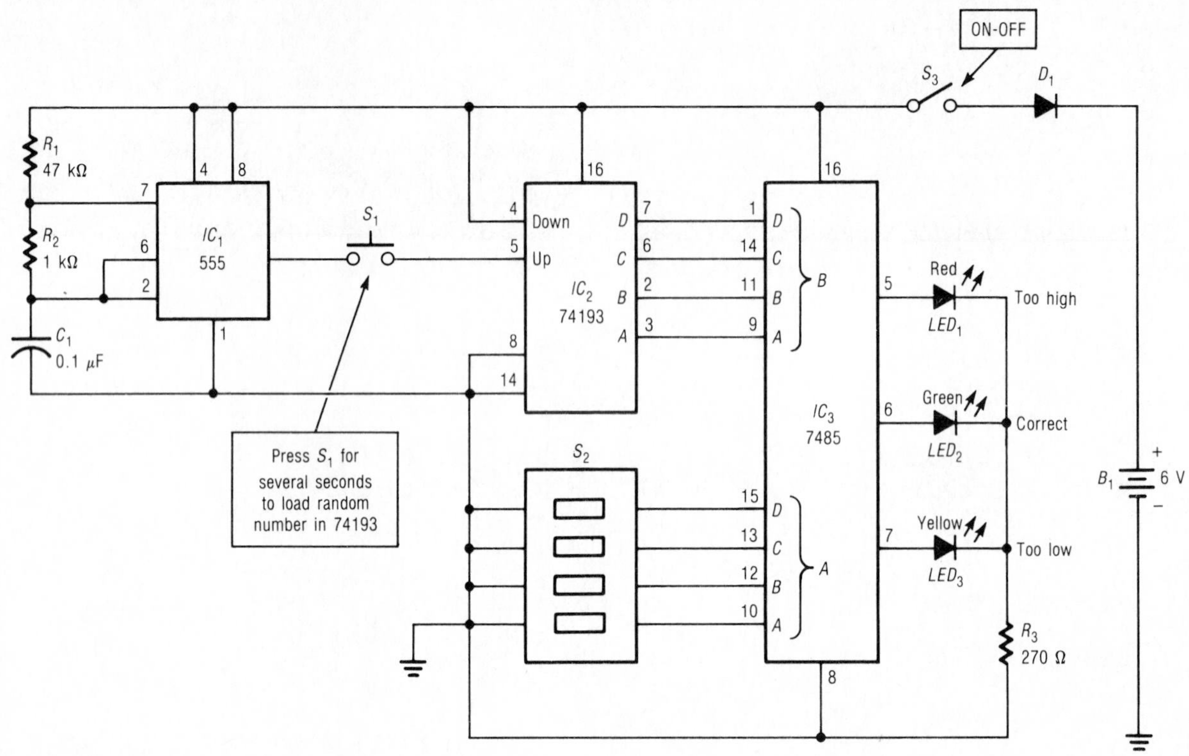

PARTS LIST

B_1	6-V battery (or four 1.5-V cells)	LED_3	Yellow light-emitting diode
C_1	0.1-μF, 50-V disk capacitor	R_1	47-kΩ, ½-W resistor
D_1	1N4001 silicon diode, 1 A, 50 PIV	R_2	1-kΩ, ½-W resistor
IC_1	555 timer IC	R_3	270-Ω, ½-W resistor
IC_2	74193 counter IC	S_1	Normally open push-button switch
IC_3	7485 4-bit magnitude comparator IC	S_2	SPST 4-position DIP switch
LED_1	Red light-emitting diode	S_3	SPST switch
LED_2	Green light-emitting diode		

FIG. 8-1 Binary hi-lo game. (Forrest Mims, *The Forrest Mims Circuit Scrapbook*, McGraw-Hill, New York, 1983, p. 121. Used by permission of Forrest M. Mims, III.)

reach its firing voltage (about 65 to 70 V) first. When the first neon lamp fires, the voltage at the top of the lamp goes positive, charging the capacitor C_2. When the capacitor has charged for a short time, it fires the other neon lamp and turns off the first lamp. The lamps alternately flash. When switch S_1 is closed, the lamp that was lit last will continue to glow while the other will not light. When the switch is closed, capacitor C_2 is short-circuited.

The high-voltage wiring must be done carefully to make sure that it is completely isolated from the enclosure. The green ground wire on the three-pronged plug must be attached to the metal chassis. Use proper strain relief techniques where cords enter and exit the case.

DIGITAL DICE

The circuit for a digital dice game is shown in Fig. 8-3. As the "roll" button is pressed, a random number between 1 and 6 will appear on the seven-segment LED display simulating the roll of a single die. The circuit may be doubled to get a pair of digital dice.

The circuit may be divided into four sections: free-running multivibrator, counter, decoder/driver, and display. Gates 1a and 1b along with capacitor C_1 and resistor R_1 form a free-running multivibrator or clock which generates a 1-kHz square wave. When switch S_2 is closed, the clock drives the counter (IC_2 and gates 1c and 1d). The clock counts from 1

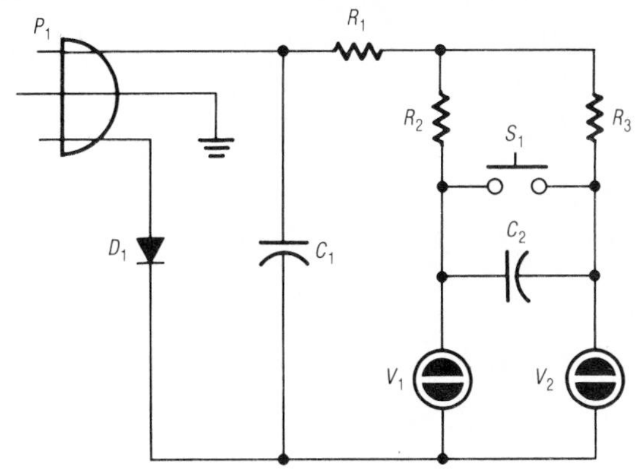

PARTS LIST

C_1,C_2	0.047-μF, 250-V polyester capacitor
D_1	1N4003 silicon diode, 1 A, 200 PIV
P_1	AC line cord
R_1,R_2,R_3	100-kΩ, ½-W resistor
S_1	Normally open push-button switch
V_1,V_2	NE-2 neon lamp

FIG. 8-2 Decision maker. (Used by permission of Graymark International, Inc.)

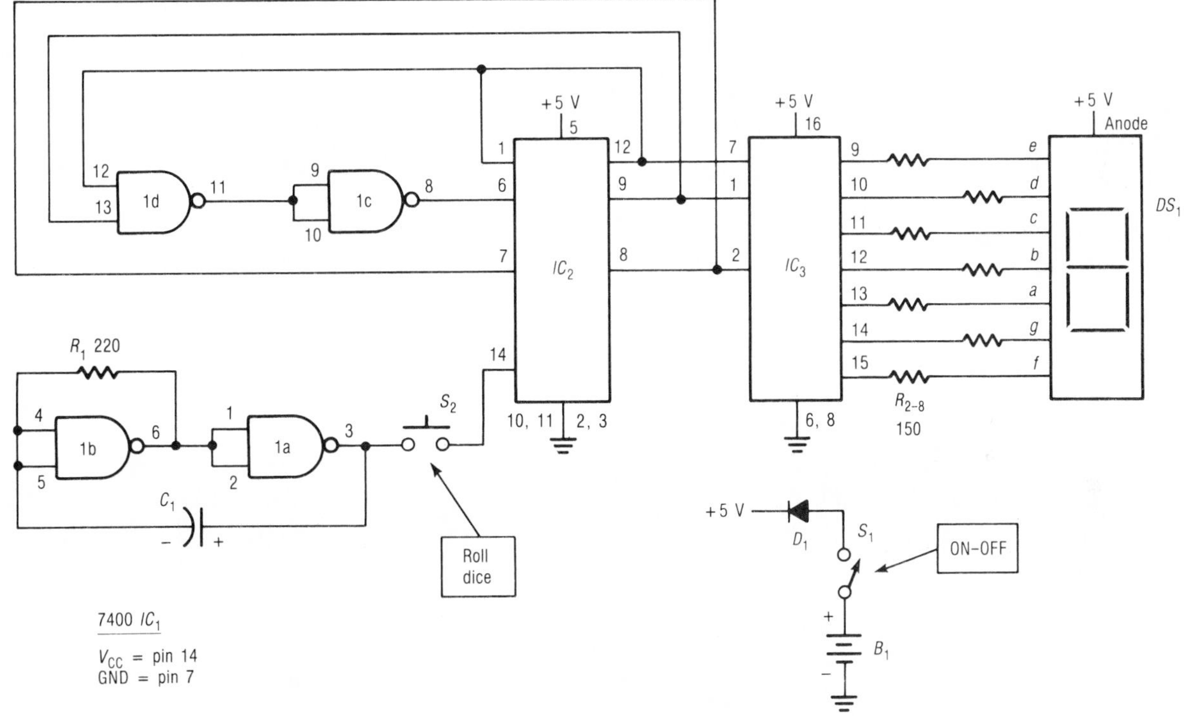

PARTS LIST

B_1	6-V battery (or four 1.5-V cells)
C_1	1-μF, 16-V electrolytic capacitor
D_1	1N4001 silicon diode, 1 A, 50 PIV
DS_1	Common-anode seven-segment LED display (Radio Shack 276-053, or TIL321, or FND507)
IC_1	7400 quad 2-input NAND gate IC
IC_2	7490 decade counter IC
IC_3	7447 seven-segment decoder-driver IC
R_1	220-Ω, ¼-W resistor
R_2 through R_8	150-Ω, ¼-W resistor
S_1	SPST switch
S_2	Normally open push-button switch

FIG. 8-3 Digital dice. (Used by permission of Robert Delp Electronics.)

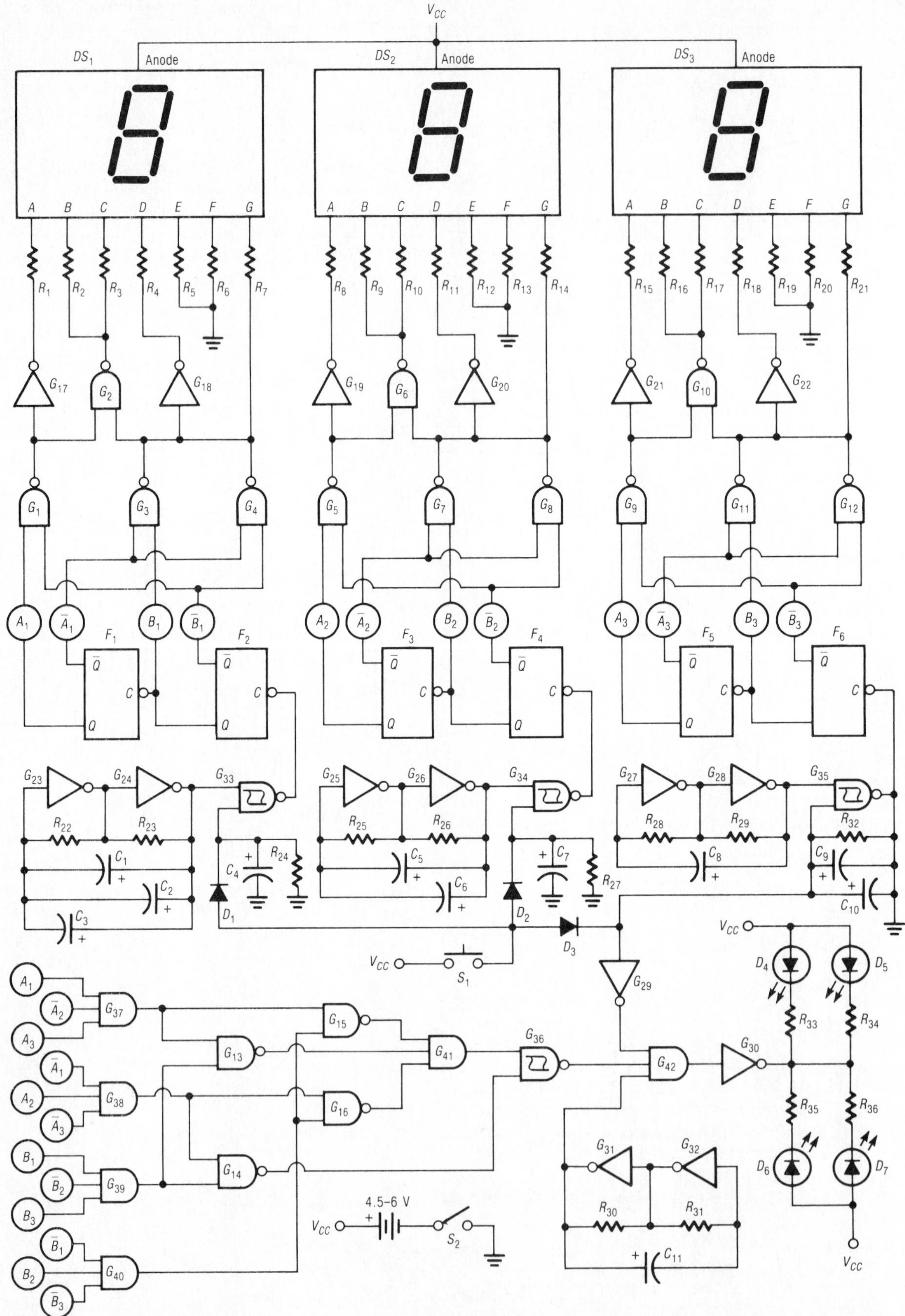

FIG. 8-4 Digital slot machine. See parts list on facing page. (Used by permission of Electronic Kits International, Inc.)

through 6 and recirculates back to 1. When switch S_2 is opened, IC_2 stops and holds at its last binary count. The decoder/driver IC_3 decodes the binary and drives the seven-segment LED display with the equivalent decimal number. The unit is powered with a 6-V battery or four 1.5-V cells. Diode D_1 drops the voltage down to about +5 V to drive the TTL ICs.

DIGITAL SLOT MACHINE

A fairly complex circuit for an electronic one-armed bandit or slot machine is detailed in Fig. 8-4. The block diagram of the digital slot machine game is sketched in Fig. 8-5. The game is played by pressing the "roll" switch (S_1). The three 7-segment LEDs rapidly display C, L, O, and A (for cherry, lemon, orange, and apple). Releasing the roll switch will cause the left display I to stop almost immediately and then display II, and finally the right display III will stop. If they match (CCC, for example), you win and the four LEDs flash on the win display. If you do not win, roll again. The odds of getting all readouts to display the same letter and win the jackpot are 1 in 64.

The schematic diagram for the slot machine game appears to be quite complex. However, the block diagram in Fig. 8-5 suggests that many of the circuits are repetitive. Each display circuit is the same except that each oscillator is set at a slightly different frequency. The display oscillators drive two flip-flops whose outputs are decoded by the logic block. The output of the logic block drives the segments of the seven-segment LED, giving displays of either C, L, O, or A. After roll switch S_1 is released, display I stops almost immediately while the other two continue to sequence rapidly through C, L, O, and A for a time. This time delay is built into the gating circuit of the display oscillators. If each seven-segment LED displays the same letter, the win logic block will detect this and activate the four single LEDs called the "win" display. The win oscillator causes the win display to flash on and off several times per second.

The circuit has a fairly heavy current draw of about 100 mA. Therefore, either a +5-V dc regulated power supply or ni-cad or alkaline dry cells should be used. A parts kit and a double-sided pc board are currently available for the digital slot machine game from Electronic Kits International, Inc.

PARTS LIST FOR FIG. 8-4

B_1	6-V battery (four 1.5-V ni-cad or alkaline cells)
C_1, C_5, C_8	10-μF, 25-V electrolytic capacitor
C_2, C_3, C_6	1-μF, 25-V electrolytic capacitor
C_4, C_{11}	100-μF, 25-V electrolytic capacitor
C_7	500-μF, 25-V electrolytic capacitor
C_9, C_{10}	1000-μF, 16-V electrolytic capacitor
D_1, D_2, D_3	1N4003 silicon diode, 1 A, 200 PIV
D_4 through D_7	Light-emitting diodes
DS_1, DS_2, DS_3	FND-510 or TIL321 common-anode seven-segment LED display
F_1 through F_6	(3) 7473 *J-K* flip-flop IC
G_1 through G_{16}	(4) 7400 quad NAND gate IC
G_{17} through G_{32}	(3) 7404 Hex inverter IC
G_{33} through G_{36}	(1) 74132 Schmitt-trigger NAND gate IC
G_{37} through G_{42}	(2) triple 3-input AND gate IC
R_1 through R_{21}	220-Ω, ½-W resistor
R_{22} through R_{31}	1-kΩ, ½-W resistor
R_{32} through R_{36}	560-Ω, ½-W resistor
S_1	Normally open push-button switch
S_2	SPST switch

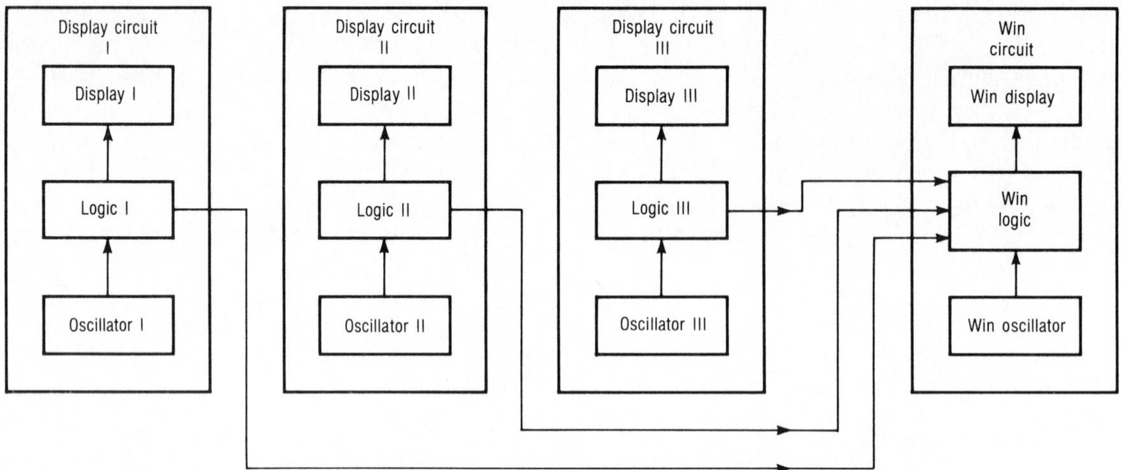

FIG. 8-5 Block diagram of digital slot machine circuit. (Used by permission of Electronic Kits International, Inc.)

ELECTRONIC SHOOT-OUT

The electronic shoot-out game circuit shown in Fig. 8-6 contains easy-to-find parts. The game consists of three LEDs and two fire switches. As the game is turned on, both players are ready to press their fire switches as they watch the green LED. After about 5 seconds the green LED lights, both players fire as quickly as possible. The red LED of the player who fired first lights and that person is the winner. If either player fires before the green LED lights, the red LED of the overanxious player will light and the green LED flashes at about 4 Hz. Firing before the green LED lights means that that player loses. After each shoot-out, the game is turned off and on again to reset for the next duel.

When the ON-OFF switch is turned on, all LEDs are off. As S_1 is moved up, time-delay capacitor C_5 begins to charge. In about 5 seconds it turns on Q_1 and Q_2, causing the green LED D_3 to light. The players now press one of the fire switches on the left. Currently both pins 1 and 13 of IC_1 are low. If the left fire switch S_2 were closed, pin 13 is forced high. This latches a low at pins 9, 10, 11, and 1 of IC_1, lighting the left LED D_1. If switch S_3 is closed shortly after S_2, it does not light D_2 because pin 2 has been forced low, disabling the right LED. Moving S_1 down short-circuits C_5 and resets the electronic shoot-out circuit for another game.

Suppose that shortly after S_1 is moved up, the left player presses S_2. This drives pin 13 of IC_1 high and lights D_1. Pin 5 of IC_2 is supposed to be low at this time to disable $IC_{2/2}$. When pin 13 of IC_2 goes low, pin 5 goes high and $IC_{2/2}$ passes the slow pulsations caused by C_4 and R_5 to pin 6. This causes the green LED to flash at about 4 Hz. This flashing means that one of the fire switches was pressed prematurely. In this example, the left fire switch was pressed too early and the left player loses the duel.

A parts kit and a pc board are currently available for the electronic shoot-out game from Mode Electronics.

ELECTRONIC TENNIS

The challenging electronic tennis game circuit shown in Fig. 8-7 is played without a video screen. A typical physical layout on a pc board is sketched in Fig. 8-8(a) with a block diagram in Fig. 8-8(b). To play the game, each player must operate two switches *with one finger only.* For instance, player 2 might operate the two push-button switches on the left in Fig. 8-8(a) while the opponent operates the two on the right. As the game is turned on, the four corner and two center LEDs (D_{17} through D_{19} and D_{36}

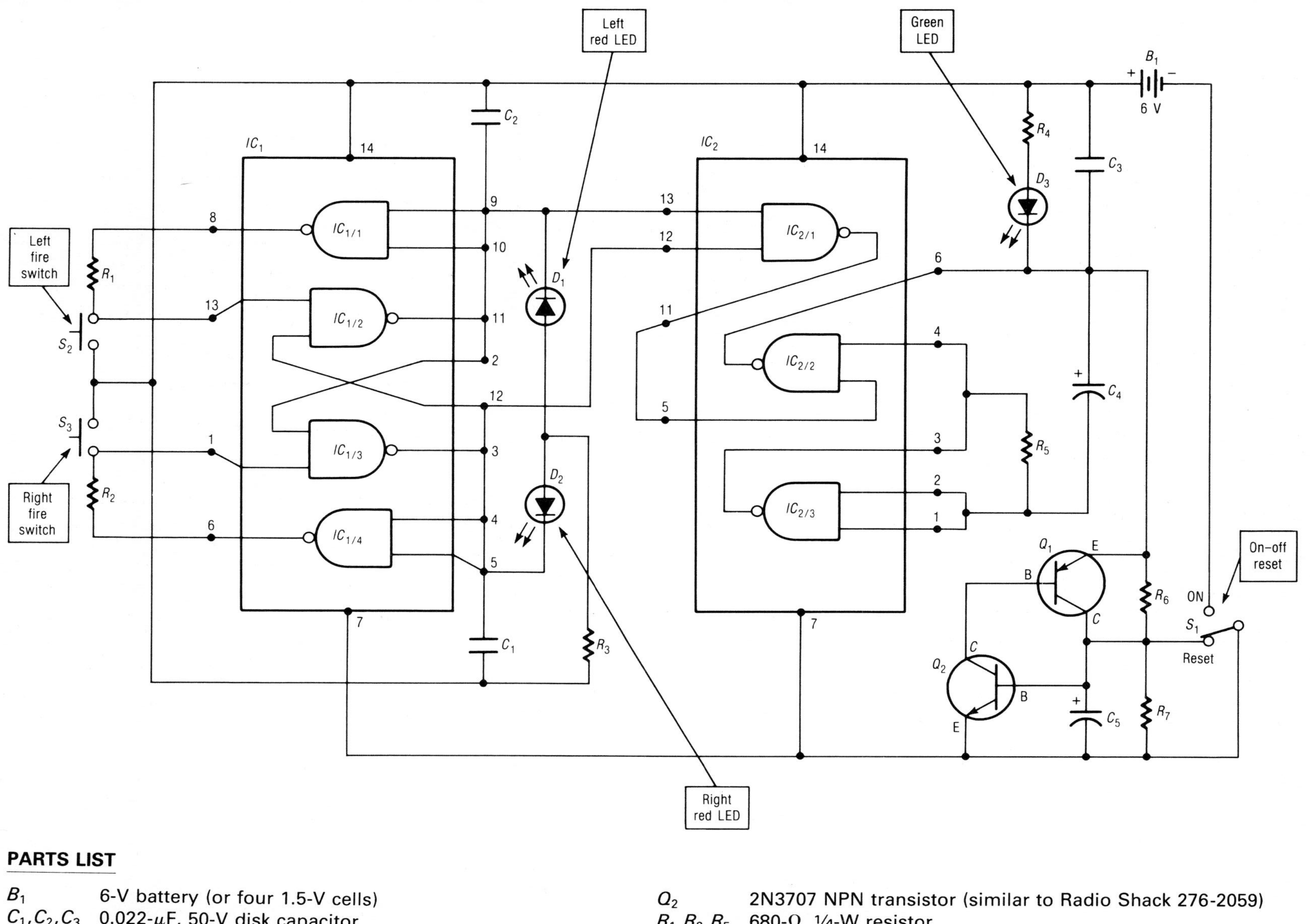

PARTS LIST

B_1	6-V battery (or four 1.5-V cells)
C_1,C_2,C_3	0.022-μF, 50-V disk capacitor
C_4	100-μF, 25-V electrolytic capacitor
C_5	47-μF, 25-V electrolytic capacitor
D_1,D_2	Red light-emitting diode
D_3	Green light-emitting diode
IC_1,IC_2	7400 quad 2-input NAND gate IC
Q_1	2N3702 PNP transistor (similar to Radio Shack 276-2023)
Q_2	2N3707 NPN transistor (similar to Radio Shack 276-2059)
R_1,R_2,R_5	680-Ω, 1/4-W resistor
R_3,R_4	330-Ω, 1/4-W resistor
R_6	330-kΩ, 1/4-W resistor
R_7	39-kΩ, 1/4-W resistor
S_1	SPDT switch
S_2,S_3	Normally open push-button switch

FIG. 8-6 Electronic shoot-out game. (Used by permission of Mode Electronics.)

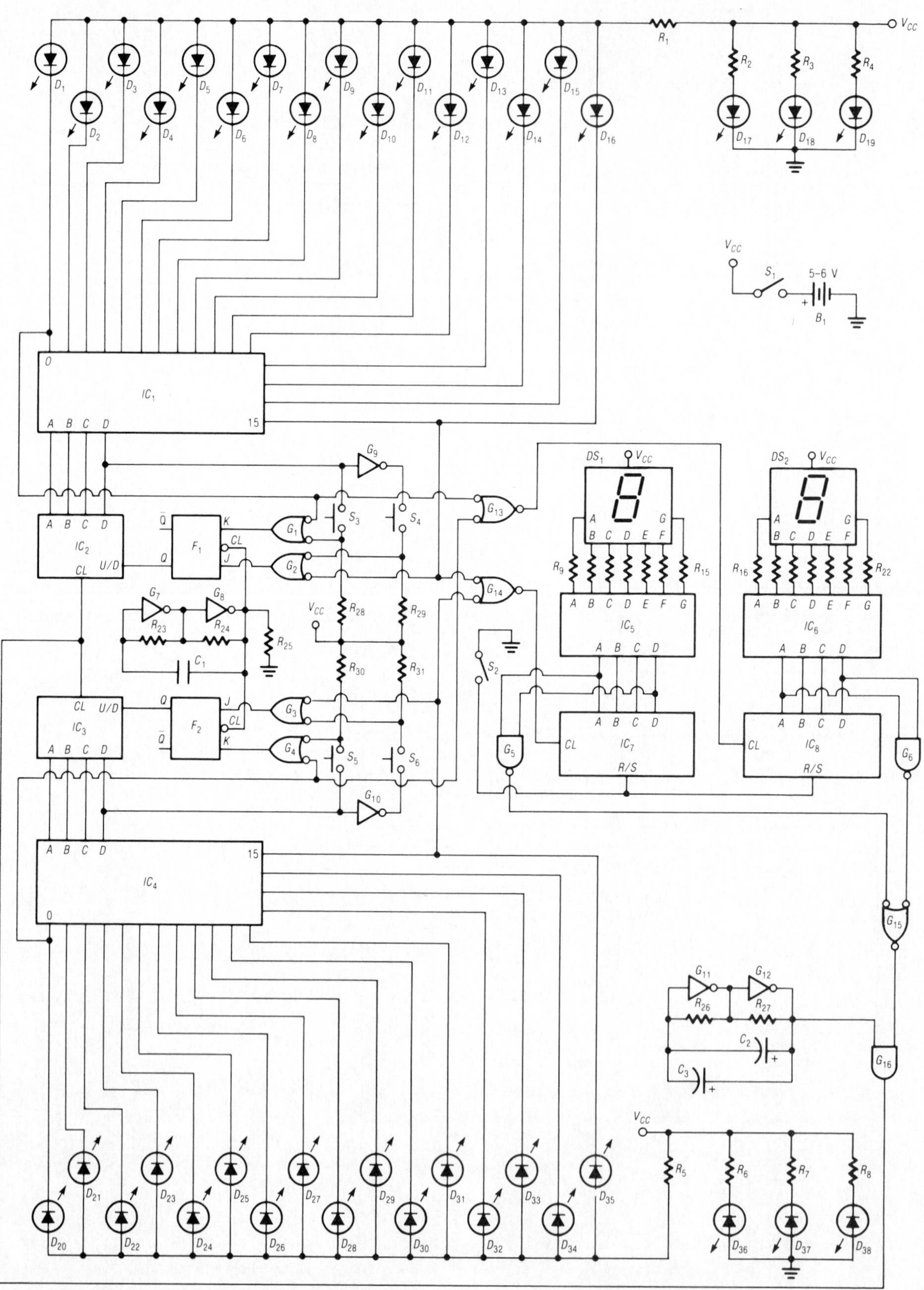

FIG. 8-7 Electronic tennis game. See parts list on facing page. (Used by permission of Electronic Kits International, Inc.)

PARTS LIST FOR FIG. 8-7

B_1	6-V battery (four 1.5-V ni-cad or alkaline cells)
C_1	0.1-μF, 50-V disk capacitor
C_2	10-μF, 25-V electrolytic capacitor
C_3	4.7-μF, 25-V electrolytic capacitor
D_1 through D_{38}	Light-emitting diode
DS_1,DS_2	FND-510 common-anode seven-segment LED display
F_1,F_2	(1) 7473 dual *J-K* flip-flop IC
G_1 through G_6	(2) 7400 quad 2-input NAND gate IC
G_7 through G_{12}	(1) 7404 Hex inverter IC
G_{13} through G_{16}	(1) 7408 quad 2-input AND gate IC
IC_1,IC_4	74154 4- to 16-line distributor IC
IC_2,IC_3	74191 up/down counter IC
IC_5,IC_6	7447 seven-segment decoder-driver IC
IC_7,IC_8	7490 counter IC
R_1 through R_8	100-Ω, ½-W resistor
R_9 through R_{22}	220-Ω, ½-W resistor
R_{23} through R_{27}	1-kΩ, ½-W resistor
R_{28} through R_{31}	10-kΩ, ½-W resistor
S_1,S_2	SPST switch
S_3 through S_6	Normally open push-button switch

through D_{38}) light continuously, marking the boundary of the net and ends of the court. Next, a single light travels back and forth on each side. It simulates two balls moving back and forth on two side-by-side tennis courts. With just one hand the player must press the button to block the moving light (the ball) from touching the end of your court. The button must be pressed after the ball comes over the net on your end of the court. If you do not block the ball before it touches the end of the court, your opponent will be given a score on the seven-segment display. The game will end when one of the scoring displays reaches 9. The reset switch S_2 will start a new game.

The block diagram in Fig. 8-8(*b*) identifies some of the functional circuits in the electronic tennis game. Notice that the two counter-distributor-display blocks are the same and are driven and controlled by the logic and clock section. The up-down counter is the key to how the light moves first one direction down a line of LEDs and then reverses. The players' control switches are fed into the control logic and the number of scores are fed to, counted, and displayed in the scoring block. Although the logic section controls the system, it does need feedback and input from the other blocks to make decisions.

The circuit has a fairly heavy current draw; therefore, either a +5-V dc regulated power supply or ni-cad or alkaline dry cells should be used. A parts kit and a double-sided pc board are currently available for the electronic tennis game from Electronic Kits International, Inc.

ELECTRONIC TUG OF WAR

An electronic tug-of-war game circuit using easy-to-find 7400 series TTL ICs is shown in Fig. 8-9. To play the game, the start switch is pressed, loading a 5 into the counter IC_5 and lighting the middle LED. The players then operate the switches S_1 and S_2 to "pull" the light toward them. The counter IC_5 will count either upward toward LED_9 or downward toward LED_1. If the light is pulled to either player's end (if LED_1 or LED_9 lights), the game automatically stops and that player wins. To play again, the start push-button must be pressed

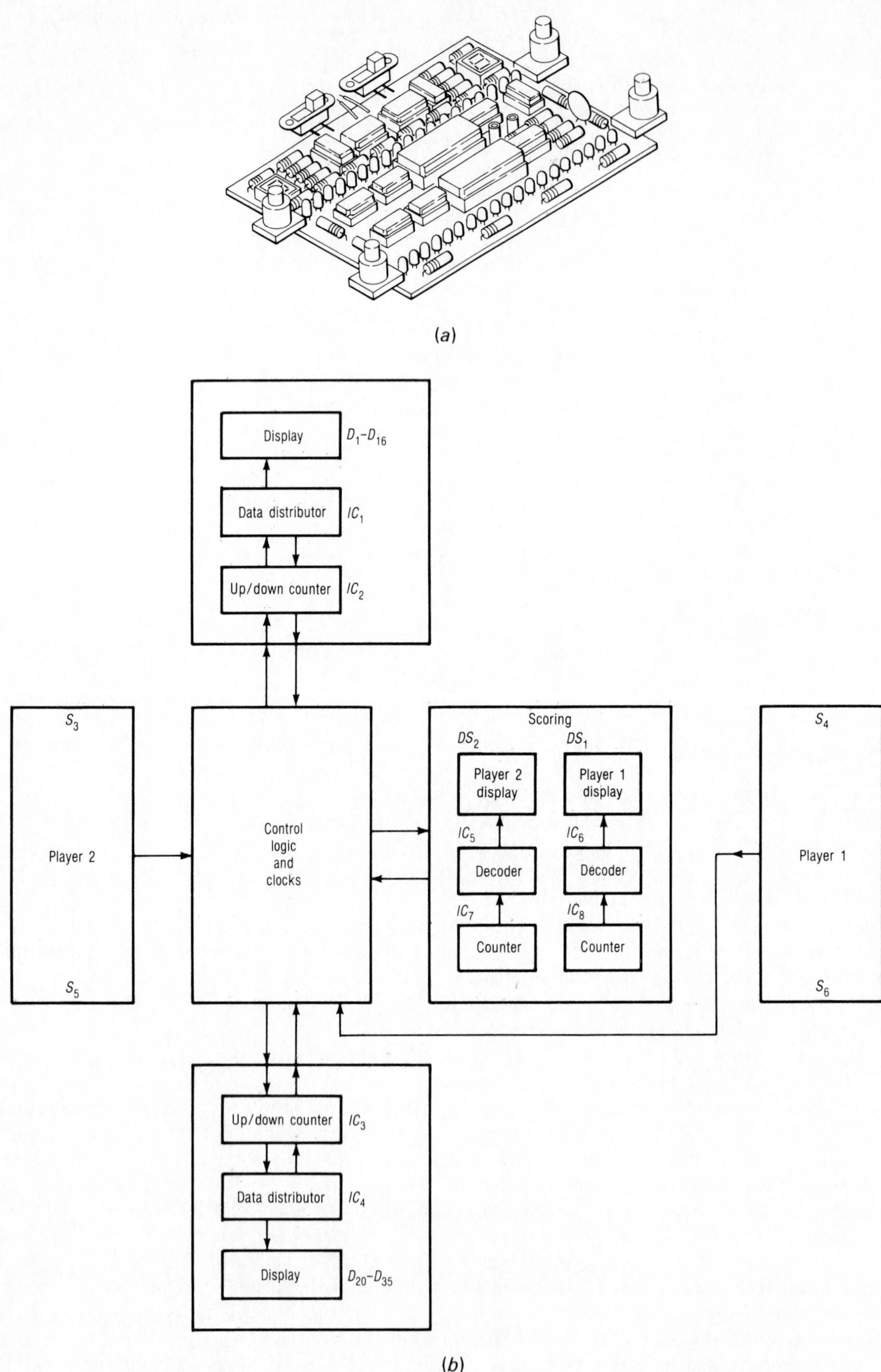

FIG. 8-8 Electronic tennis game. (Used by permission of Electronic Kits International, Inc.)

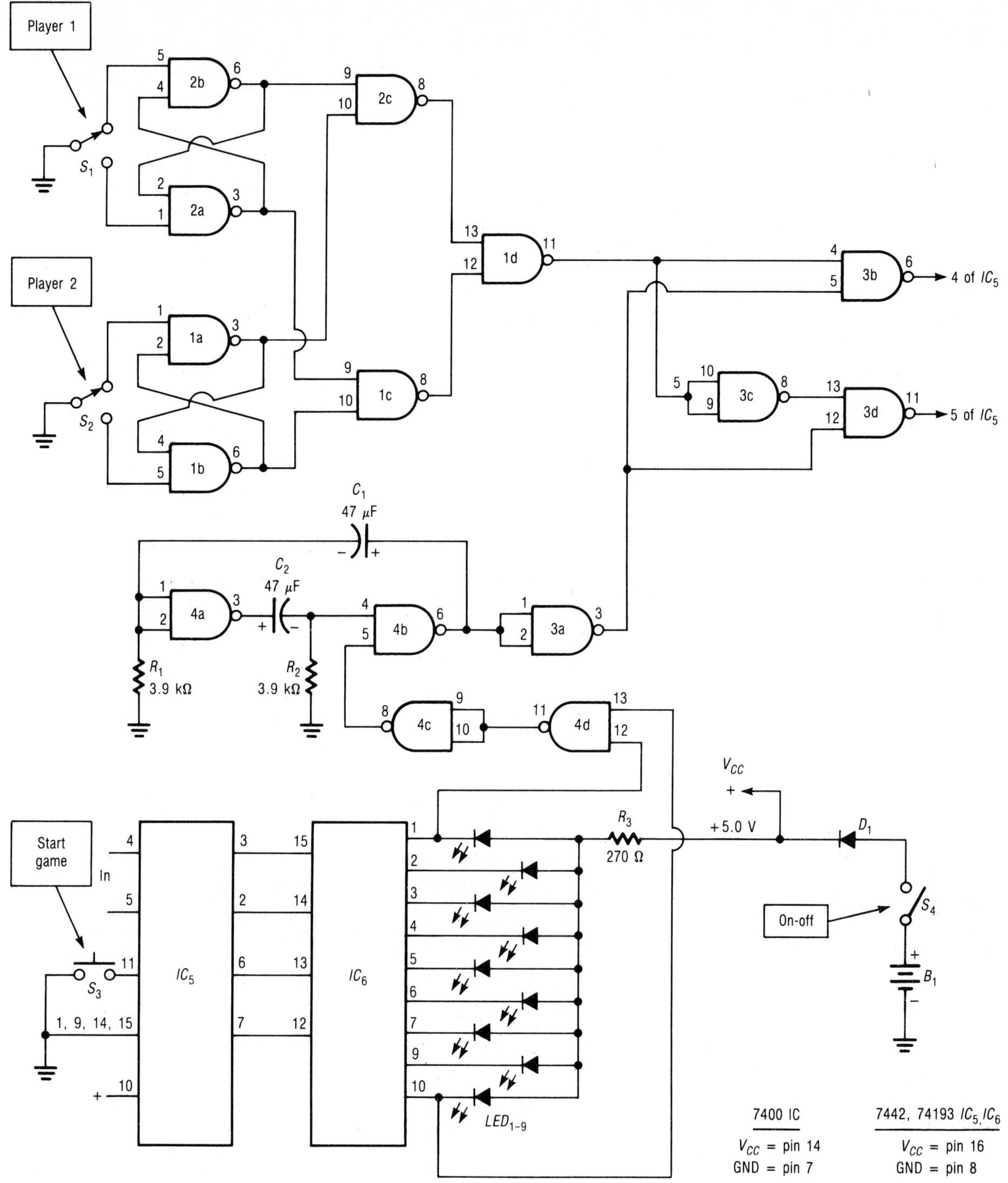

PARTS LIST

B_1	6-V battery (or four 1.5-V ni-cad or alkaline cells)	LED_1 through LED_9	Light-emitting diode
		R_1,R_2	3.9-kΩ, ¼-W resistor
C_1,C_2	47-μF, 16-V electrolytic capacitor	R_3	270-Ω, ¼-W resistor
D_1	1N4001 silicon diode, 1 A, 59 PIV	S_1,S_2	SPDT slide switch
IC_1 through IC_4	7400 quad 2-input NAND gate IC	S_3	Normally open push-button switch
IC_5	74193 up-down counter IC		
IC_6	7442 1-of-10 decoder IC	S_4	SPST switch

FIG. 8-9 Electronic tug-of-war game. (Used by permission of Robert Delp Electronics.)

again. It takes a combination of skill and luck to win the game.

The electronic tug-of-war game is based on an up-down counter (IC_5). Pressing the start button loads a 5 (binary 0101) into the counter which is decoded by IC_6 and lights the middle light-emitting diode (LED_5). The gating circuit IC_{3b}, IC_{3c}, and IC_{3d} directs clocking pulses to either pin 4 or 5 of the counter IC. If pin 4 is pulsed, IC_5 counts downward; if pin 5 receives a pulse, the count is upward. The up-down gating section is fed by the low-frequency clock IC_{4a}, IC_{4b}, and IC_{3a} and a logic circuit IC_{1c}, IC_{1d}, and IC_{2c} driven by the players' switches. The players' switches are debounced by IC_{1a} and IC_{1b} and IC_{2a} and IC_{2b}. To stop the game when either end LED has been lit, lines from LED_1 and LED_9 go to the gating circuit IC_{4b}, IC_{4c}, and IC_{4d}, causing the clock to turn off when either player wins.

Resistors R_1 and R_2 along with capacitors C_1 and C_2 are part of the clock circuit. Diode D_1 drops the voltage of the 6-V battery to about +5 V required to operate the TTL ICs. Resistor R_3 limits the current through the LEDs to a safe level.

SHOCKER

The shocker game shown in Fig. 8-10 is a favorite among students and young adults. The player grasps the probe, touching both sides of the double-sided pc board. Next, the player tries to place the probe's wire tip in the slot in the metal plate *without touching the plate.* If the plate is touched with the probe, the player receives a harmless shock. The intensity of the shock can be adjusted with potentiometer R_1.

The player must be touching both sides of the double-sided pc board to receive a shock when the metal plate is touched. As the wire on the probe makes and breaks contact with the metal plate, a pulse of current flows through the 25-V coil of the transformer. The transformer steps up the voltage and the player feels the shock across the two sides of the pc board. Potentiometer R_1 acts as a voltage divider and sends more or less voltage to the fingers of the player. The unit operates best on an alkaline or ni-cad battery.

CODE PRACTICE OSCILLATOR

A oscillator circuit using discrete components is detailed in Fig. 8-11. This circuit will be of interest to budding amateur radio operators for use as a code practice oscillator. Depression of key K_1 produces a steady tone from the speaker.

Transistors Q_1 and Q_2 are wired as an oscillator. Part of the output signal from Q_2 that drives the speaker is fed back to the base of Q_1 through feedback capacitor C_1. Try changing the value of C_1 for a different tone.

An alkaline or ni-cad battery is recommended on most oscillator circuits. A parts kit and a pc board are currently available for the code practice oscillator from Mode Electronics.

WHEEL OF FORTUNE

A wheel-of-fortune circuit that features audible "clicks" as it spins is detailed in Fig. 8-12. Depression of switch S_1 starts the "spin" accompanied by clicks. The spin starts fast and gradually slows and stops with 1 of the 10 LEDs lit. The object of the game is to guess which of the 10 LEDs will remain lit when the wheel of fortune stops its spin. The LEDs would typically be arranged in a circle. A typical spin takes about 20 to 30 seconds.

A variable-frequency oscillator or clock is formed by inverters *A*, *B*, and *C* in Fig. 8-12. Clock pulses are sent to pin 14 of the 4017 counter IC. The decade counter/decoder/driver counts from 0 to 9, lighting each LED in sequence. Each pulse from the clock advances the count by one. The clock also sends pulses through inverter *D* to be amplified by Q_1. These pulses are heard as clicks from the speaker. Capacitor C_3 affects the speed of rotation. When switch S_1 is pressed for a time, capacitor C_2 charges. After S_1 is released, the charge on C_2 powers the oscillator and keeps transistor Q_2 turned on, allowing the LEDs to light. When C_2 becomes discharged, the oscillator is turned off, leaving the single winning LED lit. Inverters *E* and *F* keep transistor Q_2 turned on, allowing the LEDs to light for about ½ minute after the spin stops. Then the output of inverter *F* goes low, turning off Q_2 and turning off the LED.

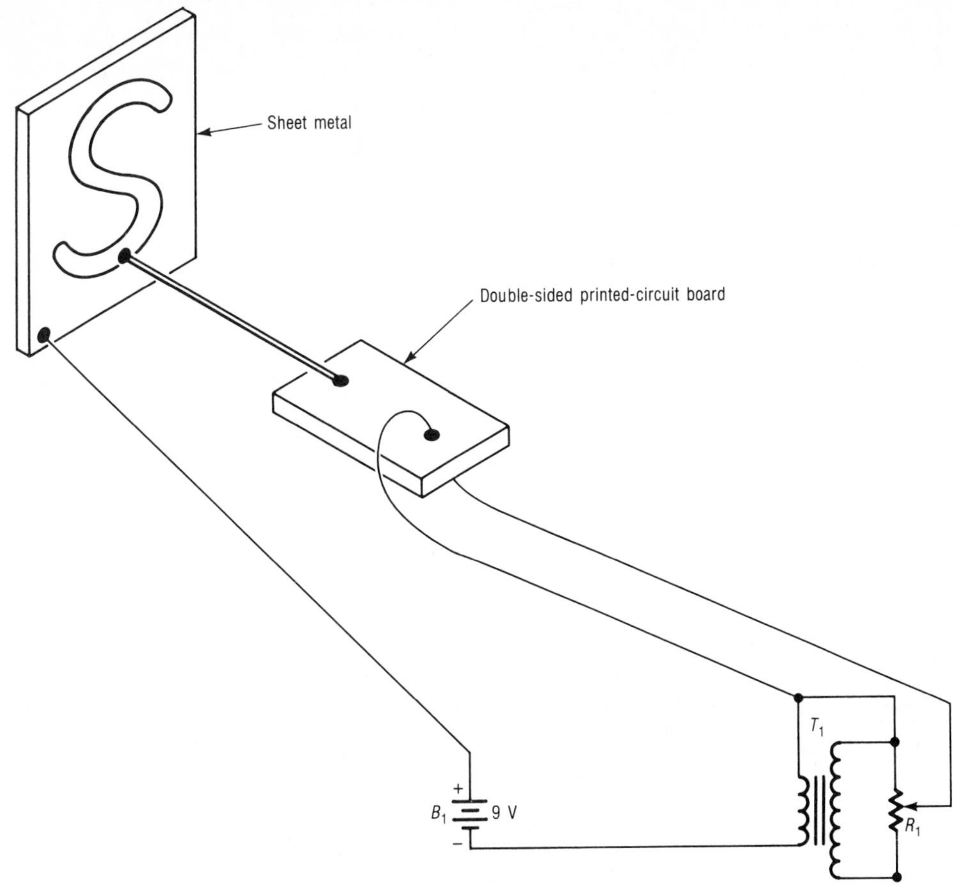

PARTS LIST

B_1 9-V battery
R_1 5-kΩ potentiometer
T_1 25.2-V miniature power transformer (Radio Shack 273-1386)
Double-sided pc board
Sheet metal

FIG. 8-10 Shocker.

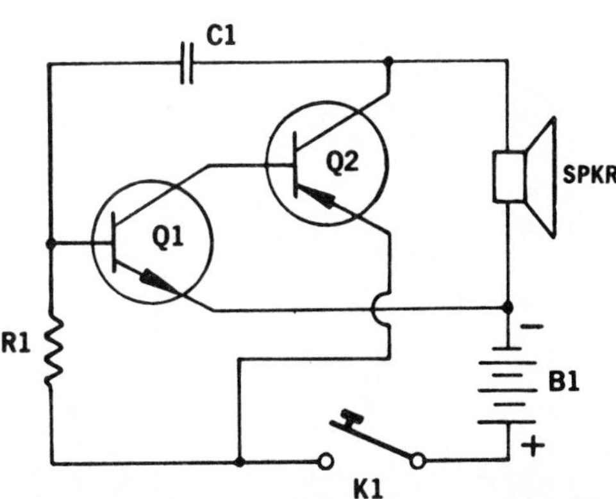

PARTS LIST

B_1	9-V battery (alkaline or ni-cad)	Q_2	2SB474 PNP power transistor (similar to ECG-226)
C_1	0.01-μF, 50-V disk capacitor	R_1	100-kΩ, ½-W resistor
K_1	Telegraph or code key	SPKR	8-Ω speaker
Q_1	BC408A NPN transistor (similar to Radio Shack 276-2016)		

FIG. 8-11 Code practice oscillator. (Used by permission of Mode Electronics.)

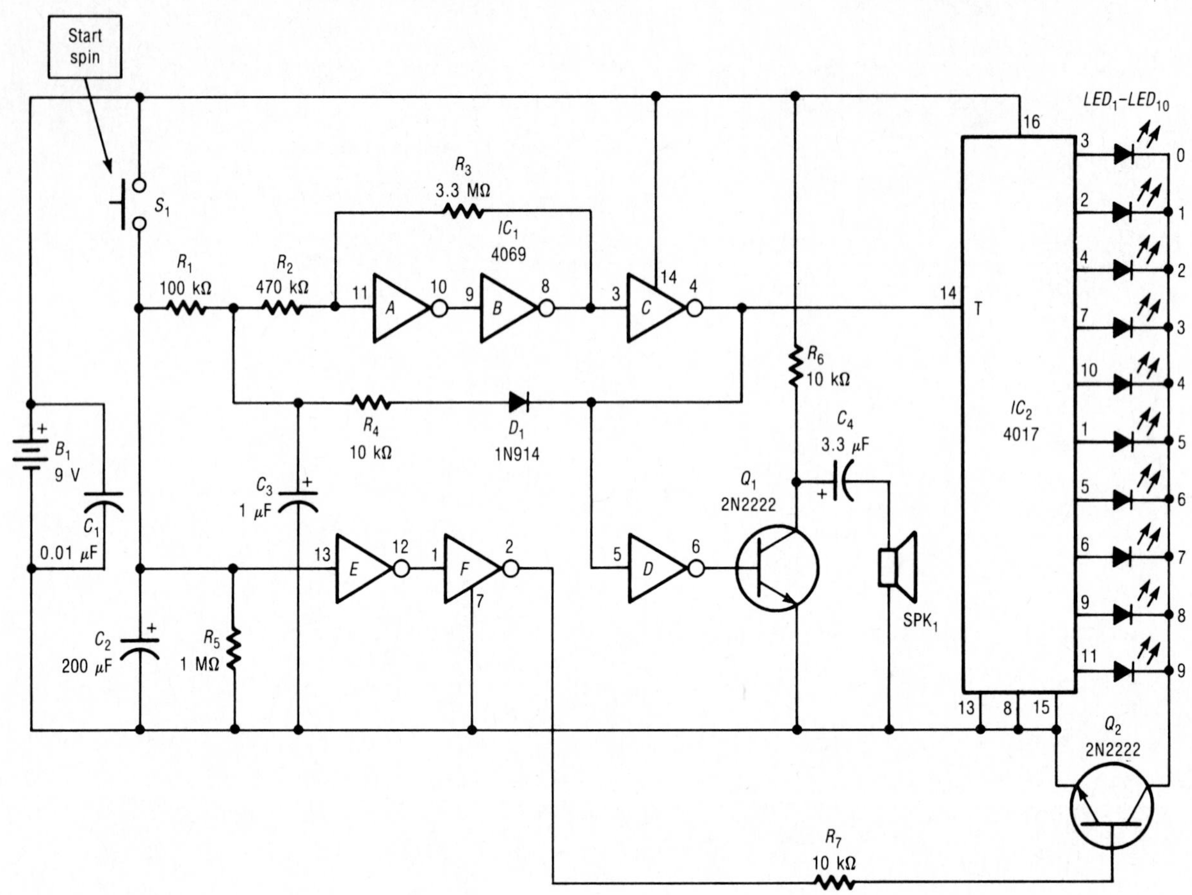

PARTS LIST

B_1	9-V battery
C_1	0.01-μF, 50-V disk capacitor
C_2	200-μF, 16-V electrolytic capacitor
C_3	1-μF, 16-V electrolytic capacitor
C_4	3.3-μF, 16-V electrolytic capacitor
D_1	1N914 silicon diode
IC_1	4069 CMOS Hex inverter IC
IC_2	4017 CMOS decade counter-decoder IC
LED_1 through LED_{10}	Light-emitting diode
Q_1,Q_2	2N2222 NPN transistor (or similar)
R_1	100-kΩ, ¼-W resistor
R_2	470-kΩ, ¼-W resistor
R_3	3.3-MΩ, ¼-W resistor
R_4,R_6,R_7	10-kΩ, ¼-W resistor
R_5	1-MΩ, ¼-W resistor
S_1	Normally open push-button switch
SPK_1	8-Ω speaker

FIG. 8-12 Wheel of fortune. (Reprinted from *Popular Electronics*. Copyright © April 1978, Ziff-Davis Publishing Company.)

chapter 9

Flasher and Display Circuits

BAR DISPLAY WITH ALARM

The bar display detailed in Fig. 9-1 is a circuit which responds like a voltmeter. As more voltage is applied to the input, more LEDs light. This forms a bar of light. If the voltage exceeds 1.2 V, all the LEDs are lit and will begin to flash on and off as an alarm of an overvoltage condition.

The heart of the circuit is the LM3914 dot/bar display driver by National Semiconductor. This IC senses analog voltage levels and drives 10 LEDs, providing a linear analog display. If pin 9 of the LM3914 IC were disconnected and left floating, a single LED will be lit at a time instead of a bar of light. This is called the *dot display mode*. The dot display mode cuts down on power consumption. Assume that 0.4 V is applied to the input; in the dot display mode, only LED_3 will light while in the bar display mode, the left three LEDs will light. Another feature of the LM3914 IC is that the current to the LEDs is regulated internally. This eliminates the need for external limiting resistors.

0- TO 5-V BAR GRAPH METER

The bar graph voltmeter circuit shown in Fig. 9-2 has a range from 0 to 5 V. As the voltage at the input increases to 0.5 V, the first light-emitting diode LED_1 on the left lights. If the voltage increases to slightly more than 2.5 V, the left five LEDs are lit, forming a bar graph indicating the input voltage.

The voltmeter circuit is based on the LM3914 dot/bar display driver IC by National Semiconductor. This IC senses analog voltage levels and drives 10 LEDs, providing an analog display. An external voltage divider could be set up to expand the range of the bar graph voltmeter.

FAST LED BLINKER

The LED blinker circuit shown in Fig. 9-3 is extremely simple and is powered by a single 1.5-V battery. The LED will flash at about 2 Hz. Current drain from the battery is very low

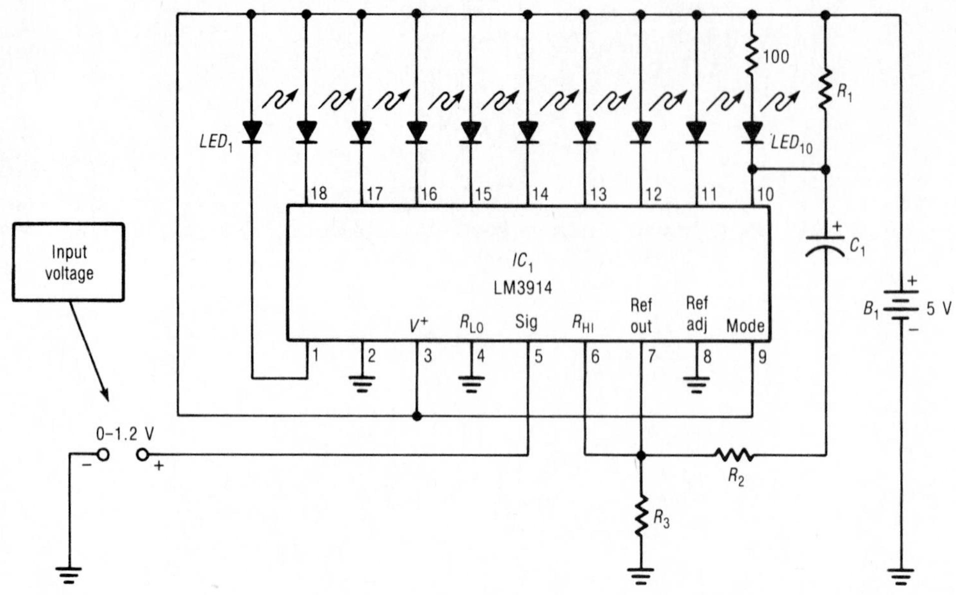

PARTS LIST

B_1	5-V dc regulated power supply (or 6-V battery)
C_1	100-μF, 16-V electrolytic capacitor
IC_1	LM3914 dot/bar display driver IC (National Semiconductor)
LED_1 through LED_{10}	Light-emitting diode (MV57164 or Radio Shack 276-081 bar graph display may be substituted for 10 LEDs)
R_1	1-kΩ, ¼-W resistor
R_2	470-Ω, ¼-W resistor
R_3	1.2-kΩ, ¼-W resistor

FIG. 9-1 Bar display with alarm. (Used by permission of National Semiconductor Corporation.)

at about 1 mA. Even an inexpensive dry cell should operate the blinker continuously for many months.

The LED blinker circuit uses the LM3909 LED flasher/oscillator IC by National Semiconductor. The IC is specifically designed to flash LEDs. By using a timing capacitor (C_1) for voltage boost, it delivers pulses of 2 V or more to the LED while operating on a supply of 1.5 V or less. Increasing the value of timing capacitor C_1 will decrease the flash rate of the blinker.

FRIENDLY FLASHER

The friendly flasher circuit detailed in Fig. 9-4 is an updated version of the older neon lamp "idiot box." The 20 LEDs flash on and off in a seemingly random fashion. The friendly flasher uses common parts.

The 555 timers IC_1, IC_2, IC_3, and IC_4 are wired as free-running multivibrators each with a slightly different low frequency. The 555 timer outputs are directed to various 7447 decoder/drivers (IC_5, IC_6, and IC_8) which light the LEDs. Resistors R_{12} through R_{32} limit the current through each LED to a safe level. The NAND gates IC_7 are used as inverters (inputs wired together). Diode D_1 drops the battery voltage to about 5 V for use with the ICs. Because of the fairly heavy current draw, alkaline or ni-cad batteries are recommended. A 5-V regulated dc power supply may be substituted to power the circuit.

INCANDESCENT BULB FLASHER

An ultrasimple incandescent bulb flasher circuit is illustrated in Fig. 9-5. With only three

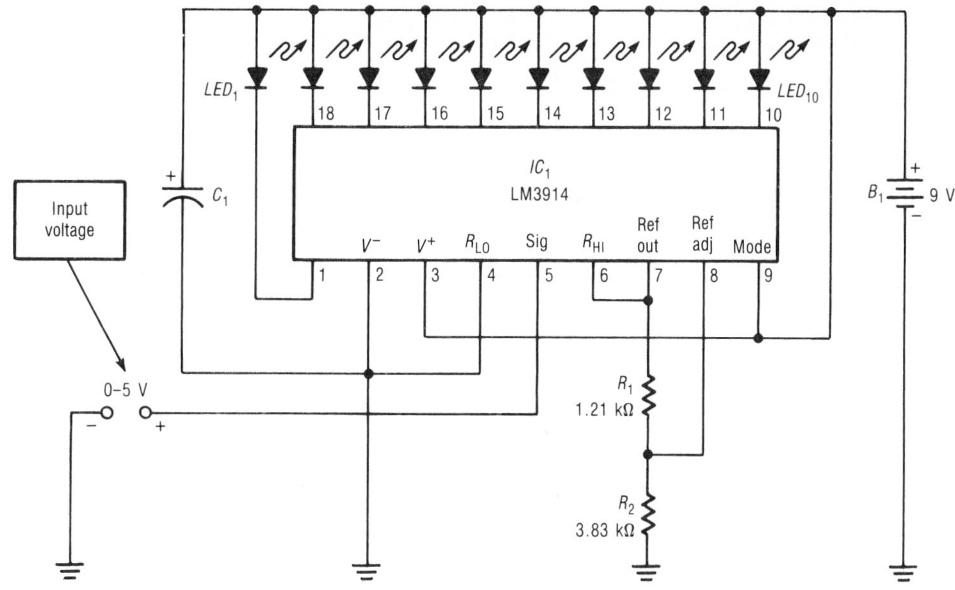

PARTS LIST

B_1	9-V battery
C_1	10-μF, 25-V electrolytic capacitor
IC_1	LM3914 dot/bar display driver IC
LED_1 through LED_{10}	Light-emitting diode (MV57164 or Radio Shack 276-081 bar graph display may be substituted for 10 LEDs)
R_1	1.21-kΩ, ¼-W resistor
R_2	3.83-kΩ, ¼-W resistor (adjust resistance for best accuracy)

FIG. 9-2 Bar graph voltmeter. (Used by permission of National Semiconductor Corporation.)

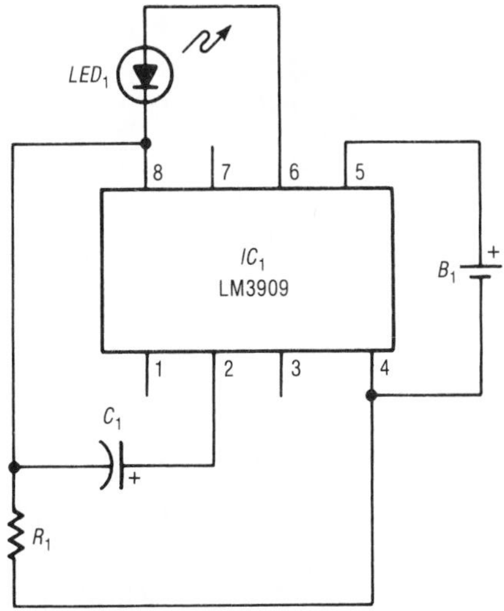

PARTS LIST

B_1	1.5-V battery
C_1	330-μF, 16-V electrolytic capacitor
IC_1	LM3909 LED flasher/oscillator IC (National Semiconductor)
LED_1	Light-emitting diode
R_1	1-kΩ, ¼-W resistor

FIG. 9-3 Simple LED blinker. (Used by permission of National Semiconductor Corporation.)

components, the circuit will cause the 6-V lamp to flash at about 1 to 2 Hz.

The bulb flasher circuit in Fig. 9-5 is based on National Semiconductor's LM3909 flasher/oscillator chip (IC_1). The rate of flashing can be adjusted by changing the value of capacitor C_1. A larger-value capacitor results in a lower flash rate. Lamp L_1 draws about 100 milliamperes (100 mA) when it lights.

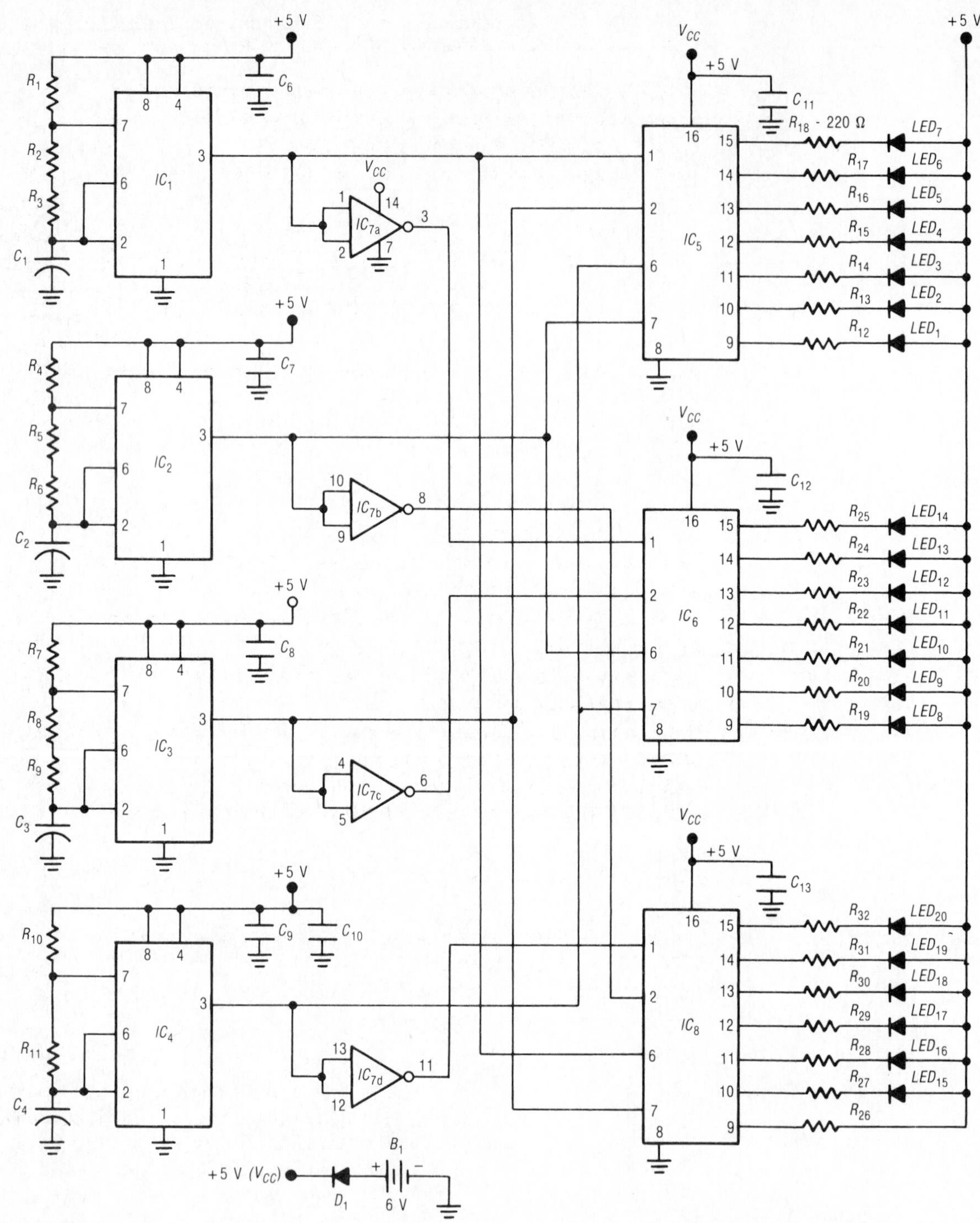

PARTS LIST

B_1	6-V battery (or four 1.5-V ni-cad or alkaline cells)	LED_1 through LED_{20}	Light-emitting diode
		R_1,R_4,R_7,R_{10}	10-kΩ, ¼-W resistor
C_1 through C_4	2.2-μF, 16-V electrolytic capacitor	R_2	220-kΩ, ¼-W resistor
C_6 through C_{13}	0.1-μF, 50-V disk capacitor	R_3,R_5,R_6	470-kΩ, ¼-W resistor
D_1	1N4001 silicon diode, 1 A, 50 PIV	R_8,R_{11}	1-MΩ, ¼-W resistor
IC_1,IC_2,IC_3,IC_4	555 timer IC	R_9	100-kΩ, ¼-W resistor
IC_5,IC_6,IC_8	7447 decoder/driver IC	R_{12} through R_{32}	220-Ω, ¼-W resistor
IC_7	7400 quad 2-input NAND gate IC		

FIG. 9-4 Friendly flasher.

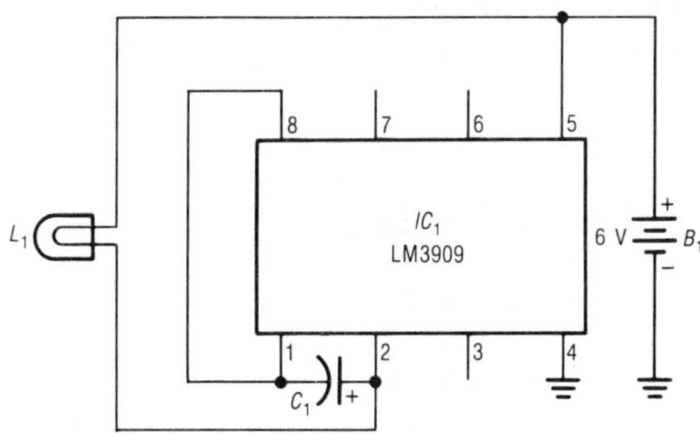

PARTS LIST

- B_1 6-V battery
- C_1 470-μF, 16-V electrolytic capacitor
- IC_1 LM3909 LED flasher/oscillator IC
- L_1 Number 47 lamp (6 V)

FIG. 9-5 Incandescent bulb flasher. (Used by permission of National Semiconductor Corporation.)

LED VOLTAGE-LEVEL INDICATOR

The circuit in Fig. 9-6 will monitor a 5-V power supply used with TTL ICs. The green light-emitting diode (LED_1) in the voltage indicator circuit will be lit only when the voltages are within acceptable limits (between 4.5 and 5.5 V). The red light-emitting diode (LED_2) will serve as a pilot light. The red LED is the power on indicator. The green LED is the in tolerance indicator. This circuit can be added to a 5-V dc power supply to indicate correct output voltage when operating TTL ICs.

Assume that the voltage to be monitored in Fig. 9-6 is 5 V. The transistor level detector Z_1, R_1, and R_2 will turn on Q_2 and light the green LED. If the input voltage drops below 4.5 V, neither Q_1 nor Q_2 conduct and LED_1 is not lit. If the input voltage rises above 5.5 V, Z_2 conducts, turning on Q_1. This causes the base of Q_2 to be grounded, turning off Q_2 and the green LED. If a different voltage range must be monitored, change to appropriate voltage zener diodes.

NEON LAMP RANDOM FLASHER

The random flasher circuit in Fig. 9-7 flashes four neon lamps in a random manner. The flashing is more interesting if a translucent piece of plastic is placed in front of the bulbs.

The neon lamp flasher circuit consists of two basic sections: the power supply and four identical relaxation oscillators. When powered by

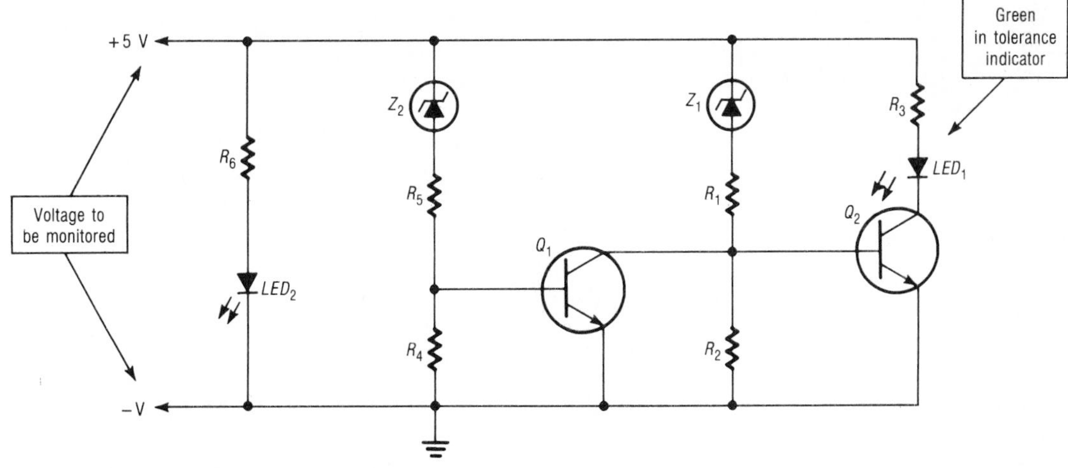

PARTS LIST

- LED_1 Green light-emitting diode
- LED_2 Red light-emitting diode
- Q_1,Q_2 2N5172 NPN transistor (similar to Radio Shack 276-2059)
- R_1,R_5 470-Ω, ¼-W resistor
- R_2,R_4 2.2-kΩ, ¼-W resistor
- R_3 270-Ω, ¼-W resistor
- R_6 220-Ω, ¼-W resistor
- Z_1 3.9-V, 5 percent, 200-mW zener diode
- Z_2 4.7-V, 5 percent, 200-mW zener diode

FIG. 9-6 LED voltage-level indicator (5 V). (R. W. Fox, *Optoelectronics Guidebook—With Tested Projects*, TAB Books, Pennsylvania, 1977, p. 93. Used by permission of TAB Books, Inc.)

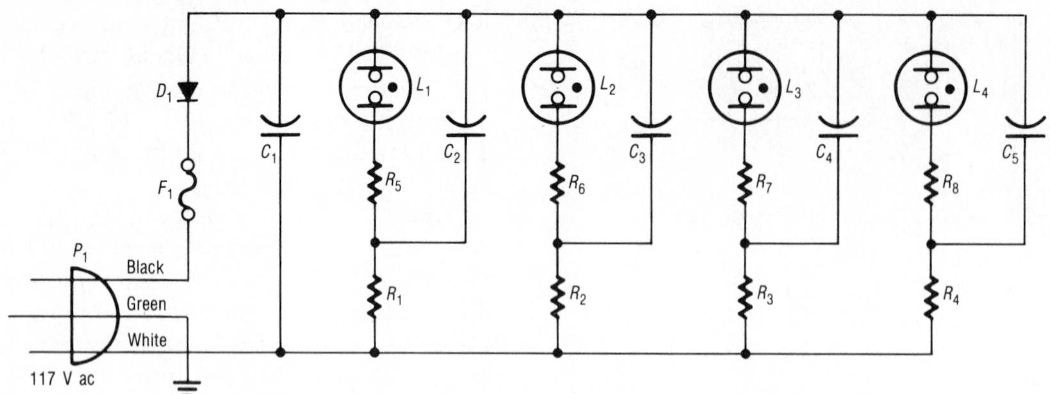

PARTS LIST

C_1	0.1-μF, 250-V capacitor	L_1 through L_4	NE-2 neon lamp
C_2 through C_5	0.47-μF, 250-V capacitor	P_1	Three-pronged ac plug
D_1	1N4004 silicon diode, 1 A, 400 PIV	R_1 through R_4	10-MΩ, ½-W resistor
F_1	0.1-A, fast-blow fuse	R_5 through R_8	1.8-kΩ, ½-W resistor

FIG. 9-7 Neon lamp random flasher.

117 V ac, D_1 and C_1 generate about 160 V dc across capacitor C_1.

Each of the four relaxation oscillators operate the same. Consider one oscillator consisting of R_1, C_2, R_5, and L_1. In the neon bulb, a voltage of 65 to 70 V is needed to ionize the neon gas and fire the lamp. Capacitor C_2 charges through R_1 until the voltage across the capacitor reaches about 70 V. The neon bulb L_1 fires, and C_2 discharges through the lamp. When the voltage across L_1 drops to about 50 V, it stops conducting and C_2 begins to charge again. The neon bulb flashes at about 2 Hz. Several more relaxation oscillators identical to R_4, R_8, C_5, and L_4 can be added to the unit to make the display more dramatic.

The power supply D_1 and C_1 produces about 160 V dc across the filter capacitor C_1. The high-voltage wiring must be done carefully to make sure that it is completely isolated from the enclosure. The green ground wire on the three-pronged plug must be attached to the metal chassis. Use proper strain relief techniques where cords enter and exit the case.

NEON LAMP RELAXATION OSCILLATOR

The single neon lamp relaxation oscillator in Fig. 9-8 is very simple and is battery operated. The current drain is extremely low, at less than 50 μA. The neon bulb flashes at about 2 Hz. Increasing the value of either R_1 or C_1 will increase the time constant of the RC circuit. This causes a decrease in the flash rate of the neon bulb.

A voltage of about 65 to 70 V is needed to ionize the neon gas and fire the neon bulb. Once lit, the neon bulb will stay lit even if the voltage is dropped to about 50 V. The relaxation oscillator makes use of this characteristic of the neon bulb. When the capacitor charges to about 70 V, the neon lamp fires (conducts) and C_1 discharges through it. The neon bulb stops conducting when the voltage drops to about 50 V. The charging time is based on the time constant of C_1 and R_1. The discharge time is set by C_1 and the "on" resistance of the neon bulb.

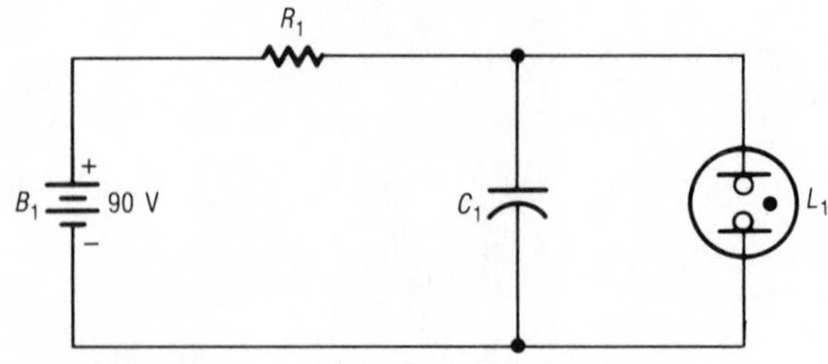

PARTS LIST

B_1	90-V battery
C_1	0.47-μF, 200-V capacitor
L_1	NE-2 neon bulb
R_1	1-MΩ, ¼-W resistor

FIG. 9-8 Neon bulb relaxation oscillator.

STROBE LIGHT

The strobe light in Fig. 9-9 is a favorite among teenagers and young adults. The flash rate can be varied from 1 to 5 per second. The strobe light features a flashtube producing high-intensity white light for short durations.

The xenon flashtube shown in Fig. 9-9 has three leads, labeled C for cathode, A for anode, and T for trigger. The tube is filled with xenon, an inert gas. To operate, the cathode-to-anode polarity must be as shown with a voltage of over 150 V. To ionize the xenon gas and cause the xenon tube to flash (conduct), a short pulse of about 4000 V is required on the trigger input to the tube.

The strobe light circuit in Fig. 9-9 can be divided into two parts: the voltage doubler power supply and the trigger and flash section. Components P_1, R_1, C_1, D_1, D_2, and C_3 make up the power supply. The diodes change ac to dc, and the voltage that appears across filter capacitor C_3 is about 300 V dc as a result of the voltage doubling action of the power supply.

The 300 V dc is applied across the cathode to anode terminals of the flashtube V_2, but a short 4000-V pulse is needed at the trigger lead to fire the xenon tube. As C_2 charges to about 70 V, the voltage fires the neon bulb V_1, causing C_2 to discharge through T_1. The pulse in the primary of trigger transformer T_1 induces a 4000-V pulse in the secondary which triggers and flashes the xenon tube. When the neon bulb V_1 stops conducting, capacitor C_2 starts to charge up again. The flash rate is controlled by the charging rate of C_2 through resistor R_2 and potentiometer R_4.

The high-voltage wiring must be done carefully to make sure that it is completely isolated from the enclosure. The green ground wire on the three-pronged plug must be attached to the metal chassis. Use proper strain relief techniques where cords enter and exit the case. A parts kit and a pc board for the strobe light are currently available from Graymark International, Inc.

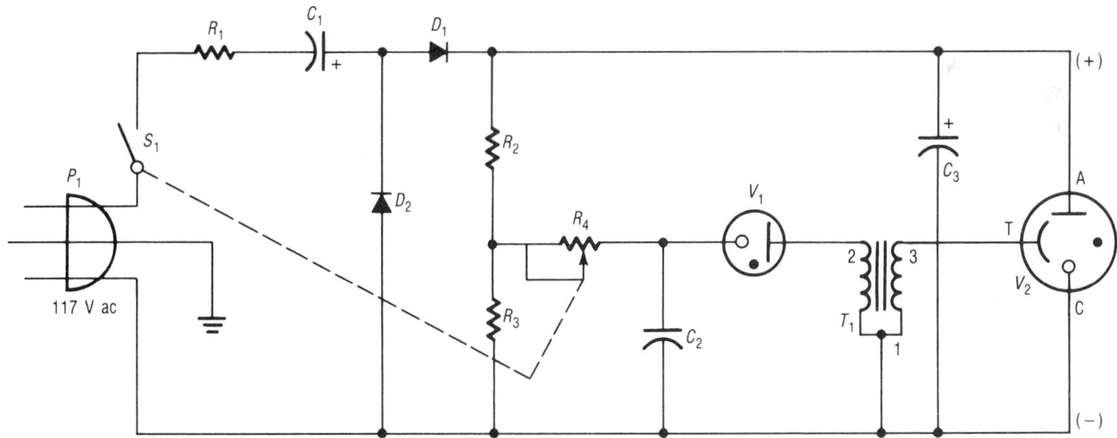

PARTS LIST

C_1	2.2-μF, 350-V electrolytic capacitor
C_2	0.22-μF, 400-V electrolytic capacitor
C_3	22-μF, 450-V capacitor
D_1,D_2	1N4004 silicon diode, 1 A, 400 PIV
P_1	Three-pronged ac plug
R_1	150-Ω, 2-W resistor
R_2	47-kΩ, ½-W resistor
R_3	2.2-MΩ, ½-W resistor
R_4	5-MΩ potentiometer with switch
S_1	SPST switch is on potentiometer
T_1	4-kV flashtube trigger transformer
V_1	Trigger tube
V_2	Xenon flashtube

FIG. 9-9 Variable strobe light. (Used by permission of Graymark International, Inc.)

2-YEAR LED FLASHER

The simple LED flasher circuit in Fig. 9-10 will operate over 2 years on a single battery. The circuit is based on the LM3909 LED flasher/oscillator IC. The flash rate is about 1 Hz. In continuous operation, the flasher circuit should flash from 3 months on a standard AA-size cell up to 2.5 years on an alkaline D-size cell.

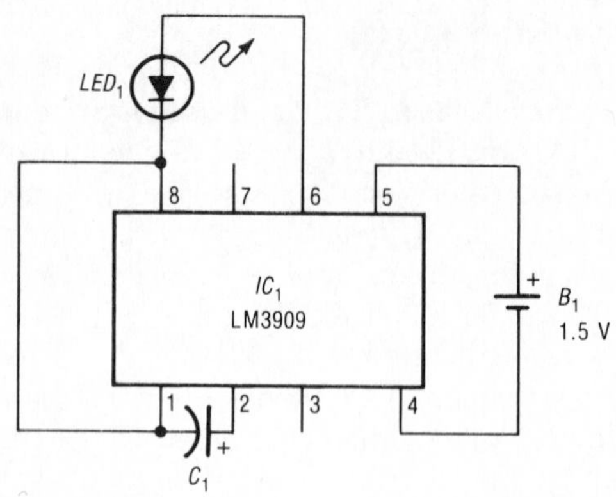

PARTS LIST

B_1	1.5-V battery
C_1	330-μF, 10-V electrolytic capacitor
IC_1	LM3909 LED flasher/oscillator IC (National Semiconductor)
LED_1	Light-emitting diode

FIG. 9-10 2-year LED flasher. (Used by permission of National Semiconductor Corporation.)

12-V SAFETY STROBE LIGHT

The 12-V strobe light in Fig. 9-11 is ideal as an emergency warning flasher to be placed on a stranded vehicle at night. Power for the 12-V strobe light is easily accessible by using the vehicle's cigarette lighter jack. The 1-Hz bright flash from the xenon tube should make the stalled vehicle visible to passing motorists.

The xenon flashtube shown in Fig. 9-11 has three leads, labeled C for cathode, A for anode,

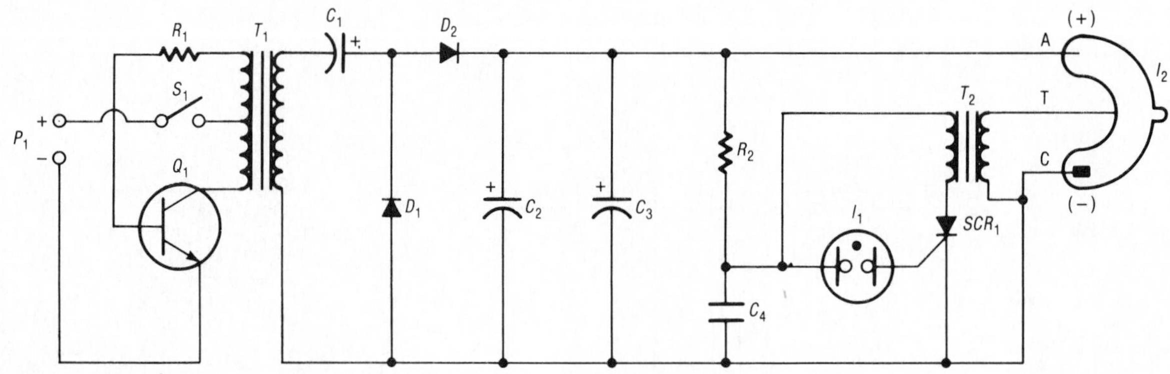

PARTS LIST

B_1	12-V automotive battery
C_1	22-μF, 350-V electrolytic capacitor
C_2,C_3	47-μF, 350-V electrolytic capacitor
C_4	0.47-μF, 250-V capacitor
D_1,D_2	1N4005 silicon diode, 1 A, 600 PIV
I_1	NE-2 neon bulb
I_2	5-W xenon flashtube
Q_1	2N3055 NPN power transistor (similar to Radio Shack 276-2041)
R_1	2.7-kΩ, ½-W resistor
R_2	1-MΩ, ½-W resistor
SCR_1	C106B1 silicon controlled rectifier (similar to Radio Shack 276-1067)
S_1	SPST switch
T_1	24-V, 400-mA, center-tapped secondary power transformer (similar to Radio Shack 273-1366)
T_2	4-kV flashtube trigger transformer

FIG. 9-11 12-V safety strobe light. (Used by permission of Electronic Kits International, Inc.)

and T for trigger. The tube is filled with xenon, an inert gas. To operate, the cathode-to-anode polarity must be as shown with a voltage of over 200 V. To ionize the xenon gas and cause the xenon tube to flash (conduct), a short pulse of about 4000 V is required on the trigger.

The 12-V strobe light in Fig. 9-11 consists of three basic circuits: the inverter, the voltage doubling power supply, and the trigger and flash section. The inverter consists of an oscillator (Q_1, R_1, and primary of T_1) and a transformer (T_1) for stepping up the voltage to more than 100 V. The voltage doubler power supply section consists of rectifiers D_1 and D_2 and capacitors C_1, C_2, and C_3 and the secondary of transformer T_1.

About 300 V dc is applied across the cathode-to-anode terminals of the flashtube I_2, but a short 4000-V pulse is needed at the trigger lead to fire the xenon tube. As C_4 charges to about 70 V, the voltage fires the neon bulb I_1, causing SCR_1 to conduct. When the silicon controlled rectifier conducts the pulse in the primary of trigger transformer T_2, a 4000-V pulse is induced in the secondary which triggers and flashes the xenon tube. When the silicon controlled rectifier is conducting, C_4 discharges through SCR_1. Neon bulb I_1 stops conducting, the SCR turns off, and C_4 starts to charge up again. The flash rate is controlled by the charging rate of C_4 through resistor R_2. Increasing the value of R_2 would increase the time duration between flashes of the strobe.

The high-voltage wiring must be done carefully to make sure that it is completely isolated from the enclosure. Use proper strain relief techniques where cords enter and exit the case. A parts kit and pc board for the 12-V strobe light are currently available from Electronic Kits International, Inc.

VARIABLE LED FLASHER

The LED flasher circuit in Fig. 9-12 features a variable flash rate. The circuit is based on the LM3909 LED flasher/oscillator IC. Adjusting potentiometer R_4 controls the flash rate from 0 to 20 Hz. The circuit draws low current (0.5 to about 10 mA) and can be operated from a single battery for a long time.

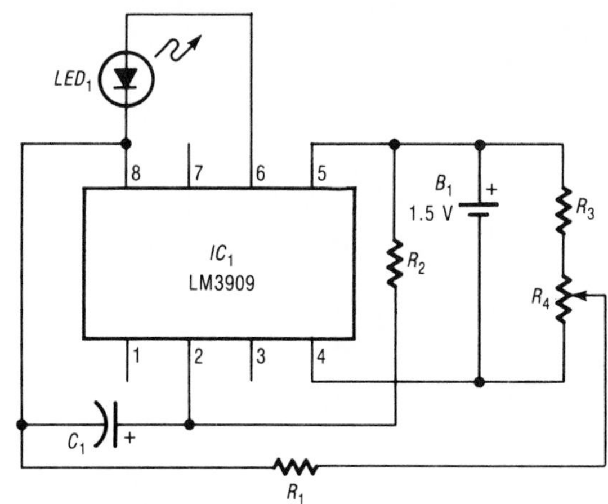

PARTS LIST

B_1	1.5-V battery
C_1	330-μF, 6-V electrolytic capacitor
IC_1	LM3909 LED flasher/oscillator IC
LED_1	Light-emitting diode
R_1,R_2	75-Ω, ¼-W resistor
R_3	2.4-kΩ, ¼-W resistor
R_4	2.5-kΩ potentiometer

FIG. 9-12 Variable LED flasher. (Used by permission of National Semiconductor Corporation.)

DIGITAL COLOR ORGAN

The digital color organ circuit in Fig. 9-13 drives 30 LEDs. When the color organ is connected to the audio output of a radio, low signal levels will light only selected red LEDs. The yellow and red LEDs will light for moderate audio signals while high audio levels light red, yellow, and green LEDs.

The color organ in Fig. 9-13 contains three basic circuits: the clock, the column decoder/driver, and the audio level sensor/LED driver. The 4011 IC_3 and associated components form the clock. The clock output feeds the CLK input of the decade counter IC_2. As the 4017 counts, one column of LEDs at a time is activated (started from left to right). At the same time, a low audio input signal at IC_1 would cause only pin 1 to go low. If pin 3 of IC_2 is activated (high) and pin 1 of IC_3 is low, the lower left red LED is the only one to light. The clock and the 4017 counter sequence the columns. If pin 11 of IC_2 were activated and a

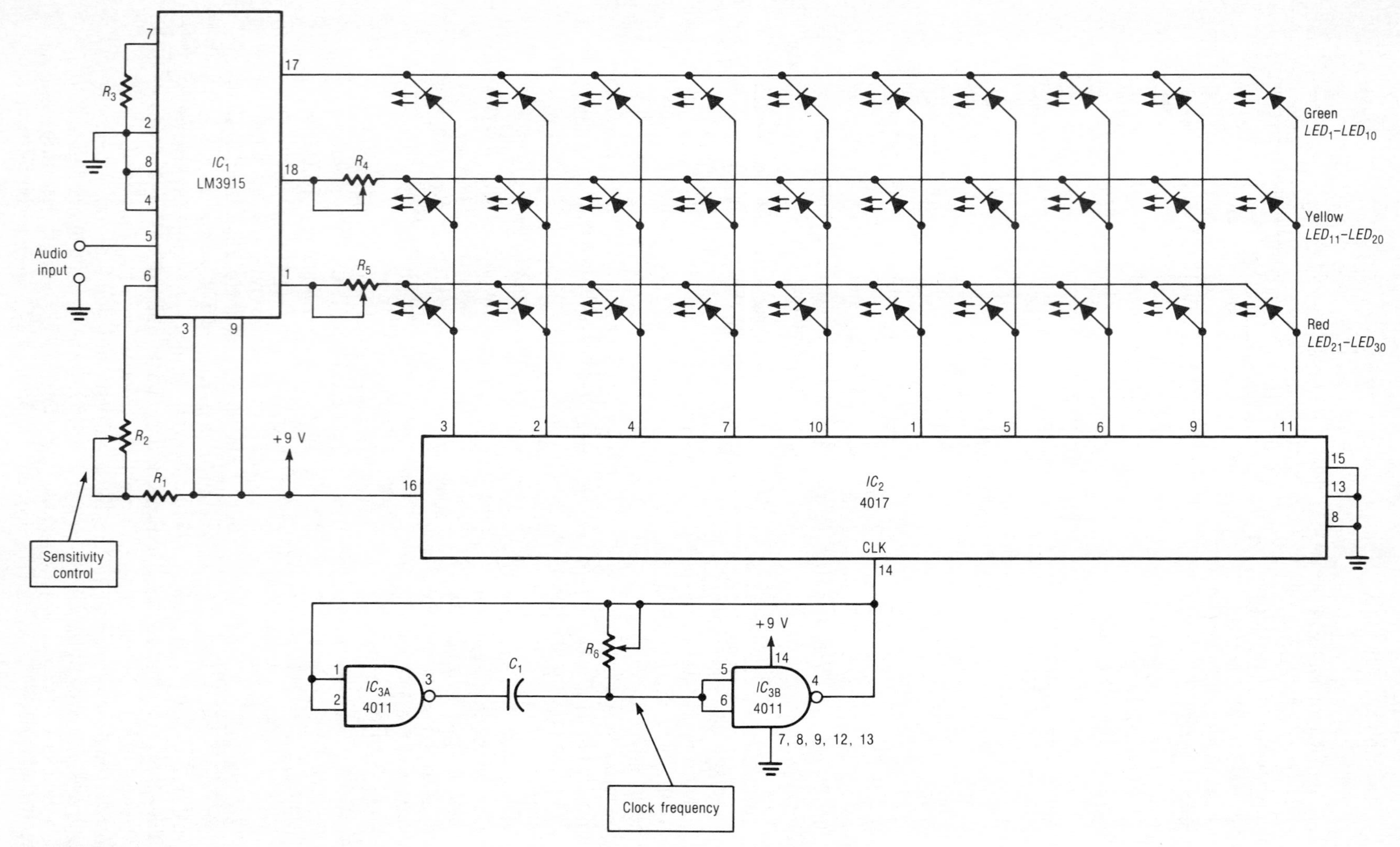

PARTS LIST

C_1	0.1-μF, 25-V disk capacitor
IC_1	LM3915 bar display driver IC
IC_2	4017 CMOS 1-of-10 output decade counter IC
IC_3	4011 CMOS quad 2-input NAND gate IC
LED_1 through LED_{10}	Green light-emitting diode
LED_{11} through LED_{20}	Yellow light-emitting diode
LED_{21} through LED_{30}	Red light-emitting diode
R_1	1.8-kΩ, ¼-W resistor
R_2,R_6	100-kΩ potentiometer
R_3	1-kΩ, ¼-W resistor
R_4,R_5	1-kΩ trimmer potentiometer

FIG. 9-13 Digital color organ. (Forrest Mims, *The Forrest Mims Circuit Scrapbook*, McGraw-Hill, New York, 1983, p. 102. Used by permission of Forrest M. Mims, III.)

high audio signal were present, all three outputs of IC_1 (pins 1, 18, and 17) would go low, causing the red, yellow, and green LEDs in the right column to light. Potentiometer R_2 adjusts the sensitivity of IC_1 to the level of audio input.

The clock frequency can be adjusted by using potentiometer R_6. The clock frequency affects the speed at which the columns are scanned. Trimmer potentiometers R_4 and R_5 adjust the brightness of the yellow and red LEDs.

chapter 10

Meters and Test Instruments

BUZZ BOX CONTINUITY AND COIL CHECKER

The buzz box tester in Fig. 10-1 is a simple continuity and coil tester. When the probes are touched to a low resistance, the speaker will buzz to indicate continuity. However, if the resistance is over 100 Ω, no sound will come from the speaker. The simple battery-operated unit is especially useful for checking continuity and relative resistance of coils, motor windings, and test leads.

The buzz box continuity checker circuit shown in Fig. 10-1 is based on the LM3909 LED flasher/oscillator IC by National Semiconductor. The circuit emits less sound from the speaker as the resistance across the probes increases. The circuit is useful for resistances up to only about 100 Ω. The difference of a few ohms in the resistance of a coil will cause the buzz box to emit a different tone.

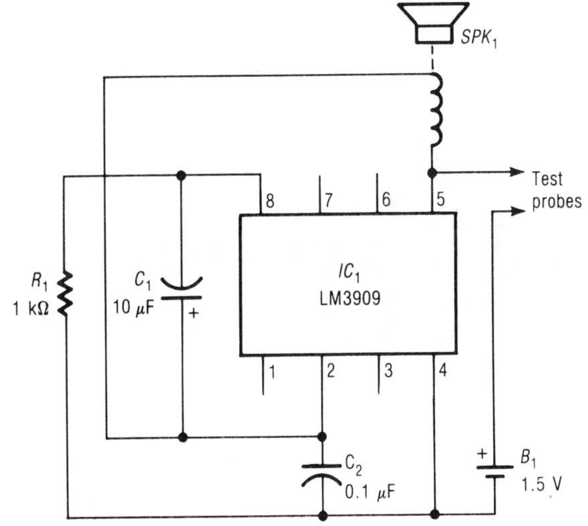

PARTS LIST

B_1	1.5-V battery
C_1	10-μF, 6-V electrolytic capacitor
C_2	0.1-μF, 50-V disk capacitor
IC_1	LM3909 LED flasher/oscillator IC (National Semiconductor)
R_1	1-kΩ, ¼-W resistor
SPK_1	12- to 16-Ω speaker

FIG. 10-1 Buzz box continuity and coil checker. (Used by permission of National Semiconductor Corporation.)

CAPACITANCE SUBSTITUTION BOX

A capacitance substitution box circuit is shown in Fig. 10-2. With a turn of the knob, the rotary switch will select one of nine possible high-voltage capacitors. Substitution boxes are used during experimenting and circuit design work. A parts kit and a pc board for the capacitance substitution box are currently available from Graymark International, Inc.

CONTINUITY TESTER

The continuity tester circuit in Fig. 10-3 combines a continuity tester with a low-voltage checker. With the mode switch in the continuity test position (down on the schematic dia-

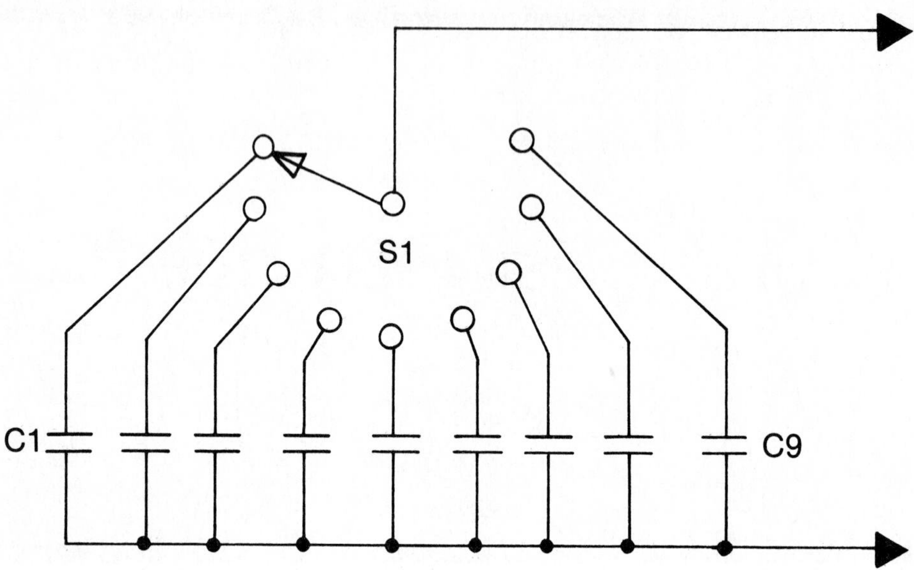

PARTS LIST

- C_1 0.0001-μF, 400-V capacitor
- C_2 0.001-μF, 400-V capacitor
- C_3 0.0022-μF, 400-V capacitor
- C_4 0.0047-μF, 400-V capacitor
- C_5 0.01-μF, 400-V capacitor
- C_6 0.022-μF, 400-V capacitor
- C_7 0.047-μF, 400-V capacitor
- C_8 0.1-μF, 400-V capacitor
- C_9 0.22-μF, 400-V capacitor
- S_1 SP12P rotary switch (similar to Radio Shack 275-1385)

FIG. 10-2 Capacitance substitution box. (Used by permission of Graymark International, Inc.)

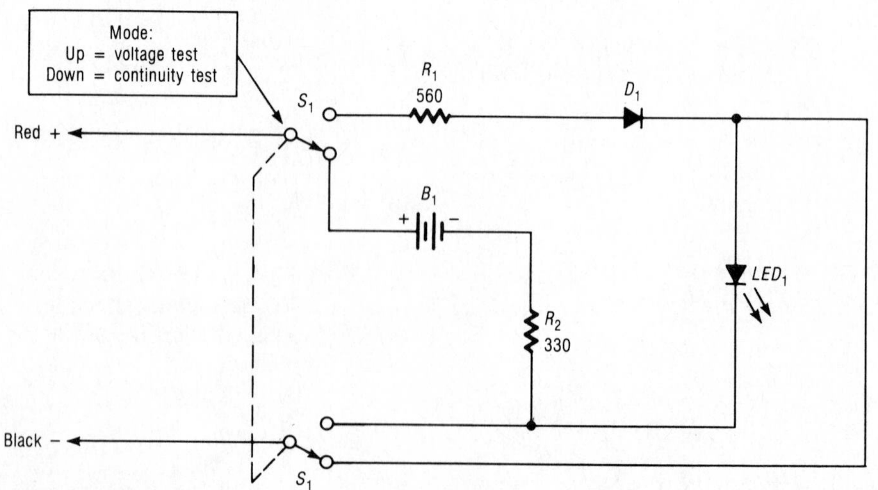

PARTS LIST

- B_1 9-V battery
- D_1 1N4001 silicon diode, 1 A, 50 PIV
- LED_1 Light-emitting diode
- R_1 560-Ω, ½-W resistor
- R_2 330-Ω, ½-W resistor
- S_1 DPDT switch

FIG. 10-3 Continuity and low-voltage tester. (Used by permission of Robert Delp Electronics.)

gram), the LED will light when low resistance is present between the test clips. With the mode switch in the low-voltage test position (up on the schematic diagram), the LED will light when a low voltage is present. The low-voltage tester is very useful in automotive work.

The continuity tester circuit includes B_1, R_2, and LED_1 in series. The LED will light brightly with no resistance between the test leads, but will just barely light with 5 kΩ of resistance at the input. Above 5 kΩ, the LED does not light and the tester indicates "no continuity." This unit is also useful in testing diodes.

The voltage tester circuit includes R_1, D_1, and LED_1 in series. The LED will light with voltages from 3 to 15 V dc or ac. The LED will light more brightly at the higher voltages. The LED can be damaged if test voltages exceed 18 V.

LED BATTERY CHECKER

The battery checker in Fig. 10-4 uses a series of LEDs to indicate the condition of either a 1.5- or a 9-V battery. It is accurate for carbon-zinc, alkaline, or ni-cad batteries. If all LEDs light, the battery under test is in excellent condition. If just the red and two yellow LEDs (D_3 through D_5) light, the battery is weak but still useful for many applications. If just the red LED (D_5) lights, the battery under test is discharged and should be replaced or recharged. The red LED will light whenever the switch is in either the 1.5- or 9-V position to signal the user that the unit is turned on.

The battery checker places a load on the battery under test while reading the terminal voltage. The LEDs in Fig. 10-4 take the place of an expensive analog panel meter used on more conventional battery checkers. Light-emitting diodes D_1 through D_4 are caused to light by voltage comparators housed in the LM324 IC. Resistors R_1 through R_4, R_{12}, and R_{13} form a voltage divider across the 9-V supply battery to set the reference voltages on pins 3, 12, 10, and 5 of the voltage comparators. The reference voltage at pin 3 would be the lowest and that at pin 5, the highest. If the voltage (from the battery under test) applied to each voltage comparator is larger (more positive) than the reference voltage, the output of that

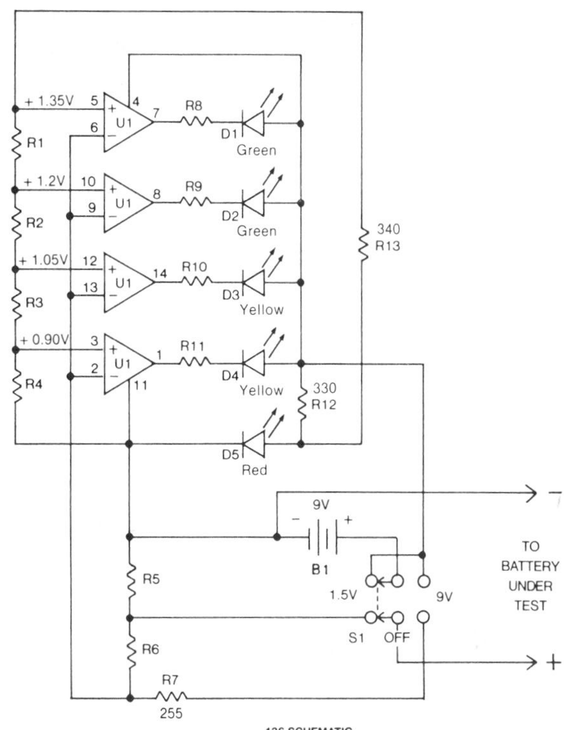

PARTS LIST

B_1	9-V battery
D_1,D_2	Green light-emitting diode
D_3,D_4	Yellow light-emitting diode
D_5	Red light-emitting diode
R_1,R_2,R_3	187-Ω, 1 percent, ¼-W resistor
R_4	1.1-kΩ, 1 percent, ¼-W resistor
R_5	6.2-Ω, ½-W resistor
R_6	45.3-Ω, 1 percent, ¼-W resistor
R_7	255-Ω, 1 percent, ¼-W resistor
R_8,R_9,R_{10},R_{11}	220-Ω, ¼-W resistor
R_{12}	330-Ω, 5 percent, ¼-W resistor
R_{13}	340-Ω, 1 percent, ¼-W resistor
S_1	DPDT, center-off slide switch
U_1	LM324 quad operational amplifier IC

FIG. 10-4 LED battery checker. (Used by permission of Graymark International, Inc.)

comparator will go low, lighting the LED. For instance, if the reference voltage at pin 10 is 1.2 V and the voltage from the battery under test is +1.3 V, the green LED (D_2) would light. In this example, four LEDs would light (D_2 through D_5). In the example, LED D_1 did not light because its reference voltage was 1.35 V while the input voltage was only +1.3 V.

The LED battery tester draws 100 to 200 mA from the 1.5-V cell under test while drawing about 20 to 30 mA from a 9-V test battery.

A parts kit and a pc board for the LED battery checker circuit are currently available from Graymark International, Inc.

LOGIC PROBE

The logic probe circuit in Fig. 10-5 can test either CMOS or TTL voltage levels. If the voltage is in the HIGH range, red LED_1 lights; however, if it is in the LOW range, green LED_2 lights. If the tip is not connected or touching something like your finger, both LEDs light. A digital signal over 30 Hz will appear to cause both LEDs to remain on. The logic probe does not need a battery because it is powered from the circuit under test. The definition of CMOS and TTL logic levels is diagrammed in Fig. 10-6.

The 555 timer IC forms the heart of the logic probe circuit in Fig. 10-5. If used to measure CMOS logic levels, P_1 goes to the positive of the power supply (from +5 to +18 V), while the GND lead P_3 is connected to the negative. Next, the input tip is touched to the point to be tested in the CMOS digital circuit. If the input is at a HIGH logic level, output pin 3 of the IC will go LOW, causing the red LED_1 to light. However, if the input is at a LOW logic level, output pin 3 of the IC will go HIGH, causing the green LED_2 to light. If the high-impedance input to the logic probe is not connected, it picks up stray 60-Hz radiation from the surroundings and alternately lights LED_1 and LED_2. The LEDs will appear to be lit continuously.

Using the circuit in Fig. 10-5 as a TTL logic probe, both P_2 and P_3 are grounded while P_1 goes to +5 V of the power supply. The diagram in Fig. 10-6 graphs the different voltage thresholds that define HIGH and LOW in TTL and CMOS circuits. Resistor R_1 adjusts the voltage thresholds of the 555 timer IC so that they agree with TTL logic levels. A parts kit and a two-sided pc board for the logic probe are currently available from Electronic Kits International, Inc.

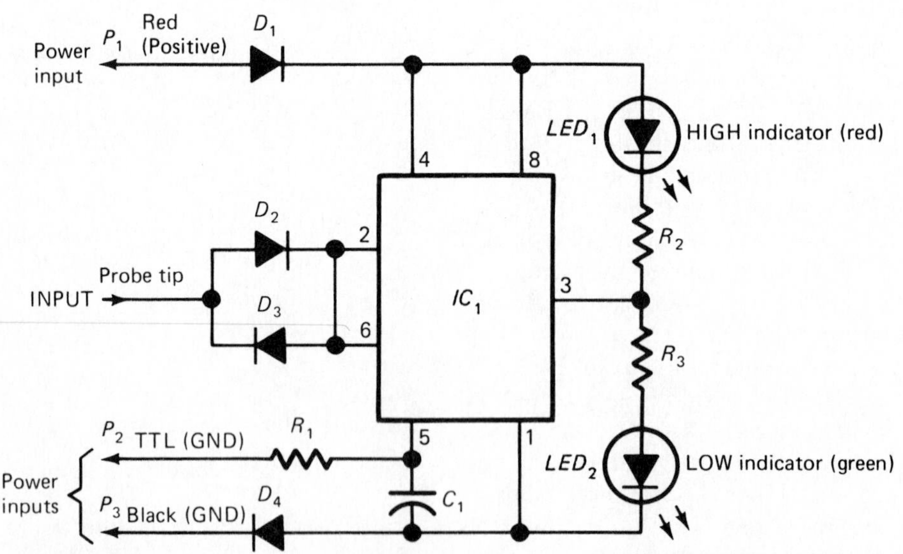

PARTS LIST

C_1	0.01-μF, 50-V disk capacitor
D_1 through D_4	1N914 silicon diode (or similar)
IC_1	555 timer IC
LED_1	Red light-emitting diode
LED_2	Green light-emitting diode
R_1	3.9-kΩ, ¼-W resistor
R_2,R_3	390-Ω, ¼-W resistor

FIG. 10-5 Logic probe. (Used by permission of Electronic Kits International, Inc.)

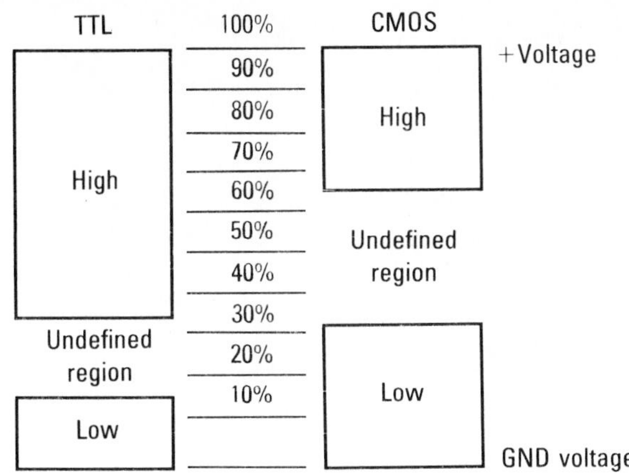

FIG. 10-6 Defining HIGH and LOW logic levels for the CMOS and TTL families of digital ICs. (Roger L. Tokheim, *Digital Electronics*, 2d ed., McGraw-Hill, New York, 1984, p. 12. Used by permission of McGraw-Hill Book Company.)

POWER LINE MONITOR

The power line monitor in Fig. 10-7 will detect a loss of 117 V ac in the power line to which it is connected. After a delay of several seconds, the "power out" alarm will sound. Then IC_1 isolates the battery-operated low-voltage alarm circuit from the power line.

The power line monitor in Fig. 10-7 is based on the General Instrument Corporation's MID400 power line monitor IC. Some of the internal details of the MID400 show that back-to-back LEDs are optically coupled to a photodetector. The photodetector amplifier drives the IC's output transistor. Resistor R_2 is the pull-up resistor for the open-collector output of the MID400 IC. If the ac input should stop, pin 6 of IC_1 goes from low to high. This triggers the

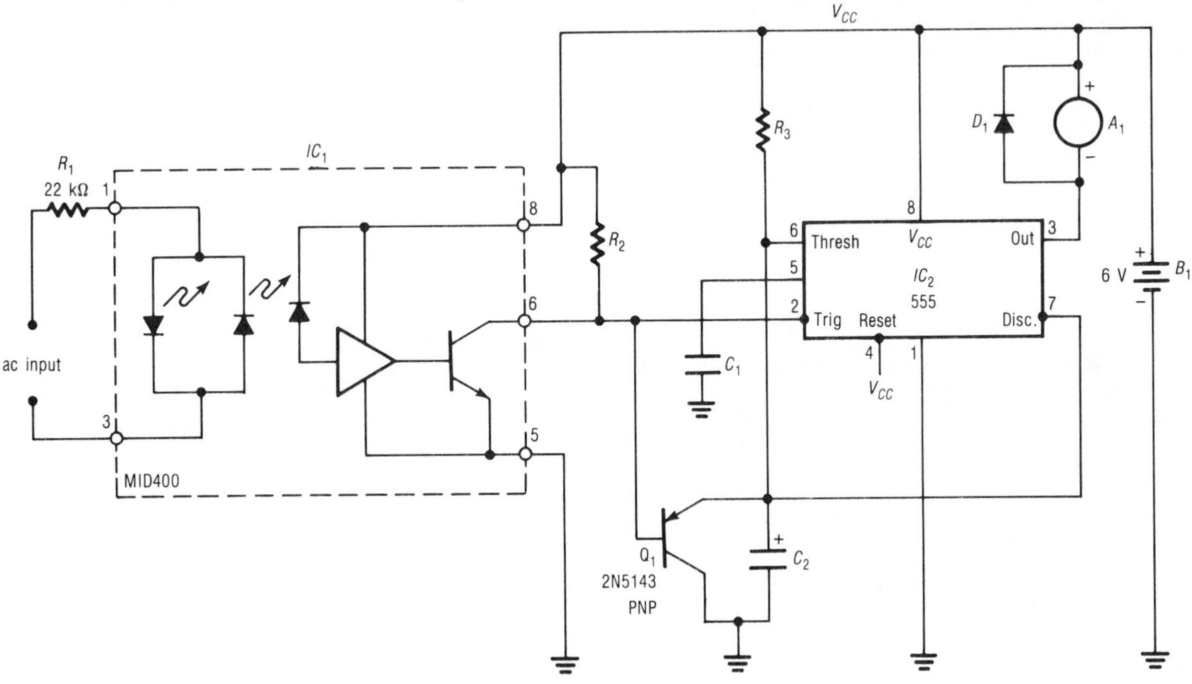

PARTS LIST

A_1 5-V piezo buzzer (Radio Shack 273-060) (or other alarm device)
B_1 6-V battery (or four 1.5-V cells)
C_1 0.01-μF, 25-V disk capacitor
C_2 10-μF, 25-V electrolytic capacitor
D_1 1N4001 silicon diode, 1 A, 50 PIV
IC_1 MID400 power line monitor IC (General Instrument)
IC_2 555 timer IC
Q_1 2N5143 PNP transistor (or similar)
R_1 22-kΩ, ¼-W resistor
R_2 4.7-kΩ, ¼-W resistor
R_3 200-kΩ, ¼-W resistor

FIG. 10-7 Power line monitor. (Used by permission of General Instrument Corporation.)

555 timer IC_2, which is used as a time delay and drives the alarm. When the 555 timer is triggered, C_2 starts to charge through R_3 until the voltage at pin 6 gets to two-thirds of V_{CC}. Then output pin 3 of the 555 timer goes from high to low, driving the alarm device A_1. The voltages at the output of the timer IC are at TTL levels and may be used to drive logic circuitry instead of a simple buzzer.

RESISTANCE SUBSTITUTION BOX

A resistance substitution box circuit is detailed in Fig. 10-8. With the use of the high/low switch and with the turn of a knob, any one of 24 different resistors can be selected. The resistance ranges from a low of 15 Ω up to 10 MΩ. Substitution boxes are used during experimenting and circuit design work. A parts kit and a pc board for the resistance substitution box are currently available from Graymark International, Inc.

RIBBON CABLE TESTER

The ribbon cable tester in Fig. 10-9 is a unit that will indicate if any of the many conductors are open. This simple tester will also indicate whether an electrical short circuit exists between adjacent wires. Typically, experimenter's IC breadboarding sockets would be used for P_1 and P_2; however, DIP IC sockets may be used if many cables are to be tested. Use of the IC breadboarding sockets gives more flexibility for testing cables with 16, 24, 32, 40, or 48 conductors. Note that the wiring pattern on the sockets repeats.

To use, insert the ribbon cable under test in sockets as suggested in Fig. 10-9. If only the green light-emitting diodes (LED_1 and LED_2) light, the cable is good. If either of the green LEDs is not lit, the cable has an open. Should one of the red light-emitting diodes (LED_3 through LED_6) light, an electrical short circuit is present in the cable.

Transformer T_1, shown in Fig. 10-9, steps the voltage from 117 V ac down to 6.3 V ac.

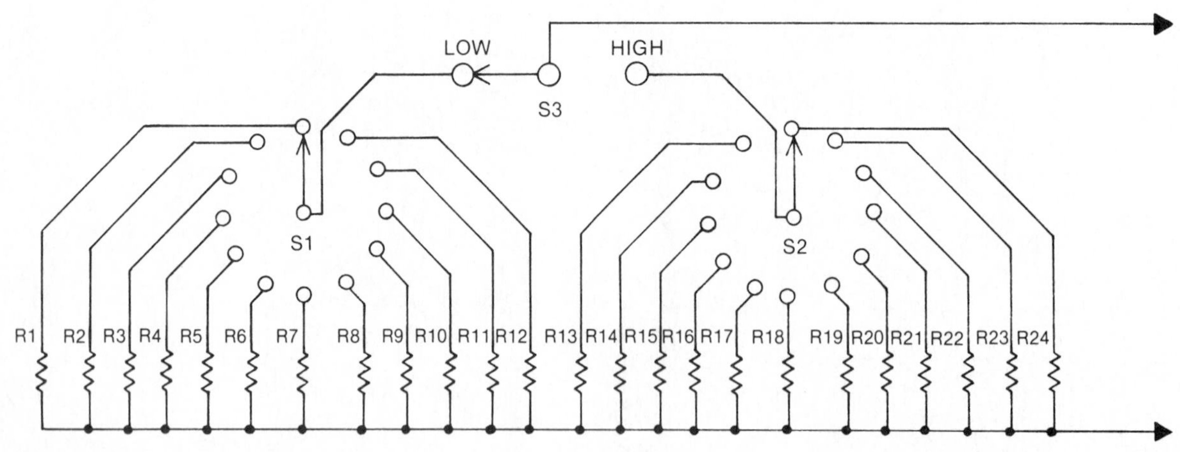

PARTS LIST

R_1	15-Ω, ½-W resistor	R_{10}	4.7-kΩ, ½-W resistor	R_{19}	1-MΩ, ½-W resistor
R_2	33-Ω, ½-W resistor	R_{11}	6.8-kΩ, ½-W resistor	R_{20}	1.5-MΩ, ½-W resistor
R_3	47-Ω, ½-W resistor	R_{12}	10-kΩ, ½-W resistor	R_{21}	2.2-MΩ, ½-W resistor
R_4	100-Ω, ½-W resistor	R_{13}	15-kΩ, ½-W resistor	R_{22}	4.7-MΩ, ½-W resistor
R_5	220-Ω, ½-W resistor	R_{14}	33-kΩ, ½-W resistor	R_{23}	6.8-MΩ, ½-W resistor
R_6	470-Ω, ½-W resistor	R_{15}	47-kΩ, ½-W resistor	R_{24}	10-MΩ, ½-W resistor
R_7	1-kΩ, ½-W resistor	R_{16}	100-kΩ, ½-W resistor	S_1, S_2	SP12P rotary switch
R_8	1.5-kΩ, ½-W resistor	R_{17}	220-kΩ, ½-W resistor	S_3	SPDT switch
R_9	2.2-kΩ, ½-W resistor	R_{18}	470-kΩ, ½-W resistor		

FIG. 10-8 Resistance substitution box. (Used by permission of Graymark International, Inc.)

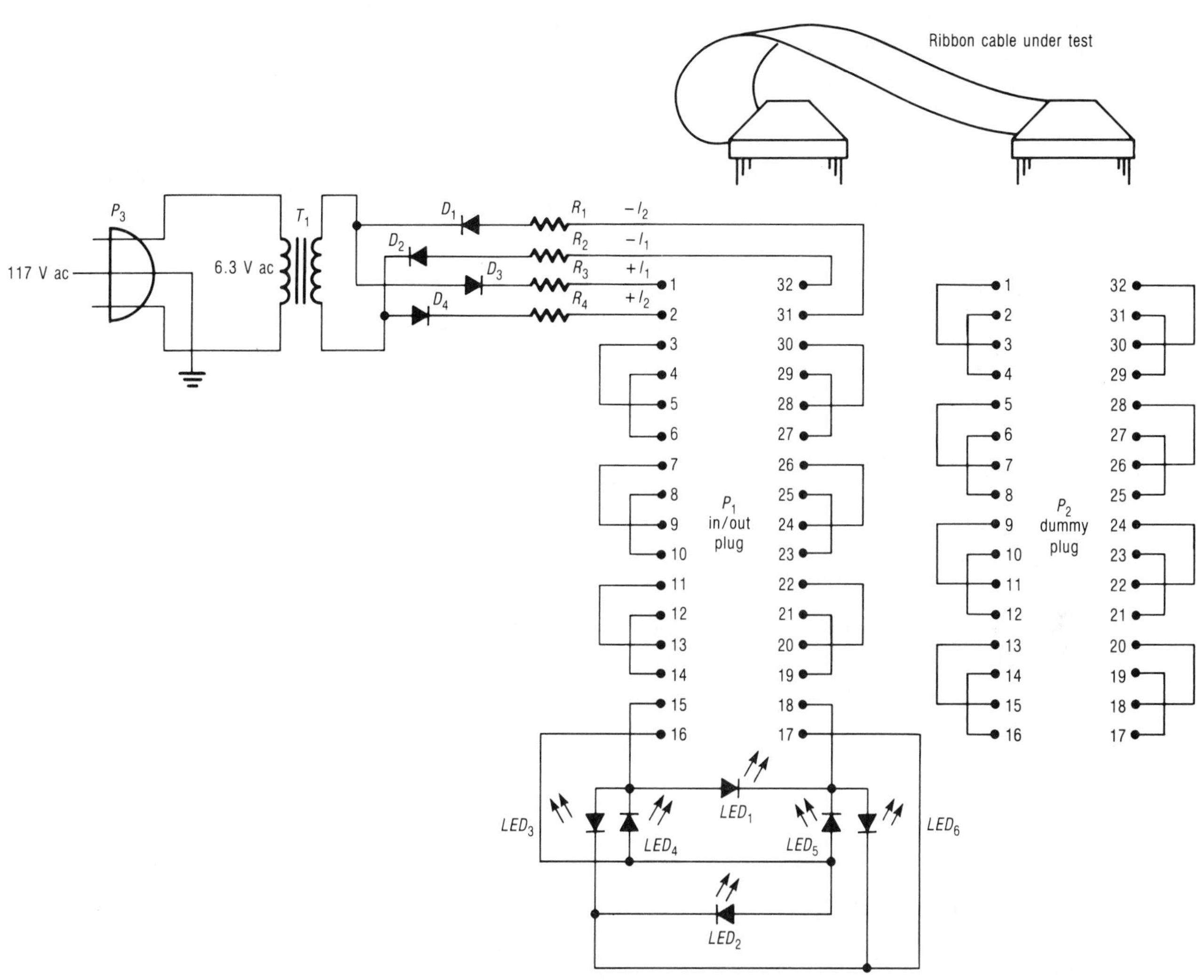

PARTS LIST

D_1 through D_4	1N4002 silicon diode, 1 A, 100 PIV
LED_1, LED_2	Green light-emitting diodes
LED_3 through LED_6	Red light-emitting diodes
P_1, P_2	Experimenter's IC breadboarding socket (or IC socket)
P_3	Three-pronged ac plug
R_1 through R_4	150-Ω, ¼-W resistor
T_1	6.3-V, 300-mA power transformer (similar to Radio Shack 273-1384)

FIG. 10-9 Ribbon cable tester. (D. E. Patrick, "Ribbon Cable Tester," *Radio-Electronics Special Projects,* Summer 1983, pp. 32–33. Used by permission of Gernsback Publications, Inc.)

When the bottom of the secondary of the transformer goes negative, a pulse of electricity follows current loop I_1. Careful tracing of this current path shows that eight lines of the ribbon cable are tested. Then the current flows through LED_1, then up into pin 15 and through another eight conductors, and back out pin 1 through R_3 and D_3 to the transformer. In summary, during the I_1 pulse, 16 conductors were tested and LED_1 lights if there are no open circuits.

When the top of the transformer secondary goes negative, current loop I_2 is used. Current will flow to pin 31 (see Fig. 10-9) through eight untested ribbon cable conductors; through LED_2, lighting it; then into pin 16, where it flows through another eight untested conductors; then out pin 2; and then back to the transformer. In summary, during the I_2 pulse, the final 16 conductors were tested and LED_2 lights if there are no open circuits. On each pulse, every other conductor in the ribbon cable is tested. If an electrical short circuit exists between adjacent wires, one of the red LEDs (LED_3 through LED_6) will light, warning of this problem.

The high-voltage wiring must be done carefully to make sure that it is completely isolated from the enclosure. The green ground wire on the three-pronged plug must be attached to the metal chassis. Use proper strain relief techniques where cords enter and exit the case.

SIGNAL TRACER

The signal tracer shown in Fig. 10-10 can be used to detect either audio or radio frequency signals in equipment such as radios or television sets. The signal tracer is an amplifier that emits sound if a signal is present at the test point. In radio frequency sections, diode D_1 demodulates the radio frequency and the resulting audio signal is amplified. Potentiometer R_3 adjusts the sensitivity of the amplifier. When troubleshooting, the signal tracer indicates whether an audio signal is present. By checking at the input and output of a stage of amplification, the amount of amplification can be estimated based on the relative volume change from the speaker.

The heart of the signal tracer is the LM388 audio amplifier IC_1. Its gain is set at about 200 in this circuit by capacitor C_3. The output of the amplifier is coupled to the speaker by capacitor C_7. Capacitor C_5 and resistor R_4 suppress amplifier oscillations. Resistors R_1 and R_2 with feedback capacitor C_4 supply base current for the output transistor inside the LM388 IC.

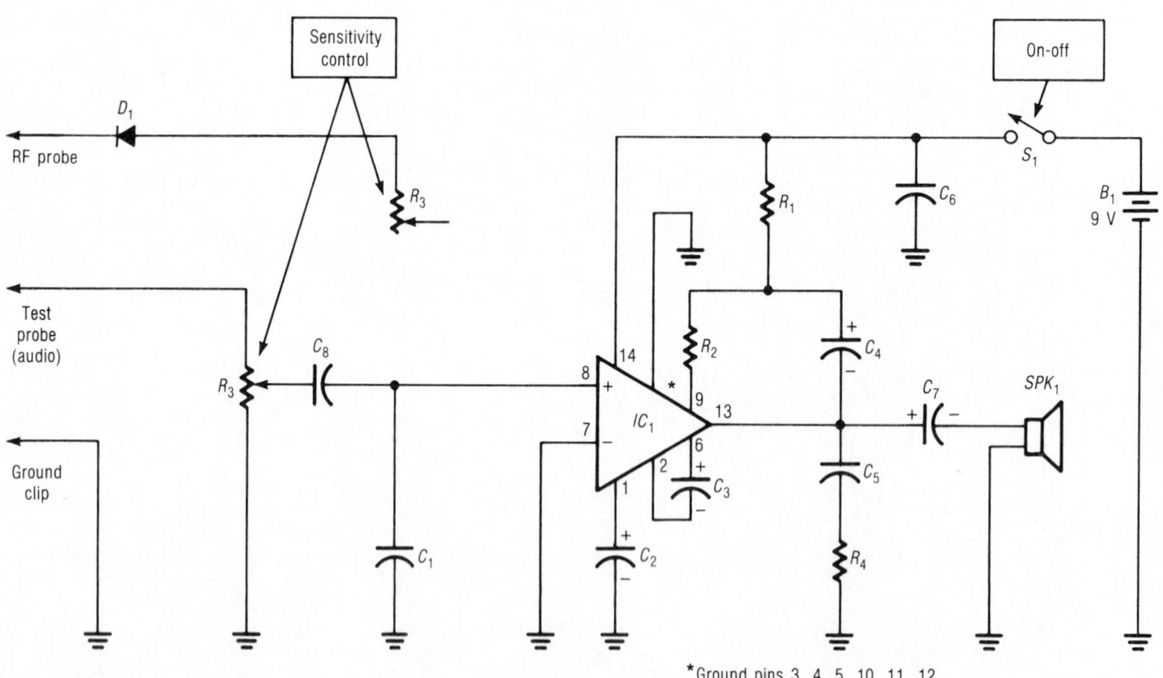

PARTS LIST

B_1	9-V battery	IC_1	LM388N power audio amplifier IC (National Semiconductor)
C_1, C_5	0.05-μF, 50-V disk capacitor	R_1, R_2	560-Ω, ½-W resistor
C_2, C_3	10-μF, 16-V electrolytic capacitor	R_3	10-kΩ potentiometer
C_4	4.7-μF, 16-V electrolytic capacitor	R_4	2.7-Ω, ½-W resistor
C_6, C_8	0.1-μF, 50-V disk capacitor	S_1	SPST switch
C_7	100-μF, 16-V electrolytic capacitor	SPK_1	8-Ω speaker
D_1	1N34A germanium signal diode		

FIG. 10-10 Signal tracer. (Used by permission of Robert Delp Electronics.)

TRANSISTOR/DIODE CHECKER

The transistor/diode checker circuit in Fig. 10-11 will test each junction of an NPN or PNP transistor. The inexpensive checker will also check diodes and LEDs for short and open circuits. The results are indicated on the four LEDs D_1 through D_4. A parts kit and a pc board for the transistor/diode checker are currently available from Graymark International, Inc.

To test a diode or LED with the checker, do the following (see Figure 10-11):

1. Connect the diode/LED to the E and C terminals of jack J_1.
2. Move switch S_2 to the Diode-EB/EC position.
3. Press switch S_1 and note which LEDs light.
4. Use Table 10-1 to determine whether the diode or LED is good, short-circuited, or open.

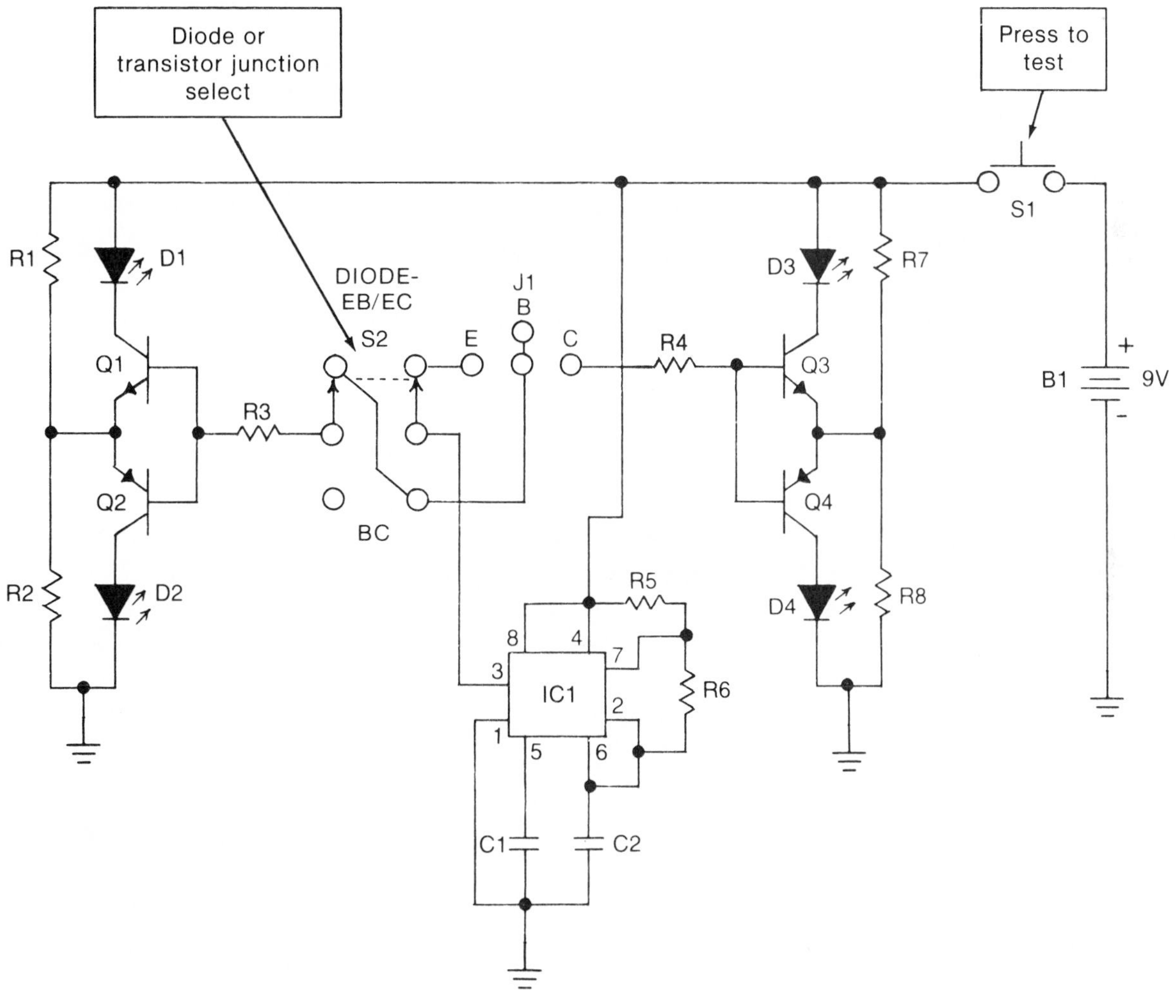

PARTS LIST

B_1	9-V battery	R_1,R_2,R_7,R_8	330-Ω, ¼-W resistor
C_1	0.01-μF, 25-V disk capacitor	R_3,R_4	8.2-kΩ, ¼-W resistor
C_2	0.047-μF, 25-V disk capacitor	R_5	1-kΩ, ¼-W resistor
D_1 through D_4	Light-emitting diode	R_6	33-kΩ, ¼-W resistor
IC_1	555 timer IC	S_1	Normally open push-button switch
Q_1,Q_3	2N3904 NPN transistor	S_2	DPST switch
Q_2,Q_4	2N3906 PNP transistor		

FIG. 10-11 Transistor/diode checker. (Used by permission of Graymark International, Inc.)

TABLE 10-1 (S_2 in Diode—EB/EC Position)

D_3	D_4	Condition	Remarks
On	Off	Good	Cathode lead in pin C of J_1
On	On	Short	Replace diode
Off	Off	Open	Replace diode
Off	On	Good	Cathode lead in pin E of J_1

To test a PNP or NPN transistor with the checker, do the following:

1. Plug the transistor into J_1. The emitter lead must be connected to E, the collector lead to C, and the base lead to B of the jack.
2. Move switch S_2 to the Diode-EB/EC position.
3. Press switch S_1 and note which LEDs light.
4. Use Table 10-2 to determine whether the transistor is either PNP or NPN and what its condition is (good, short-circuited, or open).
5. Move switch S_2 to the BC position (BC stands for base-collector).
6. Press switch S_1 and note which LEDs light.
7. Use Table 10-3 to determine whether the transistor's base-collector junction is good, short-circuited, or open.

The 555 timer IC_1 in Fig. 10-11 is wired as a free-running multivibrator with a frequency of about 500 Hz. Switch S_2 directs the output of IC_1 (pin 3) to either the emitter (E) or base (B) terminals of J_1.

TABLE 10-2 (S_2 in Diode—EB/EC Position)

D_1	D_2	D_3	D_4	Condition	Remarks
Off	On	Off	On	Good	NPN transistor
On	On	On	Off	EB short	Replace transistor
Off	On	On	On	EC short	Replace transistor
Off	Off	Off	Off	EB open	Replace transistor
Off	On	Off	Off	EC open	Replace transistor
On	Off	On	Off	Good	PNP transistor
On	On	Off	On	EB short	Replace transistor
On	Off	On	On	EC short	Replace transistor
Off	Off	Off	Off	EB open	Replace transistor
On	Off	Off	Off	EC open	Replace transistor

TABLE 10-3 (S_2 in BC Position)

D_3	D_4	Condition	Remarks
On	Off	Good	NPN transistor
On	On	BC short	Replace transistor
Off	Off	BC open	Replace transistor
Off	On	Good	PNP transistor

Assume that a good test diode is placed across J_1 with the cathode to E and anode to C. Close switch S_1, and move switch S_2 to the Diode-EB/EC position as shown in Fig. 10-11. The 500-Hz waveform from the oscillator IC_1 will be rectified by the test diode. Negative pulses are applied to the base leads of both transistors Q_3 and Q_4. Negative pulses will only turn on Q_4 causing LED D_4 to light. The current path to light D_4 would be up through D_4, Q_4, R_7, and S_1 and back to positive of the battery. Light-emitting diode D_3 does not light because transistor Q_3 needs a positive (not negative) pulse at the base to turn it on. According to Table 10-1, since D_4 is on and D_3 off, this test diode is good and the cathode is connected to E.

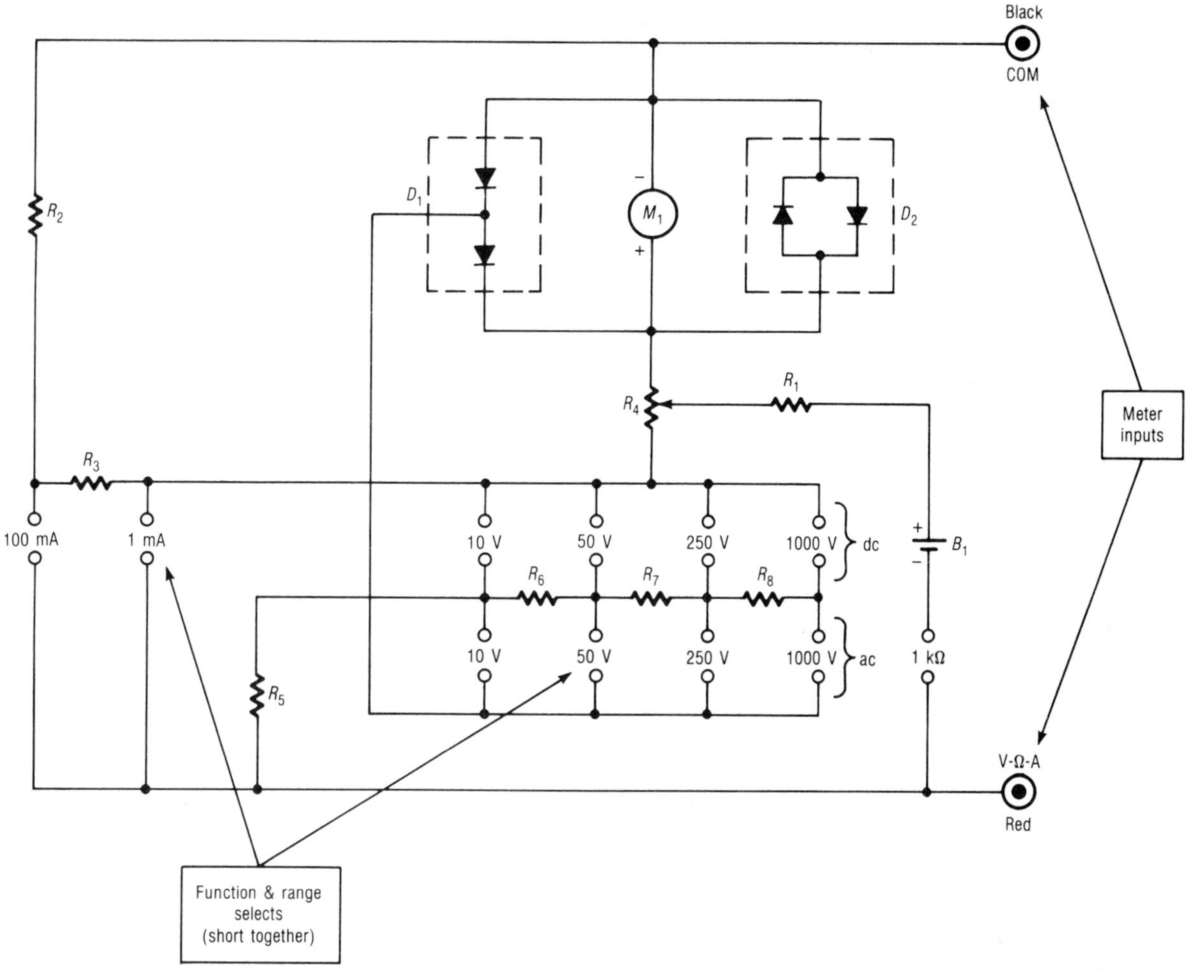

PARTS LIST

B_1	1.5-V battery	R_3	536-Ω, 1 percent, ¼-W resistor
D_1	Rectifier module (two diodes)	R_4	700-Ω potentiometer
D_2	Meter protection diode module	R_5	9.53-kΩ, 1 percent, ¼-W resistor
M_1	400-μA meter movement	R_6	40-kΩ, 1 percent, ¼-W resistor
R_1	2.1-kΩ, 1 percent, ¼-W resistor	R_7	200-kΩ, 1 percent, ¼-W resistor
R_2	5.4-Ω, 1 percent, ½-W resistor	R_8	750-kΩ, 1 percent, ¼-W resistor

FIG. 10-12 Volt-ohm milliammeter. (Used by permission of Graymark International, Inc.)

VOLT-OHM MILLIAMMETER

The volt-ohm milliammeter (VOM) in Fig. 10-12 is a useful tool for any workbench or toolbox. The VOM will measure both dc and ac voltage from 0.1 to 1000 V. It also has two current ranges of 1 mA and 100 mA and a single resistance range ($R \times 1$ kΩ). A parts kit, a pc board (including rotary switch), and complete directions for the VOM are currently available from Graymark International, Inc. Without a custom rotary switch, jumpers can be used from jack to jack to change ranges on the VOM.

On the VOM's 100-mA range, resistor R_2 is the low-resistance shunt (parallel) resistor, taking most of the current flow. The meter movement path includes M_1, R_4, and R_3. In the 1-mA range, resistors R_2 and R_3 form the shunt with M_1 and R_4, the meter movement current path. The COM input jack is negative on the current and voltage ranges.

On the VOM's ohms range, the battery B_1 causes the meter to move full scale when the input jacks are short-circuited together. Potentiometer R_4 adjusts for exact full-scale deflection of the meter or zero on the ohms scale. Resistors R_2 and R_3 are in parallel with M_1 and R_4 in the ohmmeter circuit. If a 2.2-kΩ resistor were placed across the input jacks, the ohmmeter would read half scale.

Consider the 10-V dc range on the VOM in Fig. 10-12. Resistor R_5 is placed in series with the meter movement assembly R_2, R_3, R_4, and M_1. The meter movement will deflect full scale when 10 V is applied to the input jacks. On the 10-V ac range, resistor R_5 is in series with the rectifier module D_1. When the bottom input jack goes negative, current bypasses M_1. When the top input jack becomes negative, however, current flows downward through M_1 and then upward through the bottom diode in D_1 and then through R_5.

WALL OUTLET CHECKER

The wall outlet checker in Fig. 10-13 provides a quick method of analyzing a 117-V ac household receptacle for safe operation. The checker will detect problems in a three-wire outlet such as opens, wires reversed, or no power. A handy neon test light is also included in the circuit.

The neon test light section of the wall outlet checker consists of L_4 and R_4. With about 70 V or higher, lamp L_4 will glow. The neon test lamp may be used on either ac or dc. If used on dc, only the negative post in the NE-2 bulb will glow, whereas both will glow when testing ac.

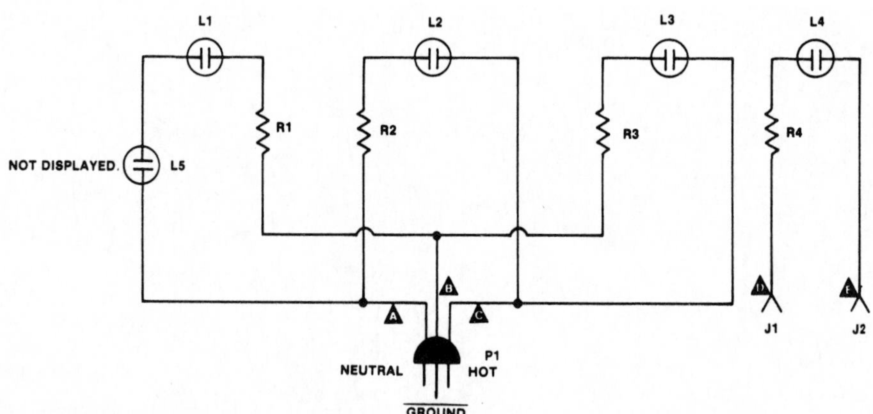

PARTS LIST

L_1 through L_5	NE-2 neon lamp
P_1	Three-pronged ac plug
R_1	15-kΩ, ¼-W resistor
R_2,R_3,R_4	56-kΩ, ¼-W resistor

FIG. 10-13 Wall outlet checker. (Carl G. Grolle and Michael B. Girosky, *Workbench Guide to Electronic Projects You Can Build in Your Spare Time,* Parker, New York, 1981, pp. 181–184. Used by permission of Parker Publishing Company, Inc.)

The outlet checker section consists of the plug, four neon lamps, and three resistors shown in Fig. 10-13. Only lamps L_1, L_2, and L_3 can be observed by the operator. The meanings of various displays are shown in Fig. 10-14. For a correctly wired outlet, L_3 lights as current flows from the hot lead to ground. Neon lamp L_2 lights as current flows from the hot to neutral of the plug. The remaining lamp L_1 does not light because both ends are grounded.

As another example, assume that the hot and neutral wires of a wall outlet were reversed. The checker would indicate this condition by lighting lamps L_1 and L_2 as shown near the bottom of Fig. 10-14. Neon lamp L_2 would light because it is directly across the hot and neutral lines. Lamp L_1 would light, as would series lamp L_5. Lamp L_3 would not light because both ends are grounded.

L1	L2	L3	CIRCUIT CONDITION
O	●	●	correct
O	●	O	open ground
O	O	O	open hot
O	O	●	open neutral
●	O	●	hot and ground reversed
●	●	O	hot and neutral reversed
light on ●			

FIG. 10-14 Interpreting the display on the wall outlet checker. (Carl G. Grolle and Michael B. Girosky, *Workbench Guide to Electronic Projects You Can Build in Your Spare Time,* Parker, New York, 1981, pp. 181–184. Used by permission of Parker Publishing Company, Inc.)

chapter 11

Microcomputer-Related Circuits

BATTERY BACKUP FOR MICROPROCESSOR MEMORY

The battery backup circuit in Fig. 11-1 may be used with static random-access memory (RAM) (such as the 2102 RAM). Such a metal oxide semiconductor RAM will lose all its data if the supply voltage falls below 2 V. This battery backup circuit is connected between the +5-V power supply and the supply input to the memory ICs. When the +5-V supply is operating normally, transistor Q_1 is conducting with a very small voltage drop from emitter to collector. Under normal operation, Q_2 is off and about 20 mA flows through the base transistor R_3 to maintain the nickel-cadmium (ni-cad) batteries in the charged condition.

If the power fails, transistor Q_1 begins to turn off while Q_2 turns on. During power failure, Q_1 isolates the standby batteries from the power supply. Transistor Q_2 drops a very small voltage from emitter to collector. Therefore, nearly the full +2.4 V is applied to the RAM chips during power failure. This is enough to prevent loss of data. A battery backup defeat switch (S_1) is included if the backup is to be turned off.

COMPUTER-CONTROLLED TRIAC DIMMER

The triac-dimmer circuit in Fig. 11-2 can be used to dim high-wattage lamps (≥200 W) or it can be used to control the speed of ac-dc universal motors. The dimmer circuit can be operated manually by R_{26} or by a computer. As the control voltage entering S_1 increases from 0 to 5 V, a load lamp plugged into SO_1 would gradually light more brightly. The dimmer/speed control circuit also features optical isolation (IC_2).

Transformer T_1, bridge rectifier D_1 through D_4, filter capacitors C_1 and C_2, and voltage regulator IC_3 form the +12-V power supply. Com-

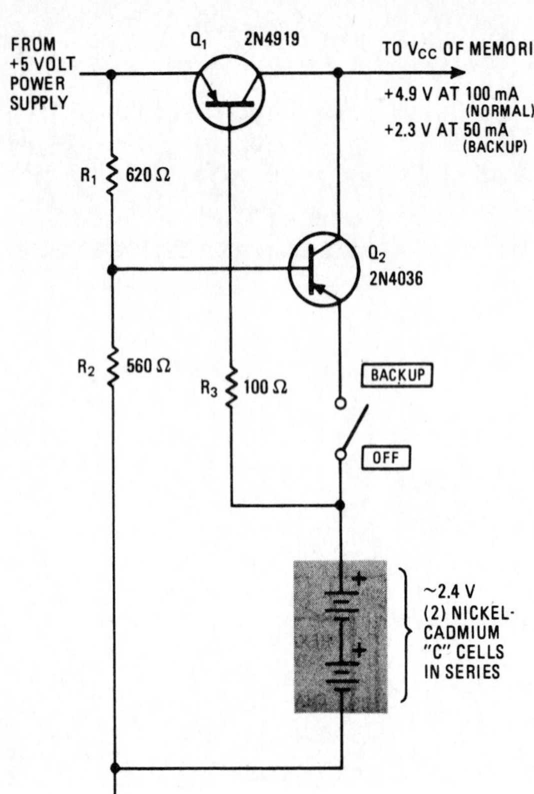

PARTS LIST

Q_1	2N4919 PNP transistor (or similar)
Q_2	2N4036 PNP transistor (or similar)
R_1	620-Ω, ½-W resistor
R_2	560-Ω, ½-W resistor
R_3	100-Ω. ½-W resistor

FIG. 11-1 Battery backup for microcomputer memory. (Raymond N. Bennett, *Circuits for Electronics Engineers,* McGraw-Hill, New York, 1977, p. 304. Used by permission of *Electronics Week.*)

ponents Q_1, Q_2, C_3, and IC_{1a} are the major parts in the ramp generator. The ramp voltage is applied to pin 10 of the voltage comparator IC_{1b}. As the ramp voltage increases above the reference voltage at pin 9, the output of the voltage comparator goes positive, turning on transistor Q_3. When Q_3 conducts, the LED inside the LASCR IC_2 lights. This fires the low-power silicon controlled rectifier in the light-activated silicon controlled rectifier, which in turn fires the power triac Q_4. When the triac conducts the load device receives power. If Q_4 conducts for a greater percentage of time, the load device receives more power. The reference voltage at pin 9 of IC_{1b} is the key to how much power is delivered to the load device. The reference voltage at pin 9 of IC_{1b} is adjusted by the input control voltage from S_1, IC_{1d}, IC_{1c}, and associated resistors. A low-level reference voltage at pin 9 of IC_{1b} causes the triac to trigger early in the sine wave and more current passes through the load. However, increasing the reference voltage at pin 9 reduces the on time of the triac, causing the load device to receive less power.

The high-voltage wiring must be done carefully to make sure that it is completely isolated

PARTS LIST FOR FIG. 11-2

D_1 through D_9	1N4004 silicon diode, 1 A, 400 PIV
C_1	470-μF, 50-V electrolytic capacitor
C_2,C_3,C_4	0.1-μF, 50-V disk capacitor
C_5	0.1-μF, 600-V capacitor
IC_1	LM324 quad op-amp IC
IC_2	MCS2400 light-activated SCR (similar to ECG-3046)
IC_3	LM340-12 voltage regulator (12 V)
L_1	100-μH choke (select current rating dependent on triac)
P_1	Three-pronged ac plug
Q_1,Q_2	2N2222 NPN transistor (or similar)
Q_3	2N3567 NPN transistor (or similar)
Q_4	2N5445 triac (40 A, 400 V) (select according to power needs)
R_1,R_2,R_4	1-kΩ, ¼-W resistor
R_3	4.7-kΩ, ¼-W resistor
R_5	220-kΩ, ¼-W resistor
R_6	10-kΩ, 1 percent, ¼-W resistor
R_7,R_{12},R_{15},R_{18} through R_{21},R_{24}	100-kΩ, ¼-W resistor
R_8	470-Ω, ¼-W resistor
R_9	1.5-kΩ, ¼-W resistor
R_{10}	2-kΩ trimmer potentiometer
R_{11}	100-Ω, ¼-W resistor
R_{13}	120-Ω, ¼-W resistor
R_{14}	560-Ω, ¼-W resistor
R_{16}	47-kΩ, ¼-W resistor
R_{17}	330-Ω, ¼-W resistor
R_{22}	100-kΩ trimmer potentiometer
R_{23}	68-kΩ, ¼-W resistor
R_{25}	15-kΩ, ¼-W resistor
R_{26}	10-kΩ linear potentiometer
S_1	SPDT switch
SO_1	Three-pronged ac convenience outlet
T_1	12.6-V, 300-mA power transformer (Radio Shack 273-1385)

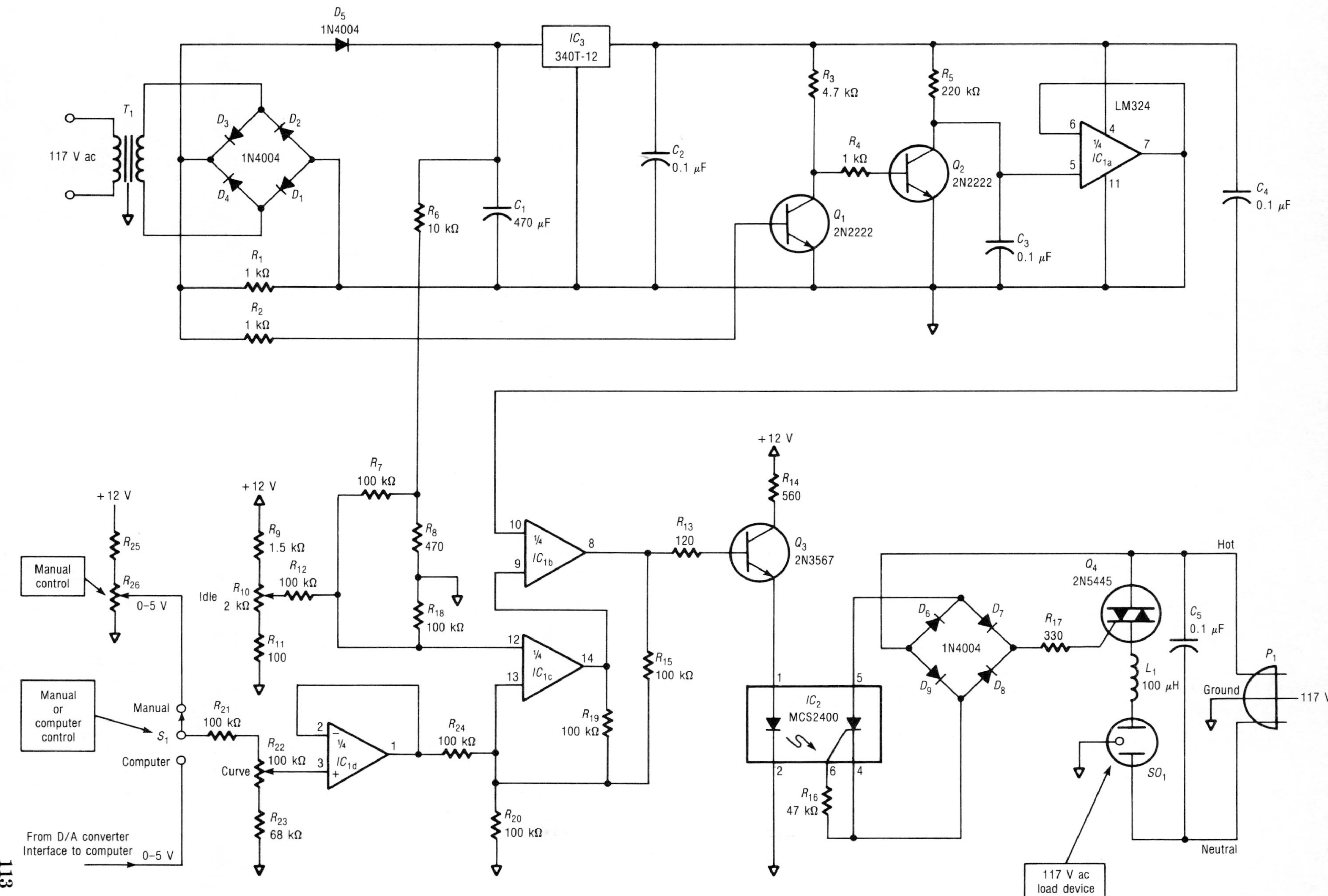

FIG. 11-2 Computer-controlled triac dimmer. See parts list on facing page. (Merrill Lessley, "Easy-to-Build Computer-Controlled Triac Dimmer," *Microcomputing,* October 1980, pp. 92–99. Used by permission of *Microcomputing.*)

from the enclosure. The green ground wire on the three-pronged plug must be attached to the metal chassis. For safety, outlet SO_1 should be wired on the neutral side of the power line. Use proper strain relief techniques where cords enter and exit the case.

OPTICAL COMPUTER INTERFACE

The optical interface in Fig. 11-3 is unique in that there is no electrical connection to the computer and no need to open the case. The simple optical computer interface will operate any device (such as lamps, alarms, and motors) that can be controlled by a relay. The optical sensor in the interface circuit is positioned in contact with the computer's monitor, and as the area under the sensor lights or is darkened, the circuit responds. The optical computer interface circuit causes the relay to snap closed when there is light and open when there is no light. This optical interface circuit can be used with any brand computer and is ideal for the beginner.

The LF353N IC wired as a voltage comparator in Fig. 11-3 is the heart of the optical computer interface circuit. When light from the cathode-ray tube (CRT) of the monitor strikes the cadmium sulfide photoresistor, its resistance drops. This causes the voltage at the noninverting input to the comparator (pin 3) to go less positive. When the voltage at pin 3 on IC_1 becomes less than the reference voltage at pin 2, the output (pin 1) goes low. Potentiometer R_1 adjusts the reference voltage on the comparator so that the circuit responds properly to light and no-light conditions on the monitor. The low-going signal from the voltage comparator (in the form of pulses due to the flicker of most CRTs) triggers the 555 timer IC_2 wired as a one-shot multivibrator. The output (pin 3) of IC_2 goes high, turning on Q_1 and causing the relay K_1 to snap closed. If less light strikes photoresistor R_2, the opposite happens. The voltage at pin 3 of IC_1 goes more positive than the reference voltage and the output of the comparator goes high, turning off the one-shot IC_2. The output of IC_2 goes low, turning off Q_1 and allowing the relay contacts to spring back into their normal positions.

The optical computer interface circuit can be operated on any power supply with a voltage between 5 and 12 V. The relay may have to be changed with a change in power supply voltage. The voltage used to power the motor, lamps, and other components would typically be 117 V ac. Care must be used when wiring 117-V ac circuits so that they are well insulated from the case.

Any computer program that will light a small section of the CRT screen can be used to turn on the computer interface. Of course, the same area on the CRT must be darkened to turn off the circuit.

JOYSTICK INTERFACE—LOW RESOLUTION

The joystick interface circuit in Fig. 11-4 has very low resolution as it converts the position of the joystick potentiometers into digital format. This interface may be adequate for an application, such as a game, where front, back, left, right, and stop might be involved. When the joystick interface is connected to a parallel input port, the joystick value can be read with a single input instruction or routine.

Each of the joystick potentiometers (R_1 and R_{14}) in Fig. 11-4 is connected as a voltage divider across 3.9 V. The output from each potentiometer is fed into an analog-to-digital (A/D) converter. The A/D converter consists of four voltage comparators set to turn on at 25, 50, 75, and 100 percent of full-scale voltage. If the voltage from the potentiometer is less than 25 percent of 3.9 V (0.975 V), all the comparator outputs will be zero. At about 1 V, the least-significant bit (LSB) of the converter will go to a logical 1. Full voltage applied to the A/D converter will cause all outputs to go high. The binary outputs of each 4-bit A/D converter will cause all outputs to go high. Five bit patterns are available from each A/D converter (0000, 0001, 0011, 0111, or 1111). The value of resistor R_x may have to be adjusted slightly to enable the MSB comparator to flip to high output near the end of the joystick's travel. This circuit will produce 25 (5×5) unique 8-bit codes that can be used by the programmer.

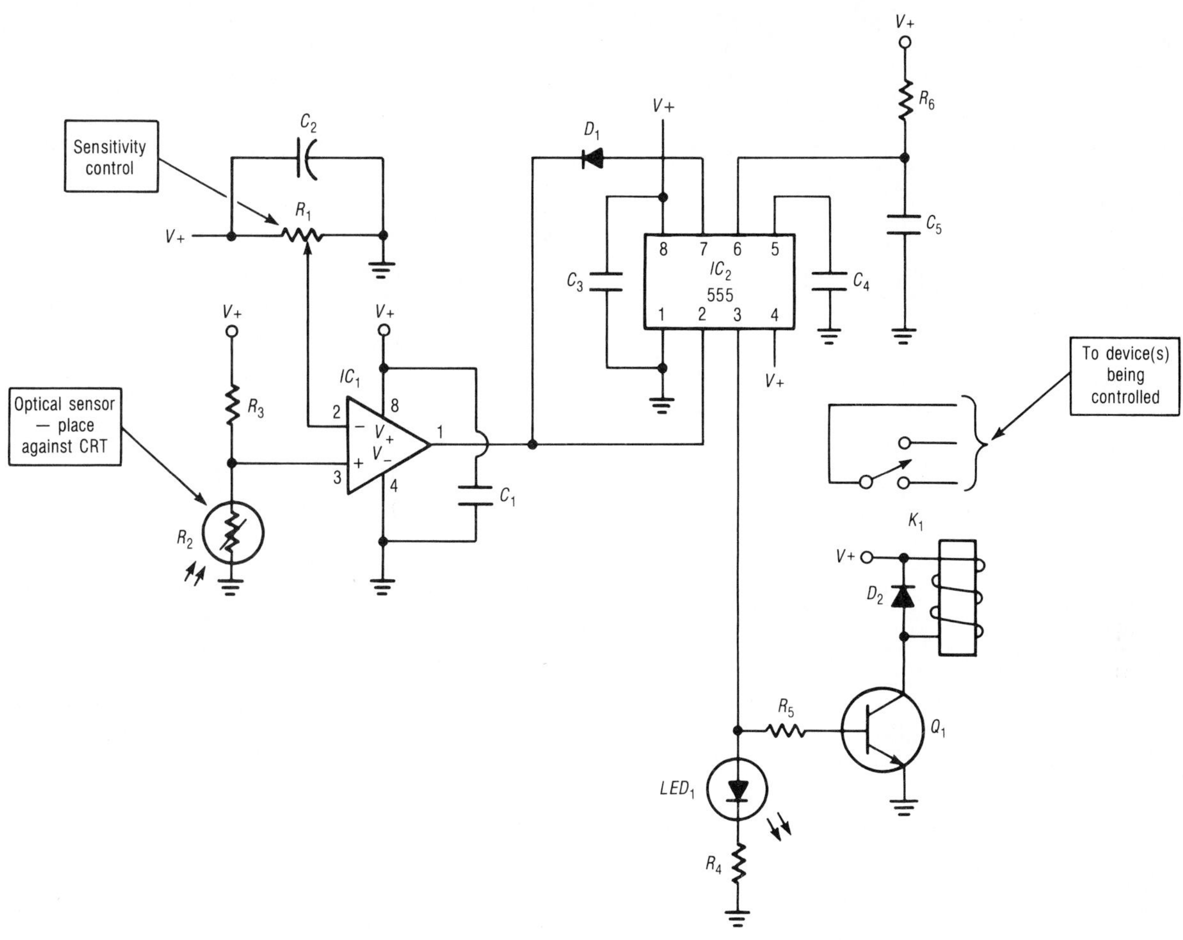

PARTS LIST

C_1, C_3, C_4, C_5	0.1-μF, 25-V disk capacitor	LED_1	Light-emitting diode
C_2	10-μF, 50-V electrolytic capacitor	Q_1	2N2222 NPN transistor (or similar)
D_1	1N914 silicon diode (or similar)	R_1	100-kΩ potentiometer
D_2	PTC205 diode (or 1N4007)	R_2	Cadmium sulfide photoresistor
IC_1	LF353N op-amp IC (Radio Shack 276-1715)	R_3	250-Ω, ¼-W resistor
IC_2	555 timer IC	R_4	1-kΩ, ¼-W resistor
K_1	12-V dc coil, 5-A contacts SPDT relay (similar to Radio Shack 275-218)	R_5	4.7-kΩ, ¼-W resistor
		R_6	1-MΩ, ¼-W resistor

FIG. 11-3 Optical computer interface. (Dave Leithauser, "Computer Interface—Build This Practical and Inexpensive Optical Interface," *Computers and Programming*, July/August 1981, pp. 55–56, 76. Used by permission of David Leithauser.)

JOYSTICK INTERFACE—HIGH RESOLUTION

The joystick interface circuit in Fig. 11-5 is a high-resolution A/D converter using easy-to-find TTL ICs. The position of each joystick potentiometer is converted into an 8-bit binary number. The outputs are then fed into two parallel ports of the computer and may be read with simple input statements or routines.

Consider the top A/D converter and clock in Fig. 11-5. The 555 timer (IC_{15}) is wired as a

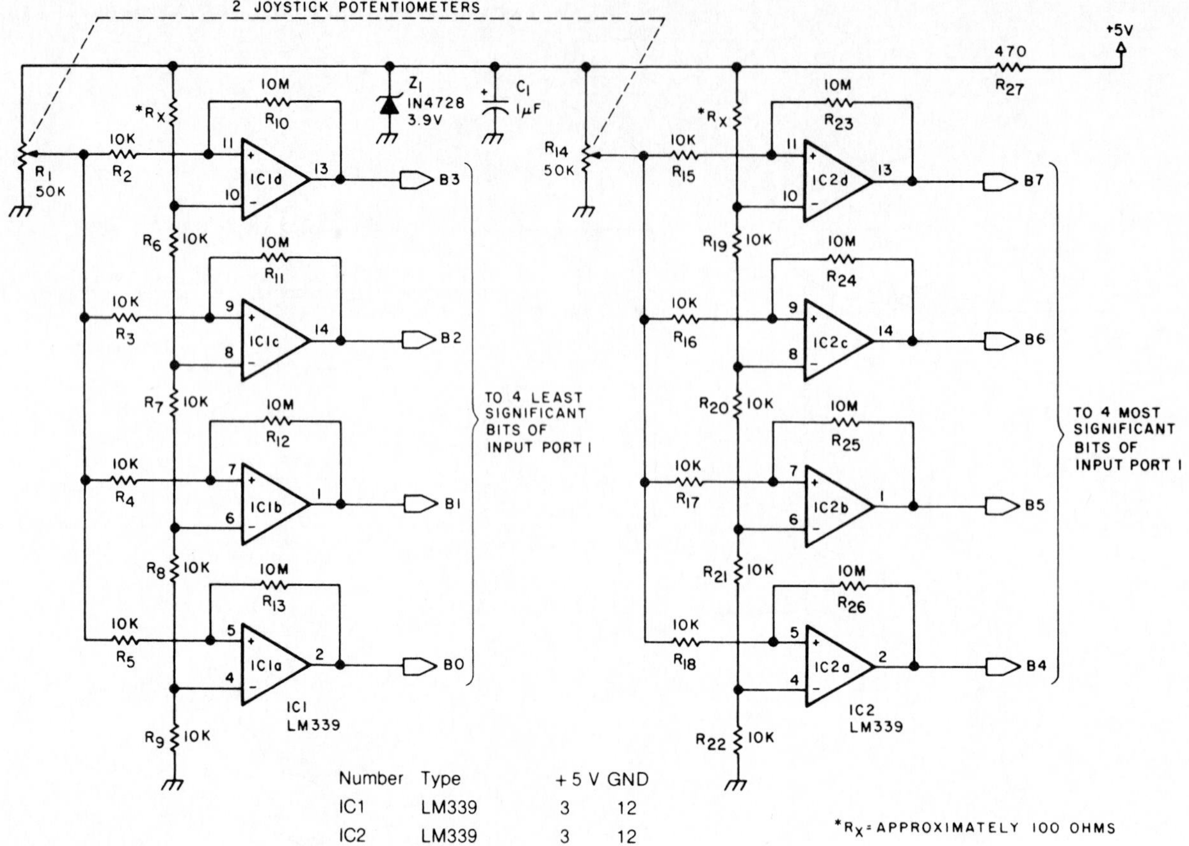

PARTS LIST

C_1	1-μF, 16-V electrolytic capacitor	R_{27}	470-Ω, ¼-W resistor
IC_1, IC_2	LM339 quad voltage comparator IC	R_x, R_x	100-Ω, ¼-W resistor (these values may have to be changed for the MSB voltage comparator to produce a high output)
R_1, R_{14}	50-kΩ joystick (with two potentiometers)		
R_2 through R_9	10-kΩ, 5 percent, ¼-W resistor		
R_{10} through R_{13}	10-MΩ, ¼-W resistor		
R_{15} through R_{22}	10-kΩ, 5 percent, ¼-W resistor		
R_{23} through R_{26}	10-MΩ, ¼-W resistor	Z_1	1N4728 zener diode (3.9 V)

FIG. 11-4 Low-resolution joystick interface. (Steve Ciarcia, *Ciarcia's Circuit Cellar,* Volume II, BYTE-McGraw-Hill, New Hampshire, 1981, pp. 103–104. Used by permission of McGraw-Hill Book Company.)

free-running multivibrator producing a clock frequency of about 7.5 kHz. The joystick potentiometer R_1 controls the pulse width of a one-shot multivibrator IC_1. The 74121 one-shot has a pulse width of 35 milliseconds when the potentiometer is at full resistance while the pulse width is a very narrow 100 microseconds or less with no resistance. When the output of the one-shot goes high, a short clear pulse (generated by IC_{3a}) resets the 7493's IC_7 and IC_8 wired as an 8-bit counter. The counter starts counting the pulses from the clock that pass through NAND gate IC_{5a}. Finally, the output of the one-shot drops low. This disables the NAND gate IC_{5a}, which stops clock pulses from reaching the 7493 counters. A short load pulse is generated by IC_{3a} and IC_{4a}, which loads the count from the 8-bit counter into the 8-bit storage register (IC_9 and IC_{10}). The 8-bit count at B_0 through B_7 represents the relative position of joystick potentiometer R_1. The clear, count, stop count, and load sequence continues to update the 8-bit output lines. The longer the pulse width generated by the 74121 one-shot, the higher the 8-bit binary count at the output of the storage register. The bottom A/D converter in Fig. 11-5 operates the same as the top unit and shares the same clock (IC_{15}).

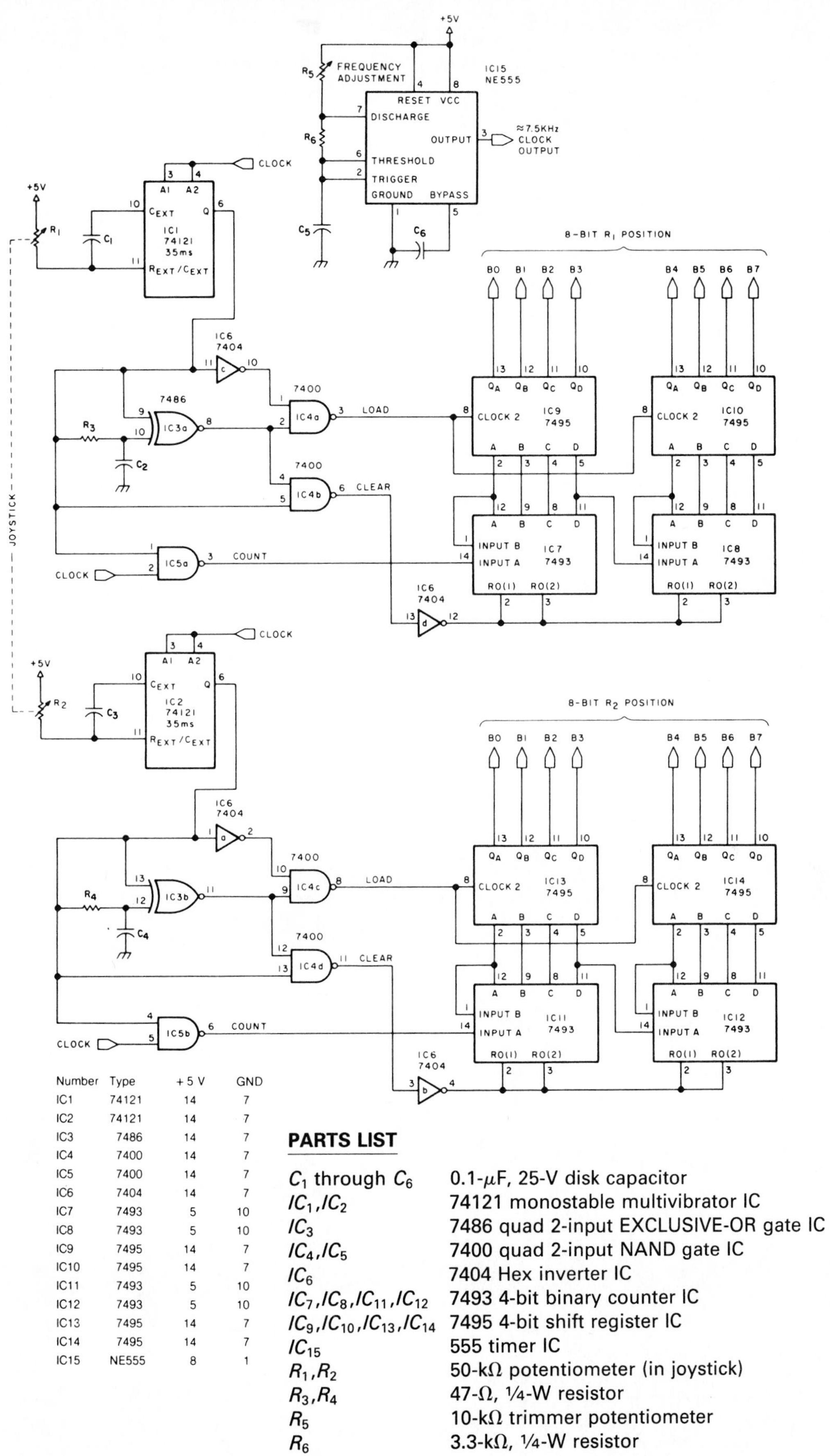

Number	Type	+5 V	GND
IC1	74121	14	7
IC2	74121	14	7
IC3	7486	14	7
IC4	7400	14	7
IC5	7400	14	7
IC6	7404	14	7
IC7	7493	5	10
IC8	7493	5	10
IC9	7495	14	7
IC10	7495	14	7
IC11	7493	5	10
IC12	7493	5	10
IC13	7495	14	7
IC14	7495	14	7
IC15	NE555	8	1

PARTS LIST

C_1 through C_6	0.1-μF, 25-V disk capacitor
IC_1, IC_2	74121 monostable multivibrator IC
IC_3	7486 quad 2-input EXCLUSIVE-OR gate IC
IC_4, IC_5	7400 quad 2-input NAND gate IC
IC_6	7404 Hex inverter IC
$IC_7, IC_8, IC_{11}, IC_{12}$	7493 4-bit binary counter IC
$IC_9, IC_{10}, IC_{13}, IC_{14}$	7495 4-bit shift register IC
IC_{15}	555 timer IC
R_1, R_2	50-kΩ potentiometer (in joystick)
R_3, R_4	47-Ω, ¼-W resistor
R_5	10-kΩ trimmer potentiometer
R_6	3.3-kΩ, ¼-W resistor

FIG. 11-5 High-resolution joystick interface. (Steve Ciarcia, *Ciarcia's Circuit Cellar,* Volume II, BYTE-McGraw-Hill, New Hampshire, 1981, pp. 105–107. Used by permission of McGraw-Hill Book Company.)

KEY-DOWN AUDIBLE SIGNAL FOR COMPUTER KEYBOARD

The key-down audible signal circuit in Fig. 11-6 emits a short beep each time a key is pressed on the computer keyboard. The circuit needs to be connected to the keystrobe signal from the keyboard encoder of the computer. (This signal is not available on all computer keyboards.) The circuit is adaptable for either a positive or negative keystrobe pulse. The computer power supply can typically be used because the key-down signal circuit draws only about 6 mA (using +5 V) and can be used on +5- to +12-V power supplies.

The key-down circuit uses two 555 timers housed in the NE556 (IC_2). One section is wired as a one-shot and the second, as a free-running multivibrator. The keystrobe pulse passing through the NOR gate (wired as an inverter) triggers the one-shot. The one-shot generates a 0.1-second high pulse which turns on and then turns off the free-running multivibrator. The oscillations from the free-running multivibrator drive the piezo buzzer.

If only IC_{1c} is used because the key-down signal circuit is connected to a computer with a positive keystrobe pulse, inputs 1 and 2 of IC_1 should be tied to V_+ and pin 4 of IC_{1b} should be left unconnected. Switch S_1 should be *left open for normal operation.* Closing S_1 will disable the audible signal circuit. Decreasing the value of either R_1 or C_1 in the one-shot will result in reducing the time constant and the length of the beep will be shortened.

COMPUTER MUSIC BOX PERIPHERAL

The music box circuit in Fig. 11-7 can be driven by a computer to generate 12 tones in each of four octaves. The computer's parallel output port (parallel printer port) passes an 8-bit con-

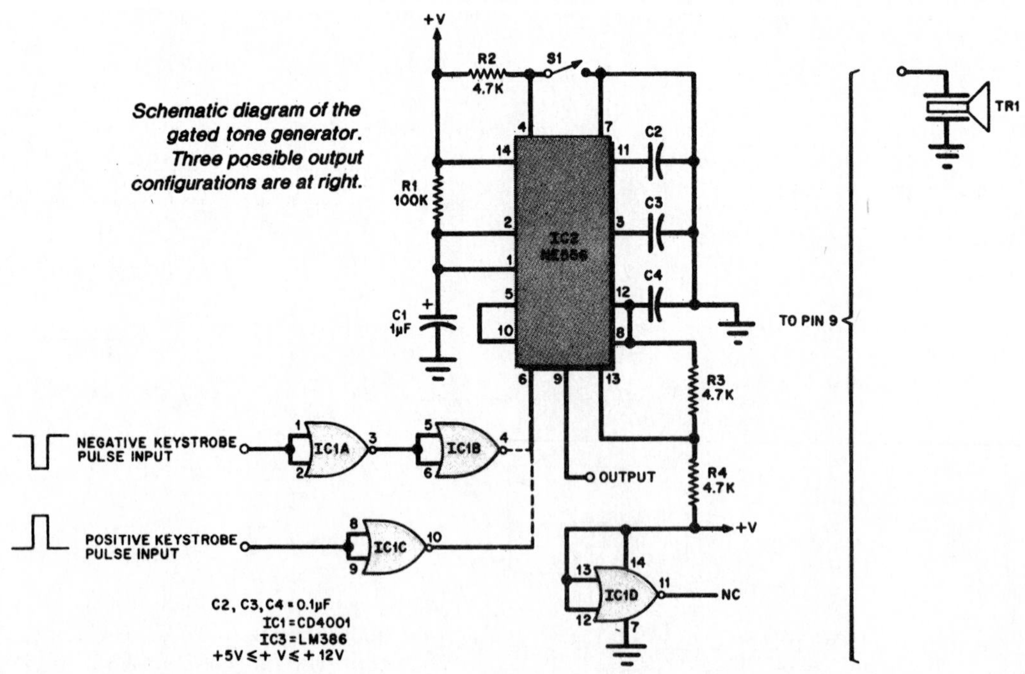

PARTS LIST

C_1	1-μF, 25-V electrolytic capacitor	R_1	100-kΩ, ¼-W resistor
C_2, C_3, C_4	0.1-μF, 25-V disk capacitor	R_2, R_3, R_4	4.7-kΩ, ¼-W resistor
IC_1	CD4001 CMOS quad NOR gate IC	S_1	SPST switch
IC_2	NE556 dual timer IC	TR_1	Piezo buzzer (Radio Shack 273-060)

FIG. 11-6 Key-down audible signal circuit can be added to most brands of microcomputers. (Reprinted from *Popular Electronics*. Copyright © October 1980, Ziff-Davis Publishing Company.)

trol word to the music box circuit. Some computers may not have the MSB (bit 7) available at their parallel printer port. According to Table 11-1, if binary 10000000 (decimal 128) is output at the computer's music box port, the frequency would be zero and the music box would be off. However, if binary 10000001 (decimal 129) appeared at the input to the music box, a 523.25-Hz square wave would be generated equal to the fifth octave C note. Table 11-2 gives the codes for all the notes in various octaves. Output to the music box peripheral is accomplished with output statements or routines written by the programmer. Calibration is done with a frequency counter using the frequencies in Table 11-1 taken at the test point shown in Fig. 11-7.

The audio output from the circuit can be used to drive an audio amplifier. A common ground should be maintained between the music box circuit and the computer. The +5-V power supply can be separate, or the computer's supply can be used if it has the extra capacity. The circuit draws about 100 mA.

The music box circuit consists of three main sections: the note decoder/selector, the voltage-controlled oscillator (VCO), and the octave decoder/selector. The note decoder/selector section consists of IC_1. One of the 12 output lines from the 74154 decoder goes low depending on the control bits (0 through 3) from the computer's output port. Because of the different settings of trimmer potentiometers R_1 through R_{12}, different voltages are applied to the VCO made up of IC_2, Q_1, and associated components. The timer IC_2 is wired as an oscillator whose frequency can be changed by varying the emitter-to-collector resistance of Q_1, which determines the charging time of capacitor C_1.

The output frequency from pin 3 of IC_2 is divided by 2, 4, 8, and 16 by four flip-flops (IC_3 and IC_4) and passed to the NAND gates (IC_5). The control bits (bits 4 through 7) from the computer's output port controls which *single gate is turned on.* If bit 7 is high, the top NAND gate will pass higher-frequency tones to the audio output. However, if bit 4 is high, the bottom NAND gate will pass low-frequency tones to the audio output.

TABLE 11-1 Control Codes and Frequencies for the Musical Scale

Control bit 7 6 5 4 3 2 1 0	Frequency, Hz	Note 5th Octave
1 0 0 0 0 0 0 0	0	Off
1 0 0 0 0 0 0 1	523.25	C
1 0 0 0 0 0 1 0	554.37	C#
1 0 0 0 0 0 1 1	587.33	D
1 0 0 0 0 1 0 0	622.25	D#
1 0 0 0 0 1 0 1	659.26	E
1 0 0 0 0 1 1 0	698.46	F
1 0 0 0 0 1 1 1	739.99	F#
1 0 0 0 1 0 0 0	783.99	G
1 0 0 0 1 0 0 1	830.61	G#
1 0 0 0 1 0 1 0	880.00	A
1 0 0 0 1 0 1 1	932.33	A#
1 0 0 0 1 1 0 0	987.77	B

8085 MICROCOMPUTER SYSTEM

The microcomputer system in Fig. 11-8 uses the popular Intel 8085 8-bit microprocessor.

TABLE 11-2 Decimal Control Codes Used for Generating Notes in Various Octaves

Note*	Number Value n
Off	0
C	1
C#	2
D	3
D#	4
E	5
F	6
F#	7
G	8
G#	9
A	10
A#	11
B	12
Octave	**Number value**
5	$n + 128$
4	$n + 64$
3	$n + 32$
2	$n + 16$

* *Note:* B_5 is the highest note ($n = 140$)
C_2 is the lowest note ($n = 17$)
C_5 is middle C
A_4 is A_{440}

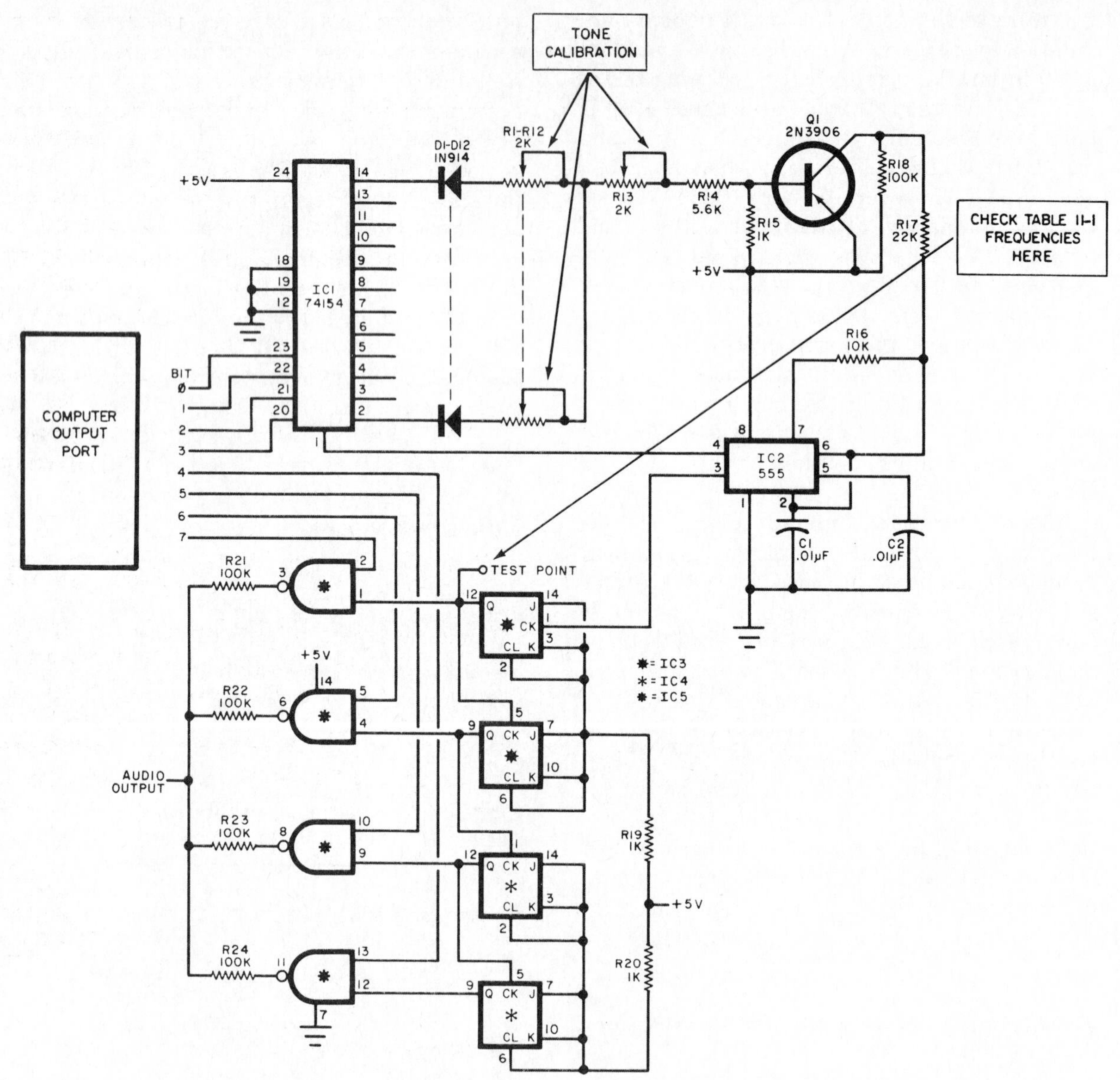

PARTS LIST

C_1,C_2	0.01-μF, 50-V Mylar capacitor
D_1 through D_{12}	1N914 diode
IC_1	74154 4- to 16-line decoder IC
IC_2	555 timer IC
IC_3,IC_4	7473 dual *J-K* flip-flop IC
IC_5	7400 quad 2-input NAND gate IC
Q_1	2N3906 PNP transistor (or similar)
R_1 through R_{13}	2-kΩ trimmer potentiometer
R_{14}	5.6-kΩ, 5 percent, ¼-W resistor
R_{15},R_{19},R_{20}	1-kΩ, 5 percent, ¼-W resistor
R_{16}	10-kΩ, 5 percent, ¼-W resistor
R_{17}	22-kΩ, 5 percent, ¼-W resistor
R_{18},R_{21} through R_{24}	100-kΩ, 5 percent, ¼-W resistor

FIG. 11-7 Computer music box peripheral. (Reprinted from *Popular Electronics*. Copyright © April 1978, Ziff-Davis Publishing Company.)

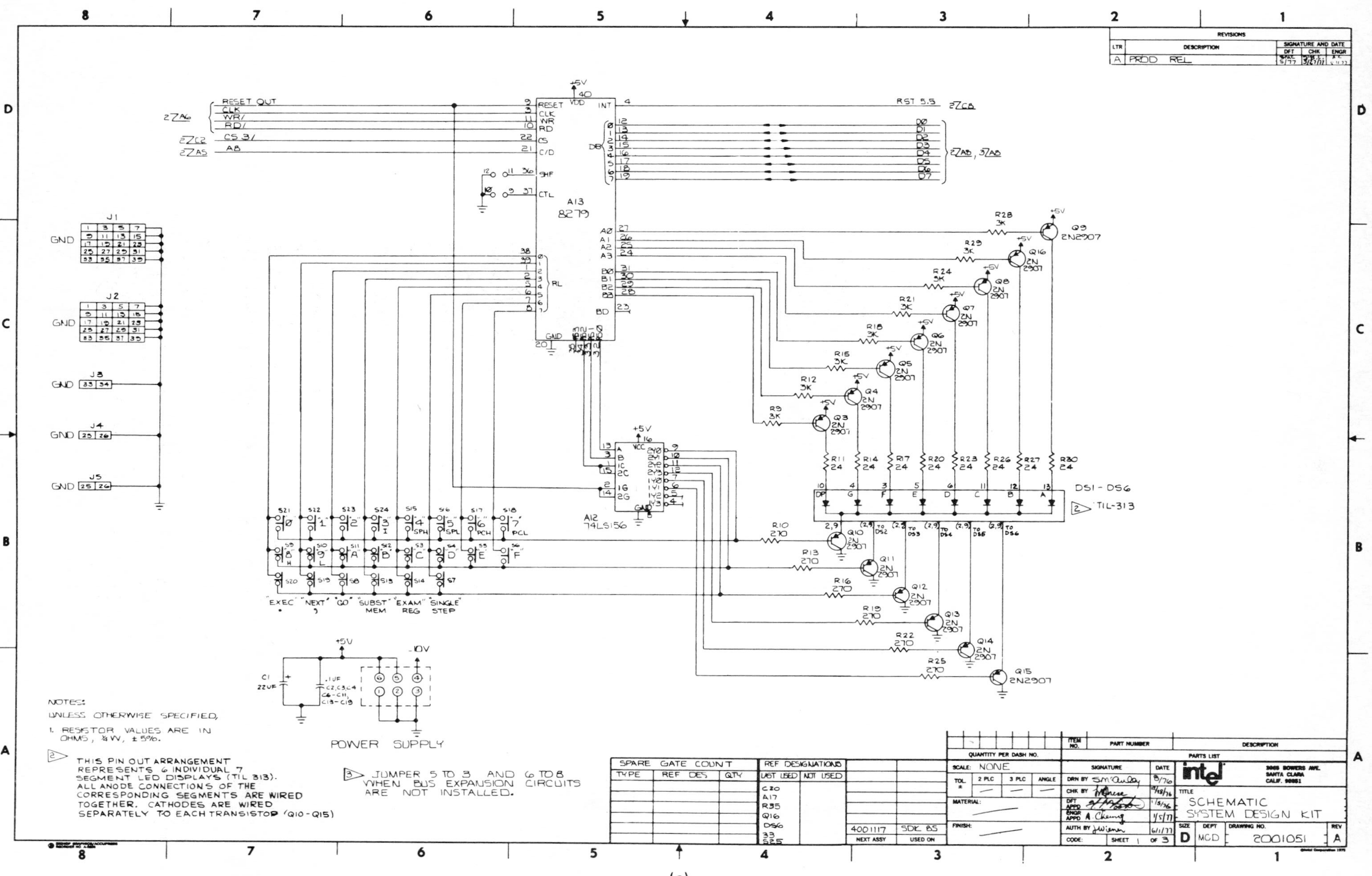

(a)

FIG. 11-8 Schematic diagram for an 8085-based microcomputer system. (Used by permission of Intel Corporation.)

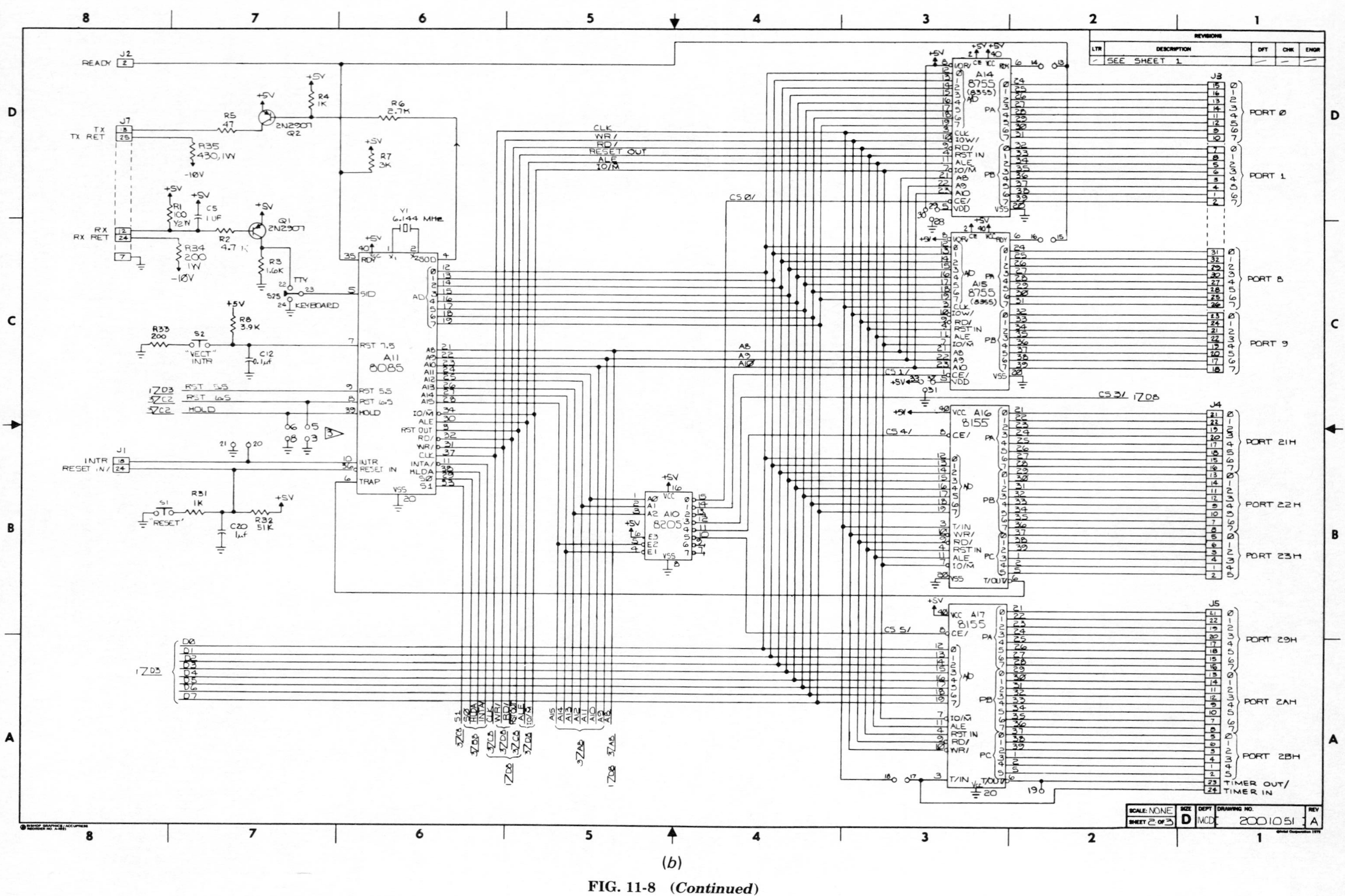

(b)

FIG. 11-8 (Continued)

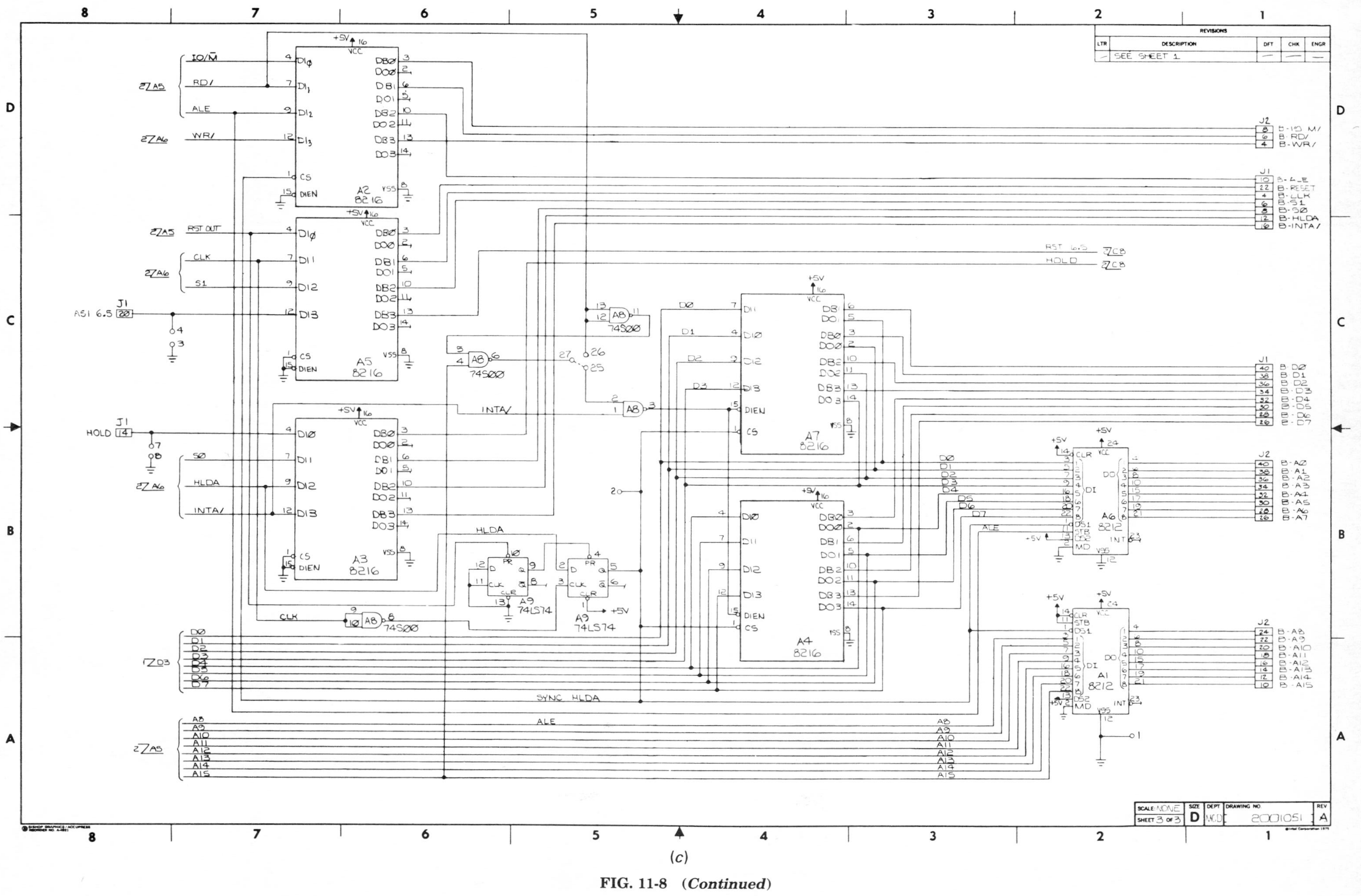

(c)

FIG. 11-8 (Continued)

The circuit is that of the MCS-85 System Development Kit packaged in kit form with a two-sided pc board and full instructions by Intel Corporation. The system design kit is an excellent way to learn about machine-level programming and interfacing the 8085 microprocessor. The minimal microcomputer system in Fig. 11-8 can be used in the development of more complex microprocessor-based systems.

The minimal microcomputer system is shown in block form in Fig. 11-9. From the block diagram it is evident that input is accomplished by using a keyboard with a six-character keyboard display for revealing address and data information. The central processing unit (CPU) is the Intel 8085 8-bit microprocessor. The system uses an 8-bit multiplexed data/address bus, an 8-bit address bus, and a control bus. The 8155 IC contains 256 bytes of static RAM, 22 programmable input-output (I/O) lines, and a 14-bit timer/counter. The 8155's RAM is available for storage of user programs and program data. The 8355 IC is a 2048-byte read-only memory (ROM) and also contains 16 I/O lines. The ROM is programmed with the system monitor. The 8279 is a keyboard/display controller chip that handles the interface between the 8085 CPU and the keyboard and LED display. The 8279 chip refreshes the display from internal memory while scanning the keyboard to detect keyboard inputs. The 8205 chip decodes the 8085's memory address bits to provide chip enables to the 8155, 8355, and 8279 ICs. The block diagram suggests that the system design kit available from Intel has space available on the pc board for system expansion.

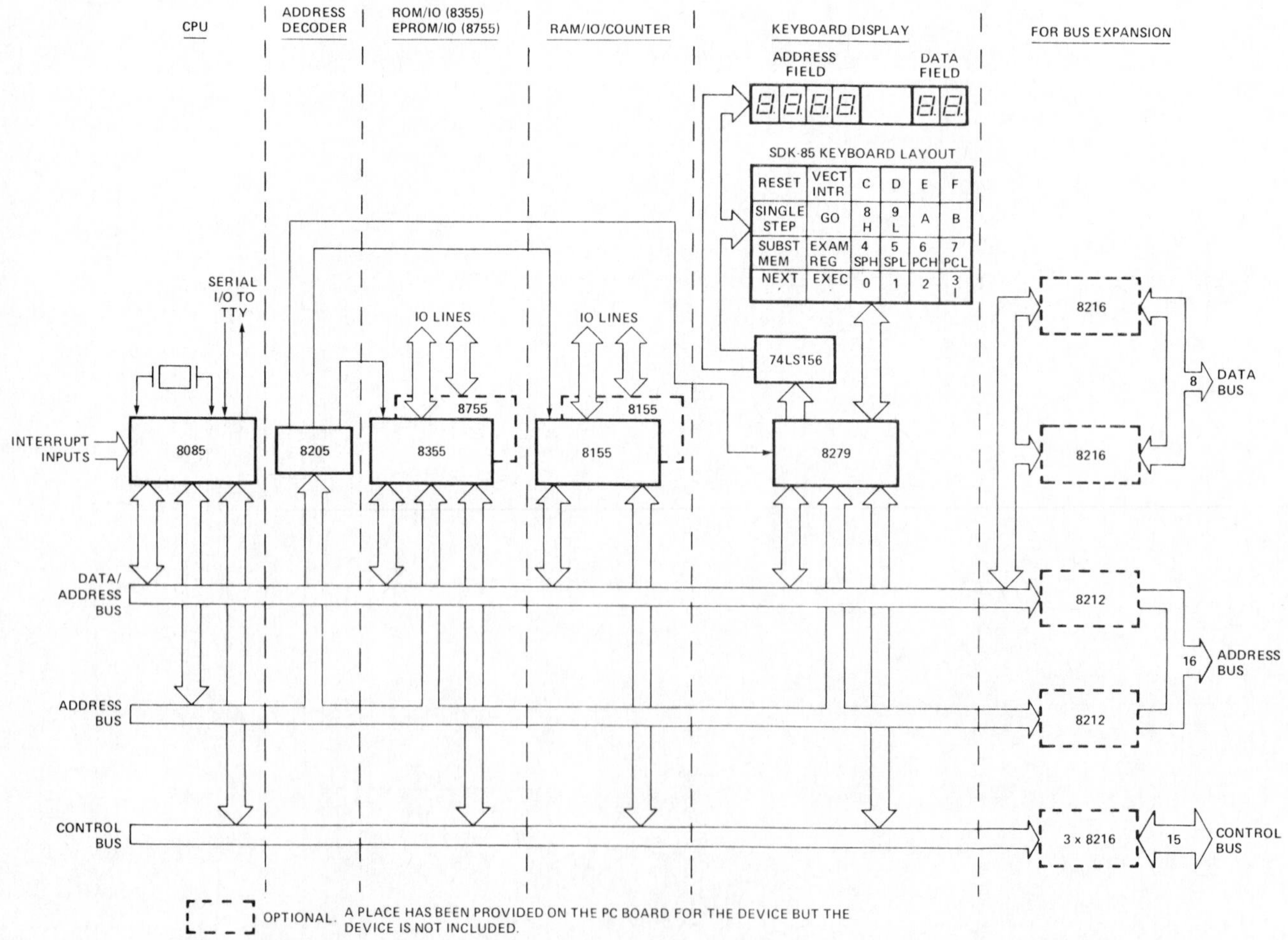

FIG. 11-9 Functional block diagram of 8085-based microcomputer system (Intel SDK-85). (Used by permission of Intel Corporation.)

chapter 12

Motor and Lamp Control Circuits

REGULATED MOTOR SPEED CONTROL

The motor speed control circuit in Fig. 12-1 can be used to vary the speed of a hand drill or other small ac/dc 117-V ac electric motor. The unit can also be used to dim incandescent lamps from off to near full brightness.

The motor speed control circuit in Fig. 12-1 is a regulated speed control or dimmer circuit. The device to be controlled is plugged into socket SO_1. The speed control is a half-wave circuit. The heart of the half-wave dimmer circuit is the silicon controlled rectifier SCR_1. When the cathode of SCR_1 goes negative (anode positive), the SCR will conduct after a positive pulse appears at the gate terminal. Potentiometer R_4 adjusts how soon in the half-cycle the SCR is turned on. If it is turned on early in the cycle, the motor speed will be greater than if it is late. Capacitor C_2 and R_3 form a feedback circuit which tends to regulate the speed of the motor under varying mechanical loads. During the half-cycle when the anode of the SCR is negative (cathode is positive), the SCR will not conduct. Closing switch S_1 bypasses the SCR and defeats the speed control (for full speed or full power).

The high-voltage wiring must be done carefully to make sure that it is completely isolated from the enclosure. The green ground wire on the three-pronged plug must be attached to the metal chassis. Use proper strain relief techniques where cords enter and exit the case. Caution must be used to ensure that only a universal (ac/dc) motor is plugged into this unit. All ac/dc motors have a wound rotor, a commutator, and brushes.

NASA MOTOR-CONTROL CIRCUIT

The motor-control circuit in Fig. 12-2 is designed to work with ac induction motors. Most

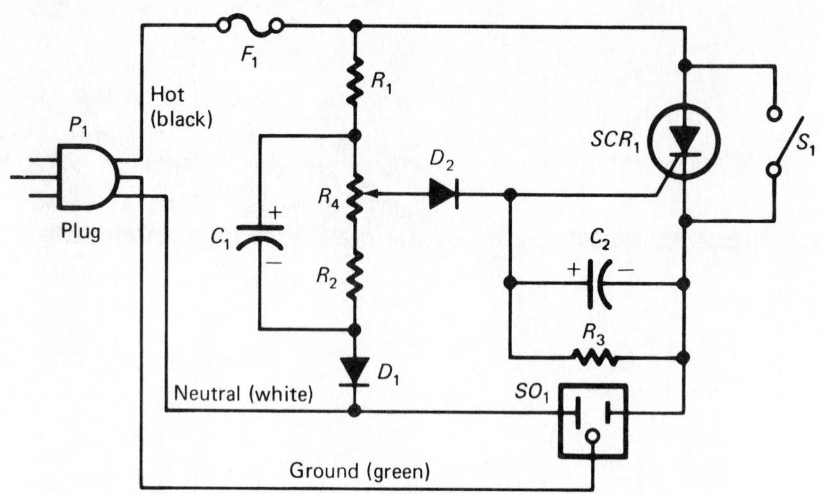

PARTS LIST

C_1	50-μF, 25-V electrolytic capacitor	R_2	100-Ω, ½-W resistor
C_2	2-μF, 25-V electrolytic capacitor	R_3	220-Ω, ½-W resistor
D_1,D_2	1N4004 silicon diode, 1 A, 400 PIV	R_4	500-Ω, 2-W linear potentiometer
F_1	5-A fuse	S_1	SPST switch
P_1	Three-pronged ac plug	SCR_1	GE-C15C silicon controlled rectifier (or similar)
R_1	2.5-kΩ, 5-W resistor	SO_1	Three-pronged ac receptacle

FIG. 12-1 Regulated motor speed control. (Charles A. Schuler, *Activities Manual for Electronics: Principles and Applications*, 2d. ed., McGraw-Hill, New York, 1984, pp. 148–149. Used by permission of McGraw-Hill Book Company.)

appliances using constant speed motors use this type of motor. The motor controller was developed by Frank J. Nola of NASA as an electric power-saving device. Power savings range up to 60 percent when an electric motor is powered with this circuit. The greatest savings occur when the motor is under light and no-load conditions. The controller circuit continually monitors the phase angle between voltage and current. When current lags too far behind voltage (such as during no-load conditions), a correction is made for better economy. The triac in the circuit may have to be placed on a heat sink if the controller circuit is used to power motors using more than 300 W.

To operate, plug the controller into 117 V ac and connect the induction motor to the circuit as shown in Fig. 12-2. Turn on both switch S_1 and the motor and adjust the level control R_{30} until a slight drop in motor speed is noticed. Leave the level control between this point and almost full speed for greatest power savings. The level adjustment must be made for each different motor.

The high-voltage wiring must be done carefully to make sure that it is completely isolated from the enclosure. The green ground wire on the three-pronged plug must be attached to the metal chassis. Use proper strain relief techniques where cords enter and exit the case. Resistor R_1 can be made from a 9-in length of 22-gauge solid copper wire.

SIMPLE DC MOTOR SPEED CONTROL

The extremely simple motor speed control circuit in Fig. 12-3 can be used with permanent magnet dc electric motors. These motors are commonly used in household cordless drills, grass shears, and children's toys. These motors are commonly powered by rechargeable batteries.

In the motor speed control circuit in Fig. 12-3, moving potentiometer R_2 gradually turns on or turns off the power transistor Q_1. When Q_1 is biased to conduct more, the motor speed will be

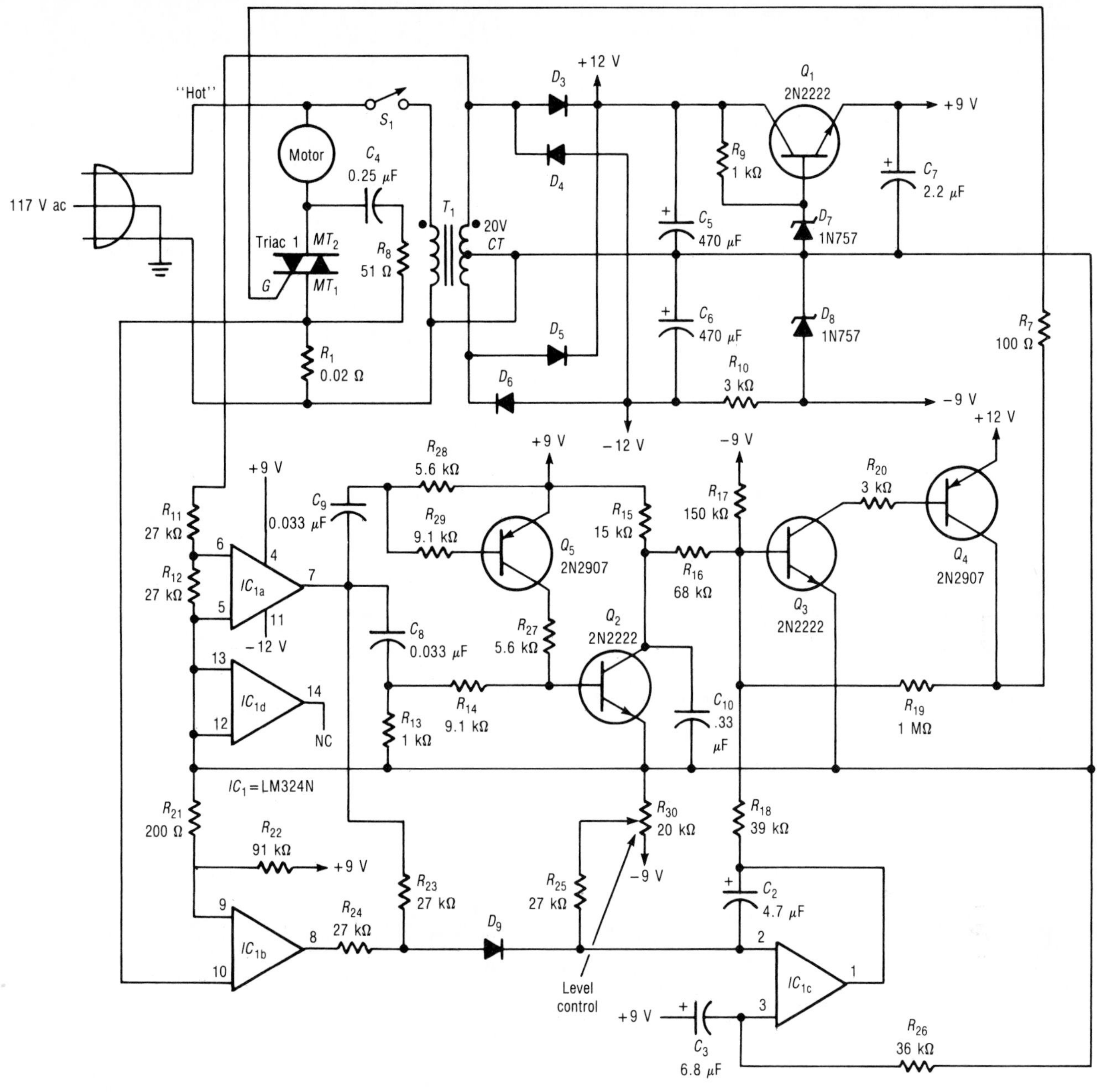

PARTS LIST

C_2	4.7-μF, 20-V electrolytic capacitor
C_3	6.8-μF, 20-V electrolytic capacitor
C_4	0.25-μF, 400-V capacitor
C_5, C_6	470-μF, 35-V electrolytic capacitor
C_7	2.2-μF, 20-V electrolytic capacitor
C_8, C_9	0.033-μF capacitor
C_{10}	0.33-μF capacitor
D_9	1N914 or 1N4148 diode
D_3 through D_6	1N4001 silicon diode, 1 A, 50 PIV
D_7, D_8	1N757, 9.1-V, 400-mW zener diode
IC_1	LM324 quad op-amp IC
Q_1, Q_2, Q_3	2N2222 NPN transistor
Q_4, Q_5	2N2907 PNP transistor
R_1	0.02-Ω, 5-W resistor
R_7	100-Ω, 2-W resistor
R_8	51-Ω, 1-W resistor
R_9, R_{13}	1-kΩ, ¼-W resistor
R_{10}, R_{20}	3-kΩ, ¼-W resistor
$R_{11}, R_{12}, R_{23}, R_{24}, R_{25}$	27-kΩ, ¼-W resistor
R_{14}, R_{29}	9.1-kΩ, ¼-W resistor
R_{15}	15-kΩ, ¼-W resistor
R_{16}	68-kΩ, ¼-W resistor
R_{17}	150-kΩ, ¼-W resistor
R_{18}	39-kΩ, ¼-W resistor
R_{19}	1-MΩ, ¼-W resistor
R_{21}	200-Ω, ¼-W resistor
R_{22}	91-kΩ, ¼-W resistor
R_{26}	36-kΩ, ¼-W resistor
R_{27}, R_{28}	5.6-kΩ, ¼-W resistor
R_{30}	20-kΩ linear potentiometer
S_1	SPST switch
T_1	20-V, center-tapped, 300-mA secondary power transformer (similar to signal DP-241-4-20)
Triac 1	200-V, 15-A triac

FIG. 12-2 NASA motor-control circuit. (Reprinted from *Electronic Experimenters Handbook—1982*. Copyright © 1981, Ziff-Davis Publishing Company.)

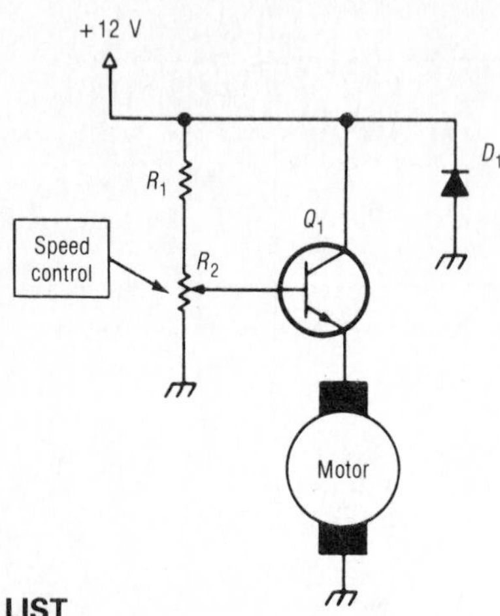

PARTS LIST

D_1	1N4002 silicon diode, 1 A, 100 PIV
Q_1	2N3055 NPN power transistor (or similar)
Motor	Permanent magnet dc low-voltage motor
R_1	47-Ω, 2-W resistor
R_2	1-kΩ, 2-W linear potentiometer

FIG. 12-3 Simple speed-control circuit for permanent magnet dc motors. (Steve Ciarcia, *Ciarcia's Circuit Cellar,* Volume III, BYTE-McGraw-Hill, New Hampshire, 1982, p. 124. Used by permission of McGraw-Hill Book Company.)

greater. The transistor must be placed on a heat sink. The power supply voltage should be varied according to the motor's voltage rating. The advantage of this circuit is simplicity. However, the transistorized speed control has the disadvantage of wasting the power dissipated by Q_1.

DC MOTOR SPEED CONTROL USING AN SCR

The dc motor speed control circuit in Fig. 12-4 uses an SCR. With the parts used in this circuit, it will handle low-voltage permanent magnet dc motors that draw up to 10 A. This circuit has a somewhat limited speed control range but maintains constant speed under varying load conditions. It is more efficient than the transistorized dc motor speed control in Fig. 12-3.

The transformer and bridge rectifier D_1 through D_4 step the voltage down and produce pulsating dc. The speed adjust potentiometer R_1 is used to control the speed of the dc motor. The SCR acts as a switch which is triggered

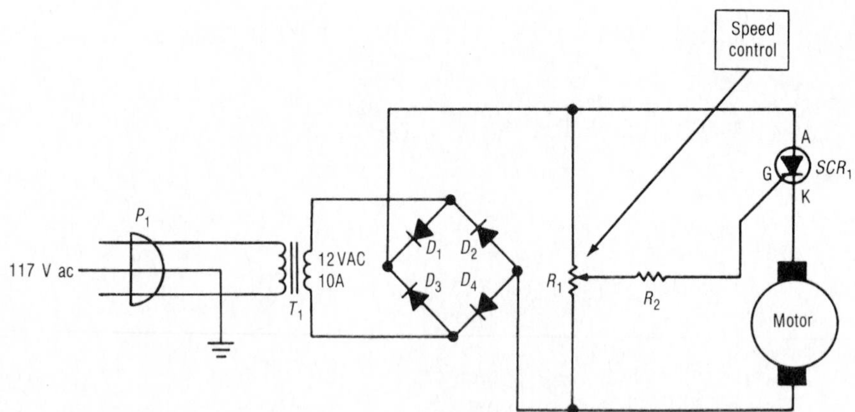

PARTS LIST

D_1 through D_4	1N1200 silicon diode, 12 A, 200 PIV (similar to ECG-5874)
Motor	12-V permanent magnet dc electric motor
P_1	Three-pronged ac plug
R_1	10-kΩ, 2-W linear potentiometer
R_2	680-Ω, ½-W resistor
SCR_1	2N688 SCR (25 A, 400 V) (similar to ECG-5527)
T_1	12-V, 10-A power transformer

FIG. 12-4 SCR speed control of permanent magnet dc motors. (Steve Ciarcia, *Ciarcia's Circuit Cellar,* Volume III, BYTE-McGraw-Hill, New Hampshire, 1982, pp. 124–126. Used by permission of McGraw-Hill Book Company.)

(turned on) by the positive voltage on the gate. Moving the wiper arm of the potentiometer up causes the SCR to fire earlier in the half-cycle, allowing more power to the motor. With the wiper arm of the potentiometer near the bottom in Fig. 12-4, the cathode (K) and gate (G) of the SCR are more nearly at the same voltage and the SCR stays off during the entire half-cycle.

The high-voltage wiring must be done carefully to make sure that it is completely isolated from the enclosure. The green ground wire on the three-pronged plug must be attached to the metal chassis. Use proper strain relief techniques where cords enter and exit the case.

PORTABLE AUTOMATIC NIGHT-LIGHT

The automatic night-light circuit in Fig. 12-5 is battery-powered and portable. The night-light circuit uses very few parts because high-gain photodarlington and darlington transistors are used. The standby current draw from the battery is very small.

As light strikes the photodarlington transistor Q_1 in Fig. 12-5, the base of Q_1 is held low. This keeps the darlington transistor Q_2 turned off during daylight hours. With nightfall, less light strikes the photodarlington transistor Q_1 and the voltage at the base of Q_2 rises. This turns on the darlington transistor Q_2, lighting the lamp.

TIME-DELAY LIGHT CONTROL

The light control circuit in Fig. 12-6 has two switches used to control a lamp plugged into socket SO_1. Switch S_2 is a regular ON-OFF switch. If push-button switch S_1 is pressed when S_2 is open, however, a lamp will light (at half brightness) for a time and then will automatically turn off. Switch S_2 is used to light the lamp at full brightness, but with no time delay. The circuit can handle up to 250 W.

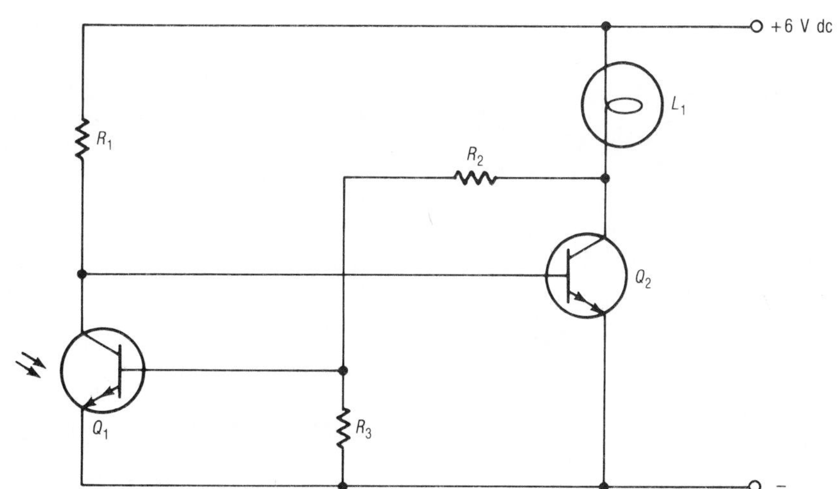

PARTS LIST

- L_1 1482 lamp (or 4512 spotlight)
- Q_1 2N5779 NPN photodarlington transistor (similar to ECG-3035)
- Q_2 D40K1 NPN darlington transistor (similar to ECG-268)
- R_1 6.8-kΩ, ¼-W resistor
- R_2 10-MΩ, ¼-W resistor
- R_3 2.2-MΩ, ¼-W resistor

FIG. 12-5 Portable automatic night light. (Used by permission of General Electric Company.)

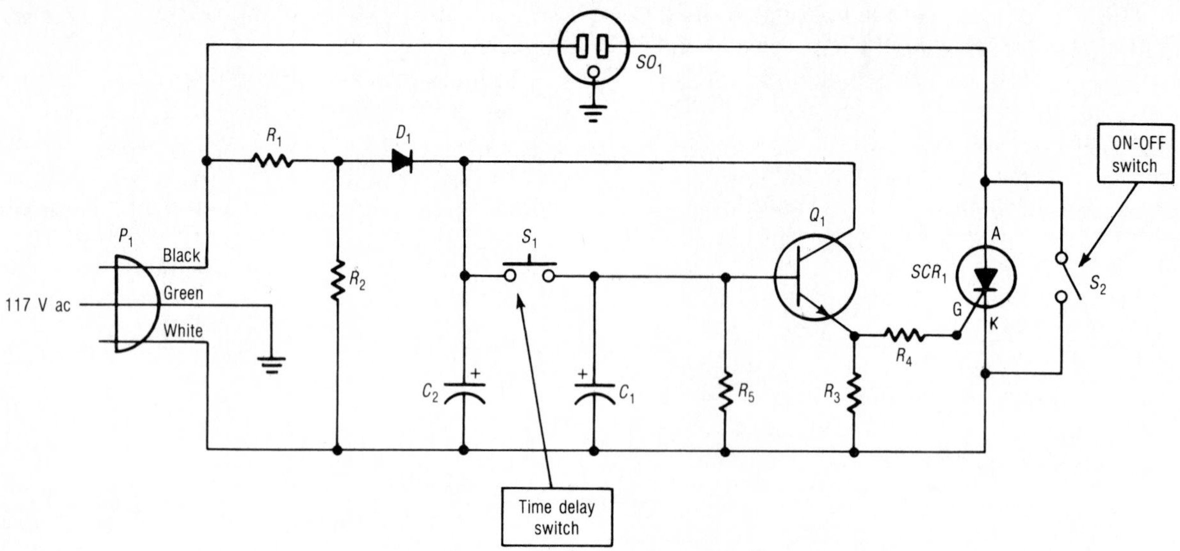

PARTS LIST

C_1	10-μF, 16-V electrolytic capacitor	R_3	4.7-kΩ, ½-W resistor
C_2	47-μF, 16-V electrolytic capacitor	R_4	22-kΩ, ½-W resistor
D_1	1N4005 silicon diode, 1 A, 600 PIV	R_5	1-MΩ, ½-W resistor
P_1	Three-pronged ac plug	S_1	Normally open push-button switch
Q_1	MPSA 14 NPN darlington transistor (similar to Radio Shack 276-2060)	S_2	SPST switch
		SCR_1	C106B1 SCR (similar to Radio Shack 276-1067)
R_1	15-kΩ, 1-W resistor	SO_1	Three-pronged ac convenience outlet
R_2	820-Ω, ½-W resistor		

FIG. 12-6 Time-delay light control switch. (Used by permission of Mode Electronics.)

With a load plugged into SO_1, the SCR is directly in series with the power line and is ready to conduct each half-cycle when the cathode (K) is negative and the anode (A) is positive. When the SCR is "ready", it needs a positive voltage at the gate (G) to cause it to conduct for the half-cycle. Components R_1, R_2, D_1, and C_2 form a rectifier-filter-voltage divider which drops the voltage down to about 6 V dc for use by the transistor circuit. When push-button switch S_1 is pressed, the positive voltage on the base of Q_2 turns on the transistor. The gate of the SCR goes positive and the SCR conducts. Capacitor C_1 is also charged when S_1 is closed. When S_1 opens, Q_2 remains on because of the positive charge remaining on the top of C_1. The SCR continues to conduct each half-cycle until C_1 becomes discharged through R_5 and the emitter-base junction of the transistor. The time delay is set at about 30 seconds in this circuit. Doubling the capacitance of C_1 will double the time delay.

The high-voltage wiring must be done carefully to ensure that it is completely isolated from the enclosure. The green ground wire on the three-pronged plug must be attached to the metal chassis. Use proper strain relief techniques where cords enter and exit the case. A parts kit and a pc board for the time-delay light control are available from Mode Electronics.

TEMPERATURE-SENSING FAN SWITCH

The temperature-sensing circuit in Fig. 12-7 is used as a fan switch. Automatic fan switches are widely used in house attics. If the temperature rises, the fan is automatically turned on. As the temperature drops, the fan is turned off.

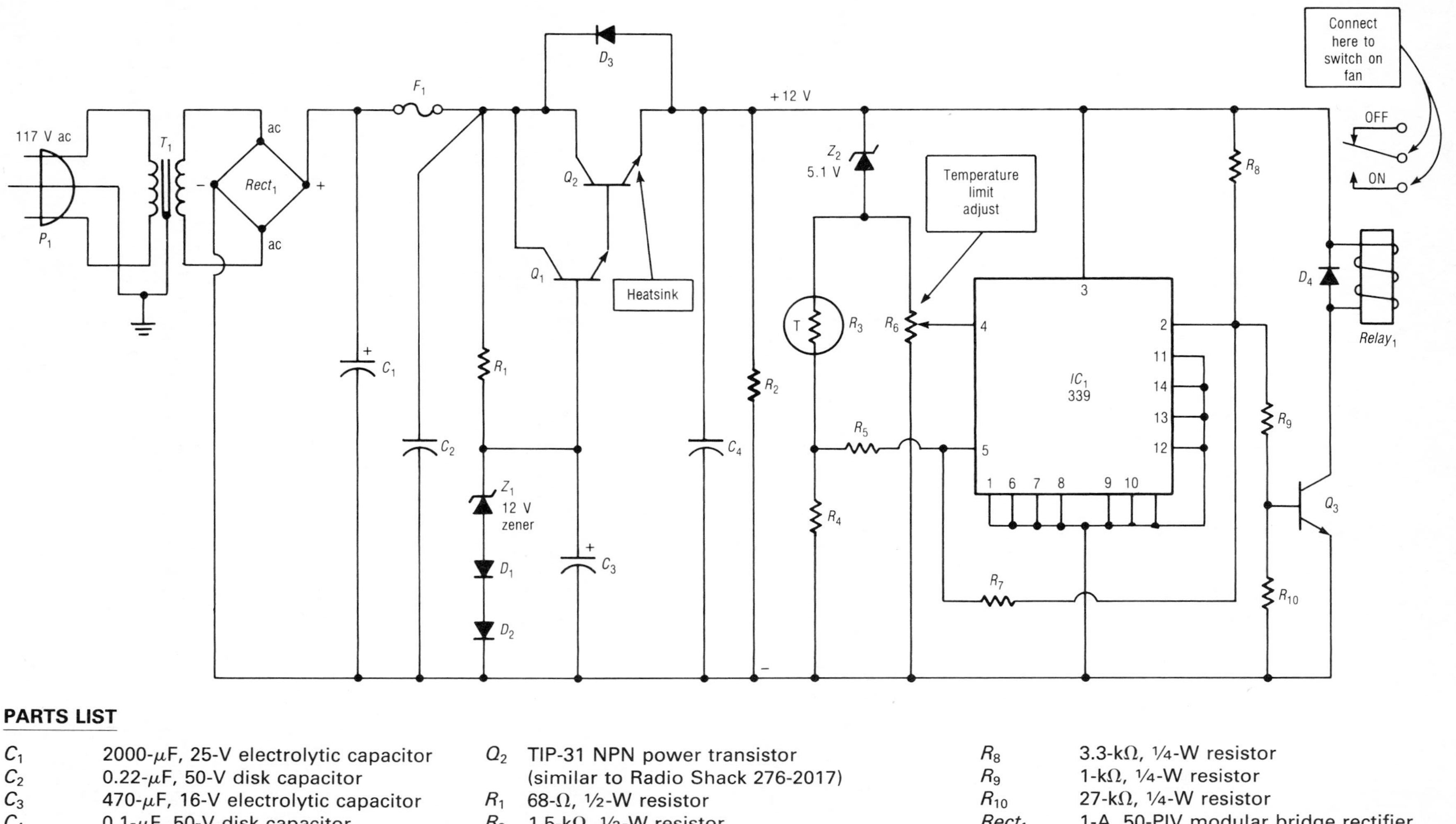

PARTS LIST

C_1	2000-μF, 25-V electrolytic capacitor
C_2	0.22-μF, 50-V disk capacitor
C_3	470-μF, 16-V electrolytic capacitor
C_4	0.1-μF, 50-V disk capacitor
D_1, D_2, D_4	1N4001 silicon diode, 1 A, 50 PIV
D_3	1N4002 silicon diode, 1 A, 100 PIV
F_1	½-A fuse
IC_1	339 quad voltage comparator IC
P_1	Three-pronged ac plug
Q_1, Q_3	2N2222 NPN transistor (or similar)
Q_2	TIP-31 NPN power transistor (similar to Radio Shack 276-2017)
R_1	68-Ω, ½-W resistor
R_2	1.5-kΩ, ½-W resistor
R_3	100- to 300-Ω "cold" resistance thermistor (with negative temperature coefficient)
R_4	220-Ω, ½-W resistor
R_5	10-kΩ, ¼-W resistor
R_6	50-kΩ linear potentiometer
R_7	10-MΩ, ¼-W resistor
R_8	3.3-kΩ, ¼-W resistor
R_9	1-kΩ, ¼-W resistor
R_{10}	27-kΩ, ¼-W resistor
$Rect_1$	1-A, 50-PIV modular bridge rectifier
$Relay_1$	12-V SPDT relay (similar to Radio Shack 275-218)
T_1	12-V, 300-mA power transformer (similar to Radio Shack 273-1385)
Z_1	1N4742 zener diode (12 V, 1 W)
Z_2	1N4733 zener diode (5.1 V)

FIG. 12-7 Temperature-sensing fan switch. (Michael Gannon, *Workbench Guide to Semiconductor Circuits and Projects,* Prentice-Hall, New Jersey, 1982, pp. 56–58, 217–218. Used by permission of Prentice-Hall, Inc.)

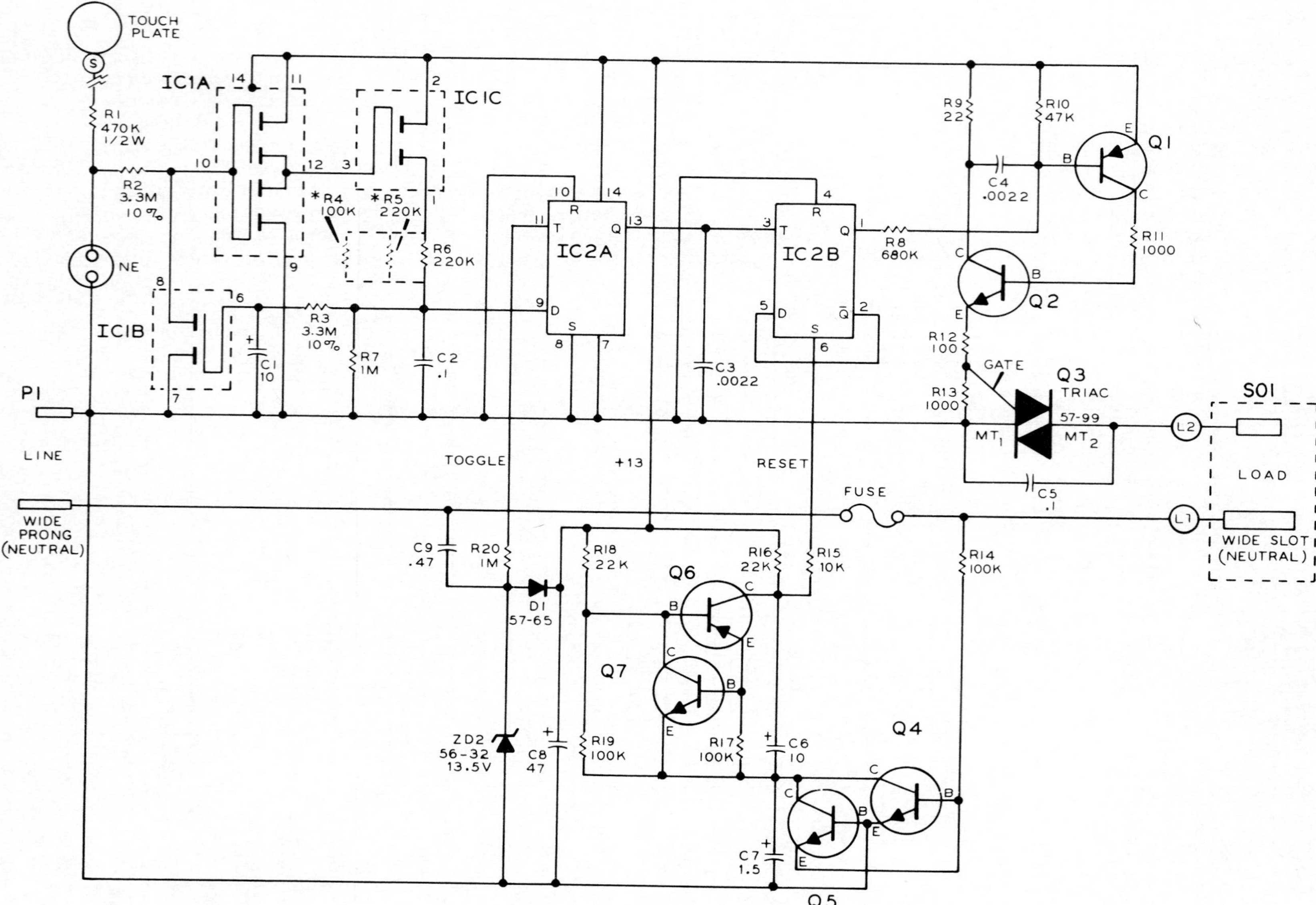

FIG. 12-8 Touch control switch. See parts list on facing page. (Reprinted by permission of Heath Company.)

The circuit can be divided into two parts: the power supply and the temperature-sensing circuit. The +12 V dc is produced by the zener-regulated power supply at the left in Fig. 12-7. Transformer T_1 drops the voltage to 12 V ac while rectifier $Rect_1$ changes it to 12 V dc with C_1 acting as a filter. Zener diode Z_1 and Q_1 and Q_2 are the major components of the voltage regulator circuit.

The temperature sensor, thermistor R_3, is typically located at a remote location. As the temperature rises, the resistance of the thermistor drops, causing an increase in positive voltage at pin 5 of the voltage comparator IC_1. When the input voltage at pin 5 becomes more positive than the reference voltage at pin 4 of the comparator, output pin 2 will go from low to high. This turns on transistor Q_3, energizing the coil of the relay. The fan would then be energized as the normally open contacts of the relay snap closed. As the temperature drops, the opposite happens in the temperature-sensing section of the circuit, and the fan turns off. Potentiometer R_6 adjusts the reference voltage on pin 4 of the voltage comparator IC_1, which sets the temperature at which the circuit will activate the relay and the fan.

The high-voltage wiring must be done carefully to make sure that it is completely isolated from the enclosure. The green ground wire on the three-pronged plug must be attached to the metal chassis. Use proper strain relief techniques where cords enter and exit the case.

TOUCH CONTROL SWITCH

The switch circuit shown in Fig. 12-8 can be used to turn lights on and off with a touch. The metal touch plate can be located on the case of the unit or at a remote location as long as it is insulated from ground. To use, plug P_1 into a

PARTS LIST FOR FIG. 12-8

C_1,C_6	10-μF, 50-V electrolytic capacitor
C_2,C_5	0.1-μF, 250-V capacitor
C_3,C_4	0.0022-μF, Mylar capacitor
C_7	1.5-μF, 50-V electrolytic capacitor
C_8	47-μF, 50-V electrolytic capacitor
C_9	0.47-μF electrolytic capacitor
D_1	1N4002 silicon diode, 1 A, 100 PIV
IC_1	4007 CMOS dual complementary pair transistor and inverter IC
IC_2	4013 CMOS dual D flip-flop IC
NE	Ne-2 neon lamp
Q_1,Q_6	2N4121 PNP transistor (similar to Radio Shack 276-2023)
Q_2,Q_4,Q_5,Q_7	MPSA20 NPN transistor (similar to Radio Shack 276-2016)
Q_3	2N6151 triac (200 V, 10 A) (similar to ECG-5614)
P_1	Three-pronged ac plug
R_1	470-kΩ, ½-W resistor
R_2,R_3	3.3-MΩ, ¼-W resistor
R_4,R_{14},R_{17},R_{19}	100-kΩ, ¼-W resistor
R_5,R_6	220-kΩ, ¼-W resistor
R_7,R_{20}	1-MΩ, ¼-W resistor
R_8	680-kΩ, ¼-W resistor
R_9	22-Ω, ¼-W resistor
R_{10}	47-kΩ, ¼-W resistor
R_{11},R_{13}	1-kΩ, ¼-W resistor
R_{12}	100-Ω, ¼-W resistor
R_{15}	10-kΩ, ¼-W resistor
R_{16},R_{18}	22-kΩ, ¼-W resistor
SO_1	Three-pronged ac convenience socket
ZD_2	VR-13.5 zener diode

117-V ac receptacle and plug the device to be controlled (such as a lamp or fan) into socket SO_1 on the touch switch. The unit does need about a 1-minute warm-up period before it will operate properly. Then the controlled lamp can be turned on or off with a touch on the touch plate. A complete kit (parts, pc board, and manual) for the touch control switch is currently available from Heath Company.

Touching the plate in Fig. 12-8 will cause the 60-Hz signal to increase. This is amplified and rectified by IC_{1a} and IC_{1c}. As the plate is touched and the finger removed, the analog voltage at the D input to IC_{2a} goes high and then low, as does output Q of the flip-flop. Output Q of the IC_{2a} toggles IC_{2b}. If output Q of IC_{2b} goes low, it turns on the astable multivibrator Q_1 and Q_2, which triggers the gate of Q_3, turning on the triac. This delivers power to output socket SO_1. If output Q of IC_{2b} goes high because someone touches the plate again, the triac driver Q_1 and Q_2 turns off, which causes the triac to turn off. When output Q of IC_{2b} is high, the touch switch is in its OFF position.

Transistors Q_4, Q_5, Q_6, and Q_7 are the major components of an automatic reset system that turns the unit to the OFF position each time power is applied or in case of momentary power interruptions. The power supply consists of C_9, D_1, ZD_2, and C_8. It powers the ICs and transistors. A 60-Hz toggle pulse is also sent to the T input of IC_{2a}. To increase the sensitivity of the unit, resistor R_5 may be added in parallel with R_6. If the unit needs to be even more sensitive, add resistor R_4 in parallel with both R_5 and R_6 as shown in Fig. 12-8.

The high-voltage wiring must be done carefully to make sure that it is completely isolated from the enclosure. The green ground wire on the three-pronged plug must be attached to the metal chassis. Use proper strain relief techniques where cords enter and exit the case.

chapter 13

Musical Instruments and Sound Effect Circuits

BOMB-BURST SYNTHESIZER

The sound synthesizer circuit in Fig. 13-1 simulates the whistle and explosion of a bomb. The circuit can be operated manually by closing switch S_1 or may be attached to a computer output port. Under computer control it could generate bomb explosions for games.

The circuit in Fig. 13-1 is based on Texas Instrument's SN76488 sound generator IC. When S_1 is pressed, there follows a high-pitched sound which falls and ends with a deep explosion. The SN76488 IC can drive the 8-Ω speaker directly through capacitor C_6. Resistor R_3 and capacitor C_3 form an RC circuit whose time constant determines the frequency range of the whistling sound. Potentiometer R_5 and capacitor C_5 form another RC circuit which connects to a one-shot in the IC to determine the length of the whistle. Resistor R_1 and capacitor C_1 determine the tone of the explosion, while potentiometer R_2 and capacitor C_2 determine the length of the explosion. Experimentation with different values in these RC circuits will give variations in sound.

ELECTRONIC DOORBELL

The electronic doorbell in Fig. 13-2 will play a short tune when the front doorbell is pressed and chimes if the back door switch is pressed. The owner can select 1 of 27 popular preprogrammed tunes for the front door. The unit also has a volume control. As suggested in Table 13-1, if the tune "Jingle Bells" is to be selected, switch 3 could be activated on A and switch 6, on B. The oscillator frequency can be adjusted by using R_2 while R_6 controls the tune speed and R_9, the sound quality. Push-button switch PB_1 is located remotely at the front door while PB_2 is the back door switch. The circuit is powered by an 8- to 16-V doorbell transformer. The electronic doorbell is currently available in kit form from Graymark International, Inc.

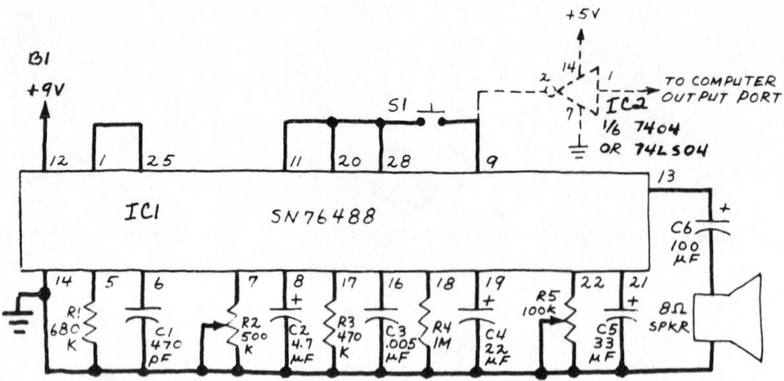

PARTS LIST

B_1	9-V battery
C_1	470-pF, 100-V disk capacitor
C_2	4.7-μF, 16-V electrolytic capacitor
C_3	0.005-μF, 50-V disk capacitor
C_4	22-μF, 16-V electrolytic capacitor
C_5	33-μF, 16-V electrolytic capacitor
C_6	100-μF, 16-V electrolytic capacitor
IC_1	SN76488 sound synthesizer IC
IC_2	74LS04 Hex inverter IC (optional—use if connected to output port of computer)
R_1	680-kΩ, ¼-W resistor
R_2	500-kΩ potentiometer
R_3	470-kΩ, ¼-W resistor
R_4	1-MΩ, ¼-W resistor
R_5	100-kΩ potentiometer
S_1	Normally open push-button switch
SPKR	8-Ω speaker

FIG. 13-1 Bomb-burst synthesizer. (Reprinted from *Popular Electronics*. Copyright © February 1982, Ziff-Davis Publishing Company.)

The AY-3-1350 melody synthesizer (IC_1) by General Instruments is a microcomputer-based synthesizer preprogrammed with 25 tunes and three chimes. The oscillator for the AY-3-1350 consists of external components R_1, R_2, R_3, C_1, and C_2. The oscillator frequency is adjustable and should be calibrated at about 250 kHz. The power supply consists of the external doorbell step-down transformer, the bridge rectifier D_1 through D_4, filter capacitors C_3 and C_4, and the +5-V voltage regulator IC_2. The attack and decay of the sound are controlled by R_9, R_{11}, R_{12}, D_5, and C_6. The field-effect transistor (FET) Q_4 drives the speaker while R_{14} is the volume control. Push-button switch PB_1 will turn on Q_3, causing the selected tune to be played. Depression of PB_2 (back door switch) causes a chime to be played.

FUZZ BOX

The fuzz box circuit in Fig. 13-3 is used between an electric guitar and its amplifier. A phone plug from the guitar pickup is plugged into J_1. The plug to the guitar amplifier fits into J_2. Using switch S_1, the musician can select either normal or fuzz (distorted) operation. With S_1 in the "normal" position, the signal is routed directly from J_1 to J_2. A parts kit and a pc board for the fuzz box are currently available from Graymark International, Inc.

The object of the fuzz box is to introduce distortion in the signal as it passes through the unit. Capacitor C_2 restricts the upper frequency response to the Q_1 amplifier. Operational amplifier U_1 clips the signal, thereby distorting the signal. The output of U_1 is di-

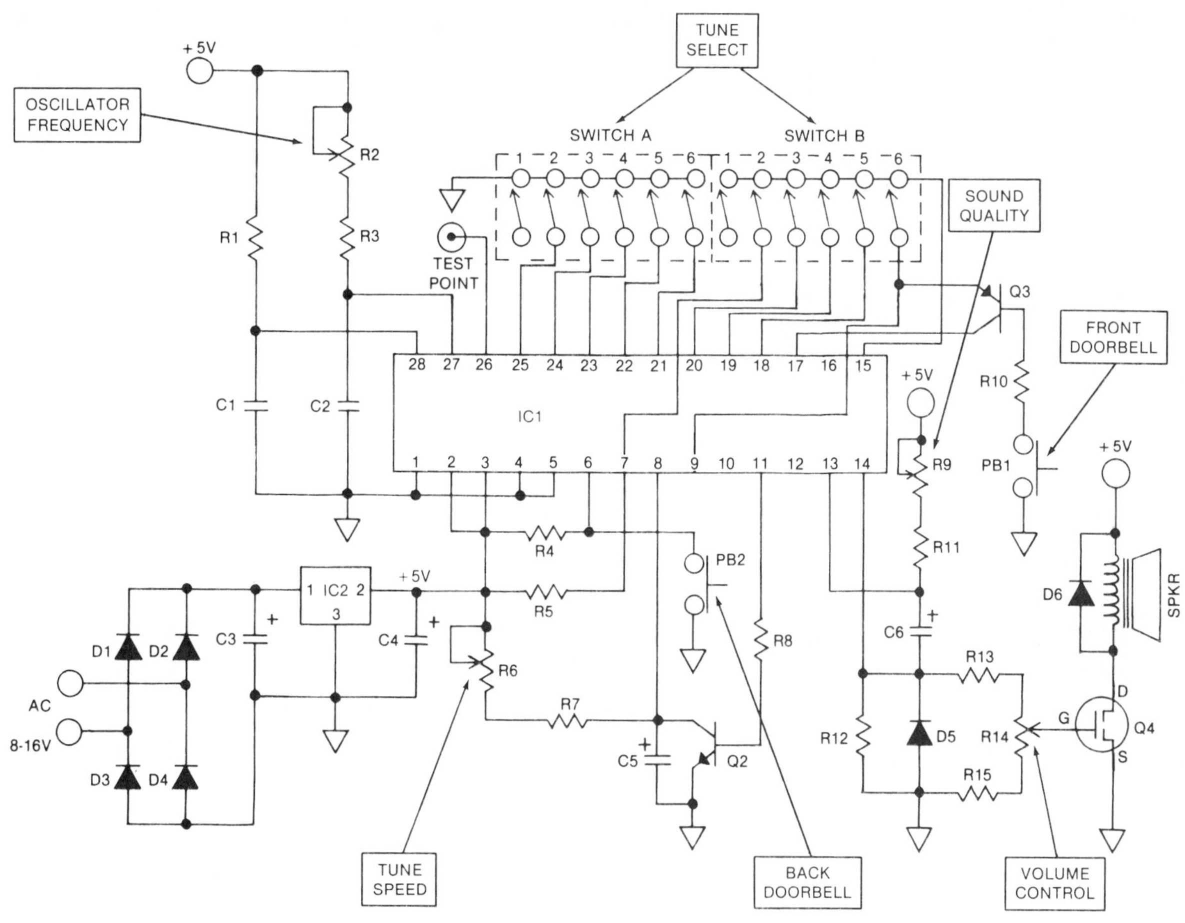

PARTS LIST

C_1	0.1-μF, 25-V disk capacitor	R_1, R_7, R_{15}	100-kΩ, ¼-W resistor
C_2	200-pF, 100-V disk capacitor	R_2	5-kΩ potentiometer
C_3	1000-μF, 25-V electrolytic capacitor	R_3	2.7-kΩ, ¼-W resistor
C_4	470-μF, 16-V electrolytic capacitor	R_4, R_5	33-kΩ, ¼-W resistor
C_5	1.0-μF tantalum capacitor	R_6	500-kΩ potentiometer
C_6	10-μF, 16-V electrolytic capacitor	R_8	10-kΩ, ¼-W resistor
D_1 through D_6	1N4002 silicon diode, 1 A, 100 PIV	R_9	10-kΩ potentiometer
IC_1	AY-3-1350 melody synthesizer IC	R_{10}	27-kΩ, ¼-W resistor
	(General Instruments)	R_{11}	1.5-kΩ, ¼-W resistor
IC_2	LM341P-5 voltage regulator IC	R_{12}	47-kΩ, ¼-W resistor
PB_1, PB_2	Normally open push-button switch	R_{13}	100-Ω, ¼-W resistor
Q_2	2N3904 NPN transistor (or similar)	R_{14}	100-kΩ potentiometer
Q_3	2N3906 PNP transistor (or similar)	SA,SB	DIP switch (6 position)
Q_4	VMOS FET, 1106	SPKR	8-Ω speaker

FIG. 13-2 Electronic doorbell. (Used by permission of Graymark International, Inc.)

rected to the guitar amplifier through S_2 and S_1, producing the FUZZ 1 sound. The FUZZ 1 sound has high frequency and amplitude distortion. With S_2 in the FUZZ 2 position, op-amp U_2 is also placed in the circuit, adding more distortion to the sound. Potentiometer R_5 is the gain control for U_1 and affects the depth and sustain characteristics of the sound. With the use of push-button switches for S_1 and S_2, the fuzz box can be foot-operated.

TABLE 13-1 DIP Switch Settings to Select 1 of 27 Tunes on the Electronic Doorbell

Switch SA	Switch SB	Selection
1	1	Westminster Chime
1	2	Simple Chime
2	2	"Toreador"
2	3	"John Brown's Body"
2	4	"America, America"
2	5	"O Sole Mio"
2	6	"Hell's Bells"
3	2	"William Tell Overture"
3	3	"Clementine"
3	4	"Deutschland Lied"
3	5	"Santa Lucia"
3	6	"Jingle Bells"
4	2	"Hallelujah Chorus"
4	3	"God Save the Queen"
4	4	"Wedding March"
4	5	"The End"
4	6	"La Vie en Rose"
5	2	"Star-Spangled Banner"
5	3	"Colonel Bogey"
5	4	"Beethoven's 5th"
5	5	"Blue Danube"
5	6	"Star Wars"
6	2	"Yankee Doodle"
6	3	"Marseillaise"
6	4	"Augustine"
6	5	"Brahms' Lullaby"
6	6	"Beethoven's 9th"

LED PENDULUM METRONOME

The metronome circuit in Fig. 13-4 features the customary audio beat and a visual LED display. The LEDs, arranged in an arc, simulate the pendulum on mechanical metronomes. The moving light on the LED display slows down at the end of each swing, acting like its mechanical counterpart. The click from the speaker is generated at the end of each swing of the light "pendulum." Potentiometer R_2 controls the tempo of the metronome. A parts kit and pc board for the LED pendulum metronome are currently available from Electronic Kits International, Inc.

A block diagram of the metronome is sketched in Fig. 13-5. Integrated circuits associated with each block are also listed on the block diagram. Block *A* is the variable-speed clock. Clock pulses are sent to block *B*, which is an up/down counter. If pin 10 of IC_4 is high, it counts upward; however, if it is low, the IC counts downward. An upward count causes the LED display to light from left to right, while a downward count causes the LEDs to sequence from right to left. The multiplexers in block *C* are driven by the counter. The 1-of-8 multiplexers light one LED at a time in sequence as IC_4 counts up and then down. Block *D* is a flip-flop which changes the count direction when the end LED on the display lights. Block *E* is a darlington amplifier that provides a click when the output of gate IC_{5d} goes high and then drops low.

COMPLEX SOUND GENERATOR

The sound circuit in Fig. 13-6 is based on the SN76477 complex sound generator IC by Texas

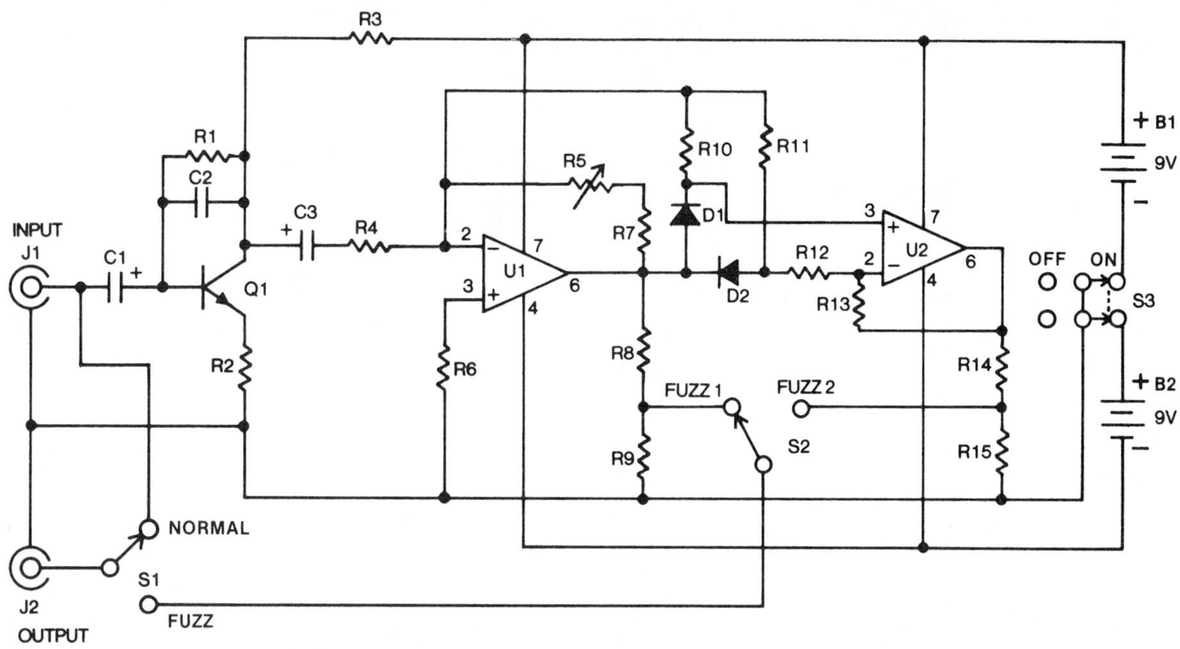

PARTS LIST

B_1,B_2	9-V battery	R_5	1-MΩ potentiometer
C_1,C_3	1-μF, 16-V electrolytic capacitor	R_6	8.2-kΩ, ½-W resistor
C_2	560-pF, 100-V disk capacitor	R_7	100-kΩ, ½-W resistor
D_1,D_2	1N4148 signal diode	R_8	39-kΩ, ½-W resistor
J_1,J_2	Phono jack (mono, ¼ in)	R_9	820-Ω, ½-W resistor
Q_1	MPS6515 NPN transistor (similar to Radio Shack 276-2009)	R_{14}	1-kΩ, ½-W resistor
		R_{15}	560-Ω, ½-W resistor
R_1	680-kΩ, ½-W resistor	S_1,S_2	SPDT switch
R_2	270-Ω, ½-W resistor	S_3	DPDT switch
R_3	6.8-kΩ, ½-W resistor	U_1	741 op-amp IC
R_4,R_{10},R_{11},R_{12},R_{13}	10-kΩ, ½-W resistor	U_2	LF356 op-amp IC

FIG. 13-3 Fuzz box for electric guitar. (Used by permission of Graymark International, Inc.)

Instruments. The SN76477 IC allows for custom sounds by changing external components. The SN76477 is a CMOS IC; therefore, it is sensitive to static electricity and must be handled with care. This circuit seems to work best with an alkaline or ni-cad battery. A parts kit and a pc board for the complex sound generator are currently available from Graymark International, Inc.

Use Table 13-2 as a guide in producing sounds of a siren, a phaser gun, a train, a propeller (prop) plane, and a space ship. To generate the siren 2 sound, for instance, switches S_3, S_4, and S_5 are all closed, and variable resistance R_6 is turned about one-half. Many other sounds can also be generated by use of the complex sound generator.

The SN76477 IC includes a noise generator, voltage-controlled oscillator, a noise filter, a super-low-frequency (SLF) oscillator, a mixer, attack/decay circuitry, an audio amplifier, and control circuitry to produce a wide variety of sounds. Switches S_1 through S_5 and potentiometer R_6 are used for programming the sound generator IC. Transistors Q_1 and Q_2 form a push-pull audio amplifier which drives the speaker.

PORTABLE ELECTRONIC ORGAN

The electronic organ in Fig. 13-7 will generate 29 notes over four octaves. For home use the circuit can be used as shown; however, for use

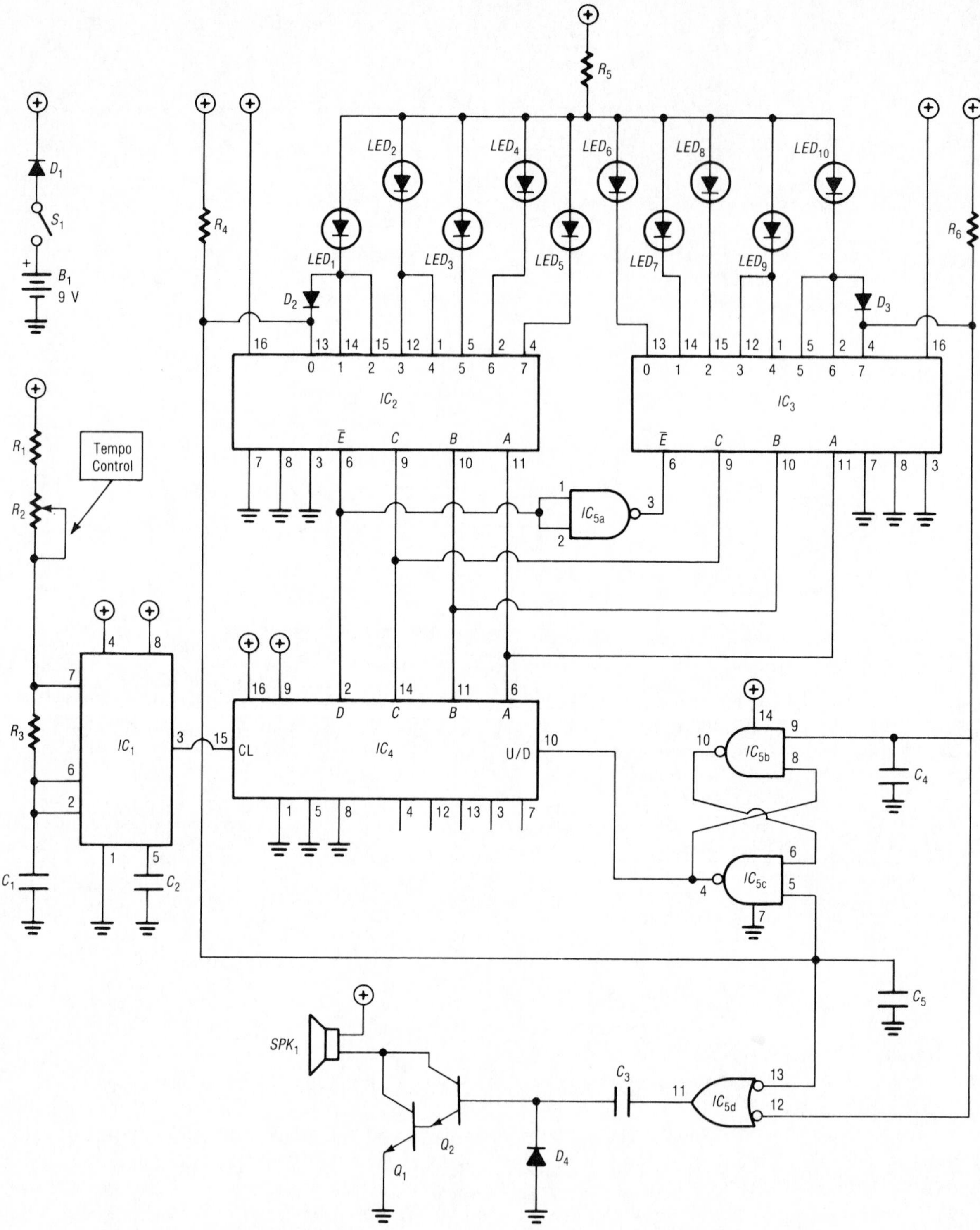

FIG. 13-4 LED pendulum metronome. (Used by permission of Electronic Kits International, Inc.)

outdoors or in larger areas, a 5- to 15-W speaker can replace SPK_1 if limiting resistor R_{11} is removed. Alkaline or ni-cad batteries should be used. To use, close power switch S_1 and press one tone switch and one octave switch at the same time. A parts kit and a pc board for the electronic organ are currently available from Electronic Kits International, Inc.

The frequency generating element of the electronic organ in Fig. 13-7 is the 555 timer (IC_1). The timer IC is wired as a free-running

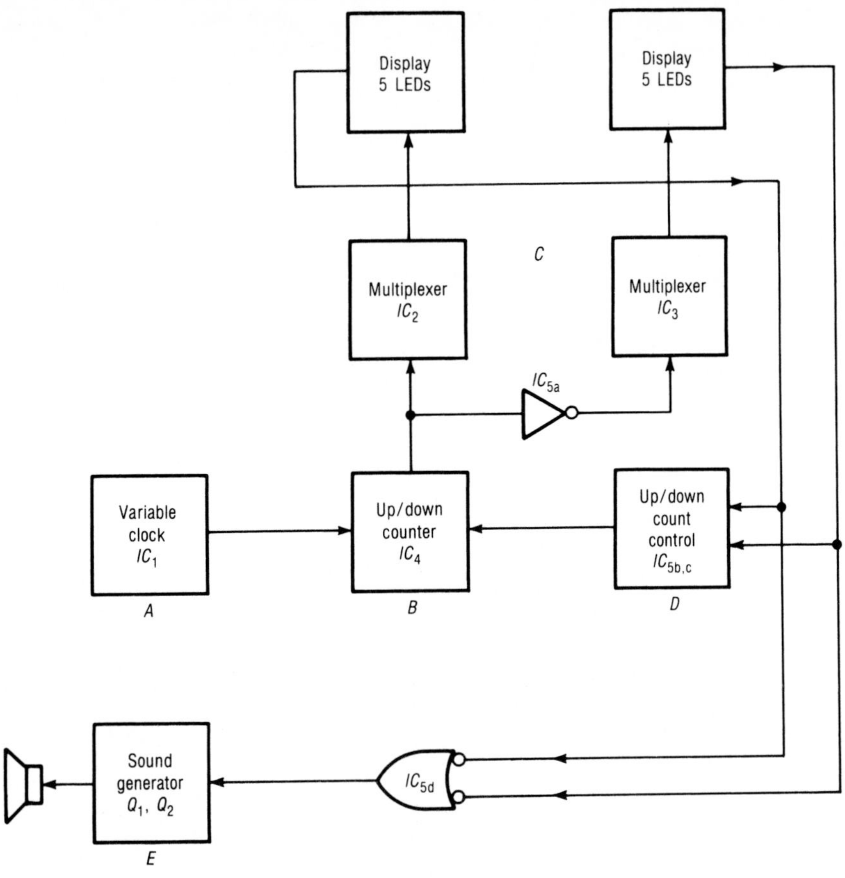

PARTS LIST

B_1	9-V battery
C_1,C_3	0.1-μF, 25-V disk capacitor
C_2,C_4,C_5	0.01-μF, 25-V disk capacitor
D_1 through D_4	1N914 diode
IC_1	555 timer IC
IC_2,IC_3	4051 CMOS analog multiplexer IC
IC_4	4029 CMOS up-down counter IC
IC_5	4011 CMOS quad 2-input NAND gate IC
LED_1 through LED_{10}	Light-emitting diode
Q_1,Q_2	2N3904 NPN transistor (or similar)
R_1,R_4,R_6	10-kΩ, ¼-W resistor
R_2	1.5-MΩ potentiometer
R_3	120-kΩ, ¼-W resistor
R_5	220-Ω, ¼-W resistor
S_1	SPST switch
SPK_1	8-Ω speaker

FIG. 13-5 Block diagram of LED pendulum metronome.

TABLE 13-2 Programming the Complex Sound Generator

	Siren 1	Siren 2	Phaser 1	Phaser 2	Train	Prop Plane	Space Ship
S_1							
S_2	On				On	On	
S_3		On		On	On	On	
S_4		On	On				
S_5		On		On			
R_6	½ turn	½ turn	½ turn	⅓ turn	¼ turn	⅛ turn	¼ turn

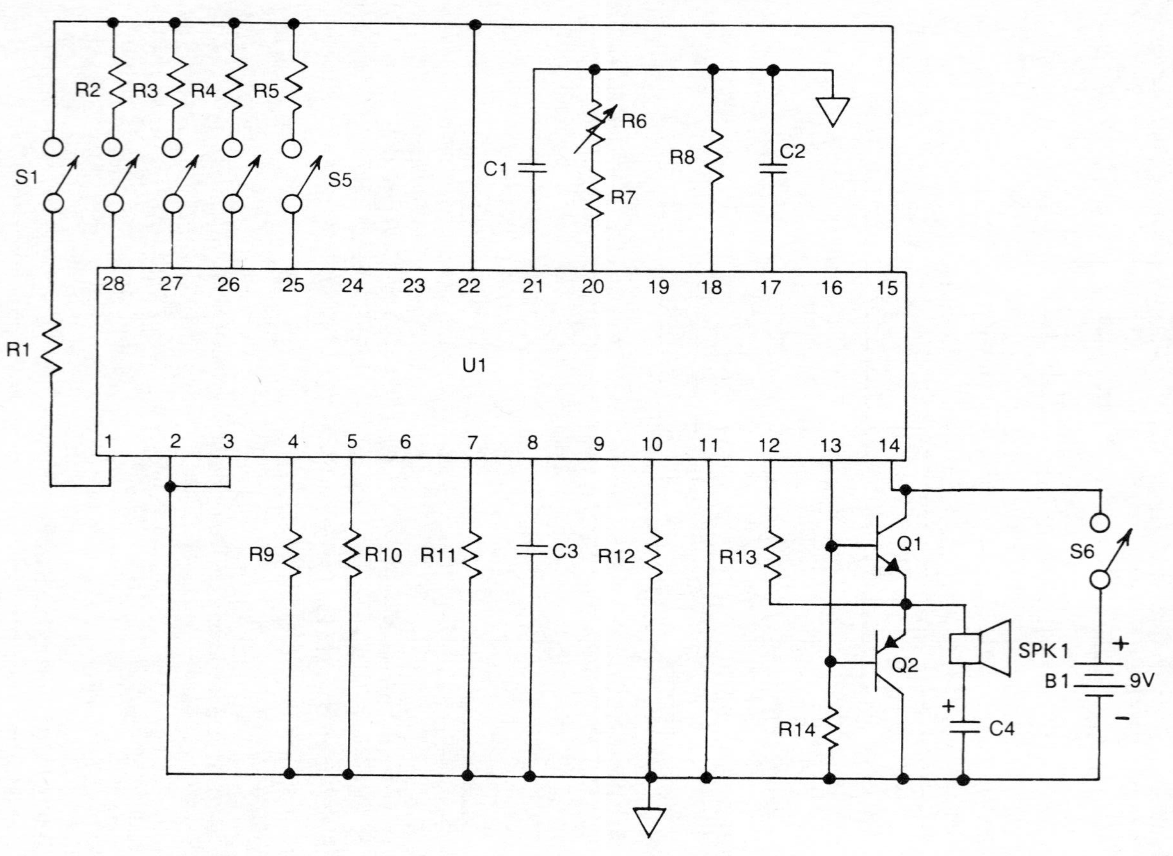

PARTS LIST

B_1	9-V battery	R_1,R_2,R_3,R_4,R_5	10-kΩ, ½-W resistor	S_1 through S_5	DIP switch
C_1,C_3	0.1-μF, 25-V disk capacitor	R_6	5-MΩ potentiometer	S_6	SPST switch
C_2	0.047-μF, 50-V disk capacitor	R_7,R_{10},R_{11},R_{12}	4.7-kΩ, ½-W resistor	SPK_1	8-Ω speaker
C_4	10-μF, 25-V electrolytic capacitor	R_8	56-kΩ, ½-W resistor	U_1	SN76477 complex sound generator IC (Texas Instruments)
Q_1	2N3904 NPN transistor (or similar)	R_9,R_{13}	47-kΩ, ½-W resistor		
Q_2	2N3906 PNP transistor (or similar)	R_{14}	3.3-kΩ, ½-W resistor		

FIG. 13-6 Complex sound generator. (Used by permission of Graymark International, Inc.)

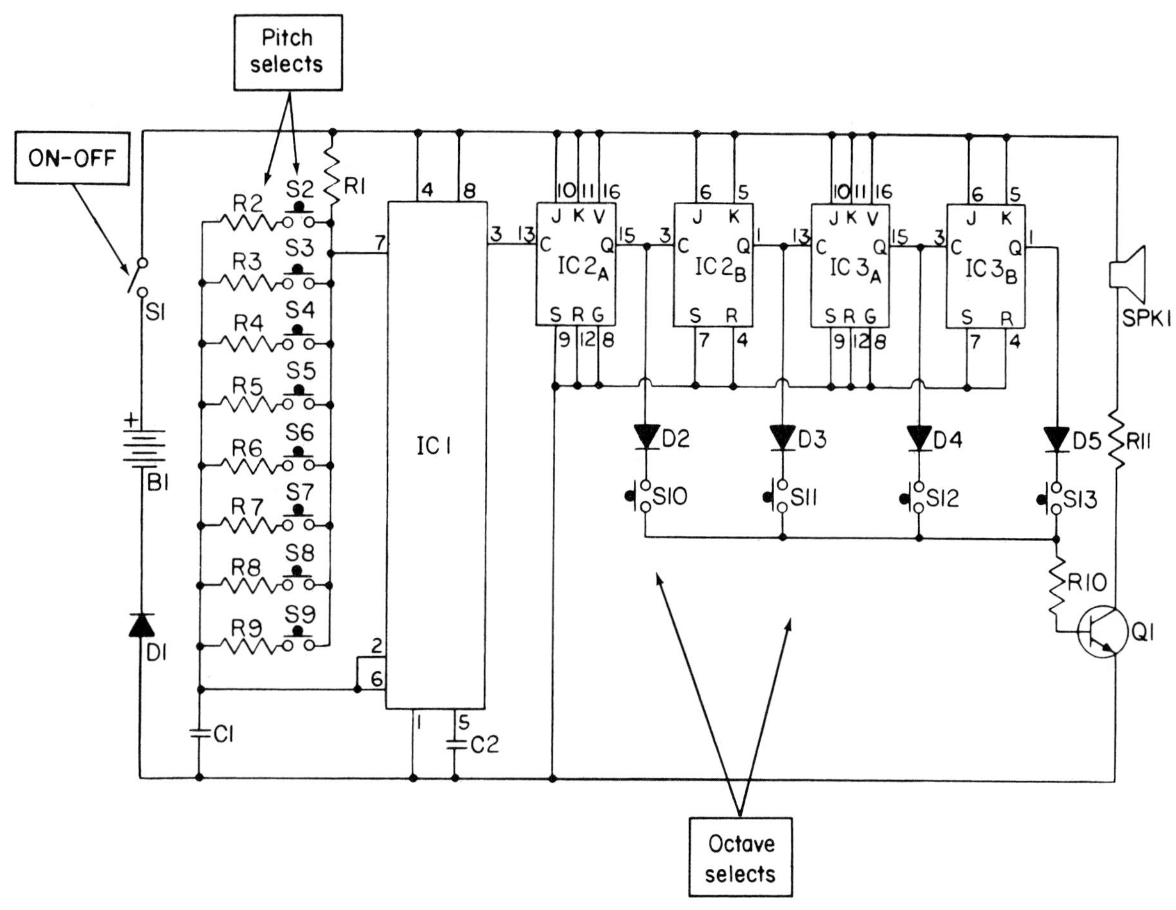

PARTS LIST

B_1	9-V battery (alkaline or ni-cad)	R_5	4.87-kΩ, 1 percent, ¼-W resistor
C_1	0.05-μF, 25-V disk capacitor	R_6	4.32-kΩ, 1 percent, ¼-W resistor
C_2	0.01-μF, 25-V disk capacitor	R_7	3.74-kΩ, 1 percent, ¼-W resistor
D_1 through D_5	1N4003 silicon diode, 1 A, 200 PIV	R_8	3.32-kΩ, 1 percent, ¼-W resistor
IC_1	555 timer IC	R_9	3.09-kΩ, 1 percent, ¼-W resistor
IC_2,IC_3	4027 CMOS dual *J-K* flip-flop IC	R_{10}	2.2-kΩ, 5 percent, ¼-W resistor
Q_1	2N3055 NPN power transistor (or similar)	R_{11}	47-Ω, 5 percent, ¼-W resistor
R_1	931-Ω, 1 percent, ¼-W resistor	S_1	SPST switch
R_2	6.65-kΩ, 1 percent, ¼-W resistor	S_2 through S_{13}	Normally open push-button switch
R_3	5.90-kΩ, 1 percent, ¼-W resistor	SPK_1	8-Ω speaker
R_4	5.23-kΩ, 1 percent, ¼-W resistor		

FIG. 13-7 Portable electronic organ. (Used by permission of Electronic Kits International, Inc.)

multivibrator. The frequency of the timer depends on the *RC* circuit containing R_1, R_2 through R_9, and C_1. For instance, if switch S_2 is closed, R_1, R_2, and C_1 determine the lowest frequency of the timer IC. The output frequency of the timer (pin 3) is then divided by 2, 4, 8, and 16 for the octaves. The square-wave signal passes through resistor R_{10} and is amplified by power transistor Q_1. Resistor R_{11} limits the current to a low-wattage speaker and may be removed if a 5- to 15-W speaker is used. Closing tone switch S_2 gives the lowest tone, while S_9 yields the highest pitch. Closing octave switch S_{10} gives the highest tones while S_{13} yields the lowest pitches.

SOUND EFFECTS GENERATOR

The sound effects circuit in Fig. 13-8 generates a "road runner" effect when used as wired. The

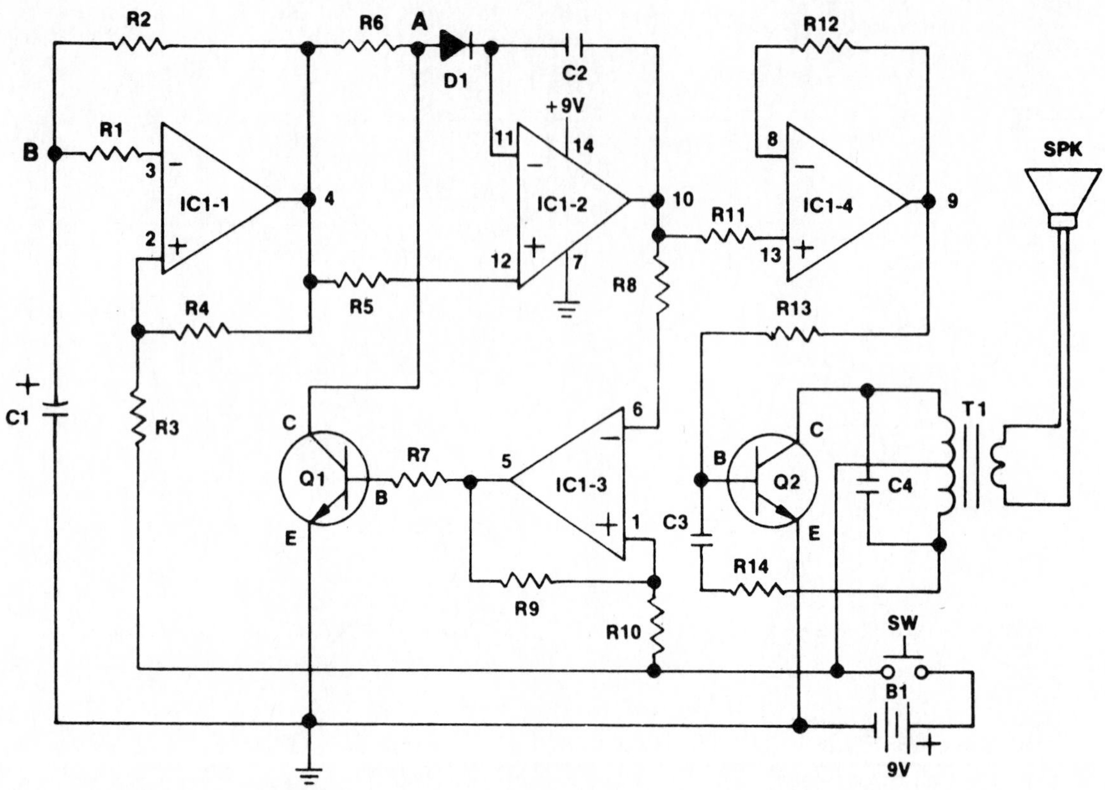

PARTS LIST

B_1	9-V battery	R_4	1.5-MΩ, ¼-W resistor
C_1	4.7-μF, 16-V electrolytic capacitor	R_5,R_{10},R_{11},R_{12}	1-MΩ, ¼-W resistor
C_2	0.1-μF, 25-V disk capacitor	R_6,R_9	510-kΩ, ¼-W resistor
C_3	0.01-μF, 25-V disk capacitor	R_8	330-kΩ, ¼-W resistor
C_4	0.022-μF, 50-V disk capacitor	R_{13}	47-kΩ, ¼-W resistor
D_1	1N914 signal diode	R_{14}	4.7-kΩ, ¼-W resistor
IC_1	3900 quad op-amp IC	S_1	Normally open push-button switch
Q_1,Q_2	2N3707 NPN transistor (or similar)	SPK	8-Ω speaker
R_1	150-kΩ, ¼-W resistor	T_1	Audio output transformer (similar to Radio Shack 273-1380)
R_2,R_7	30-kΩ, ¼-W resistor		
R_3	2.2-MΩ, ¼-W resistor		

FIG. 13-8 Road runner sound effects generator. (Used by permission of mode Electronics.)

sounds can be changed by placing jumper wires from point *A* to pins 1, 4, 8, 11, 12, and 13 of the op-amp IC_1. Other sound variations are generated by placing jumper wires from point *B* to pins 2, 5, 7, and 13 of IC_1. Placing a 0.33-microfarad (0.33-μF) capacitor across points *A* and *B* in the circuit will also change the sound. Try experimenting with this circuit. A parts kit and a pc board for the road runner sound effects generator are currently available from Mode Electronics.

STEAM ENGINE AND WHISTLE SOUND SYNTHESIZER

The sound synthesizer circuit in Fig. 13-9 simulates the sound of an antique steam locomotive. The frequency of the "puffs" can be adjusted by using potentiometer R_2. Depression of switch S_1 activates the whistle sound commonly associated with steam engines.

The SN76477 complex sound generator (IC_1) by Texas Instruments is the central component

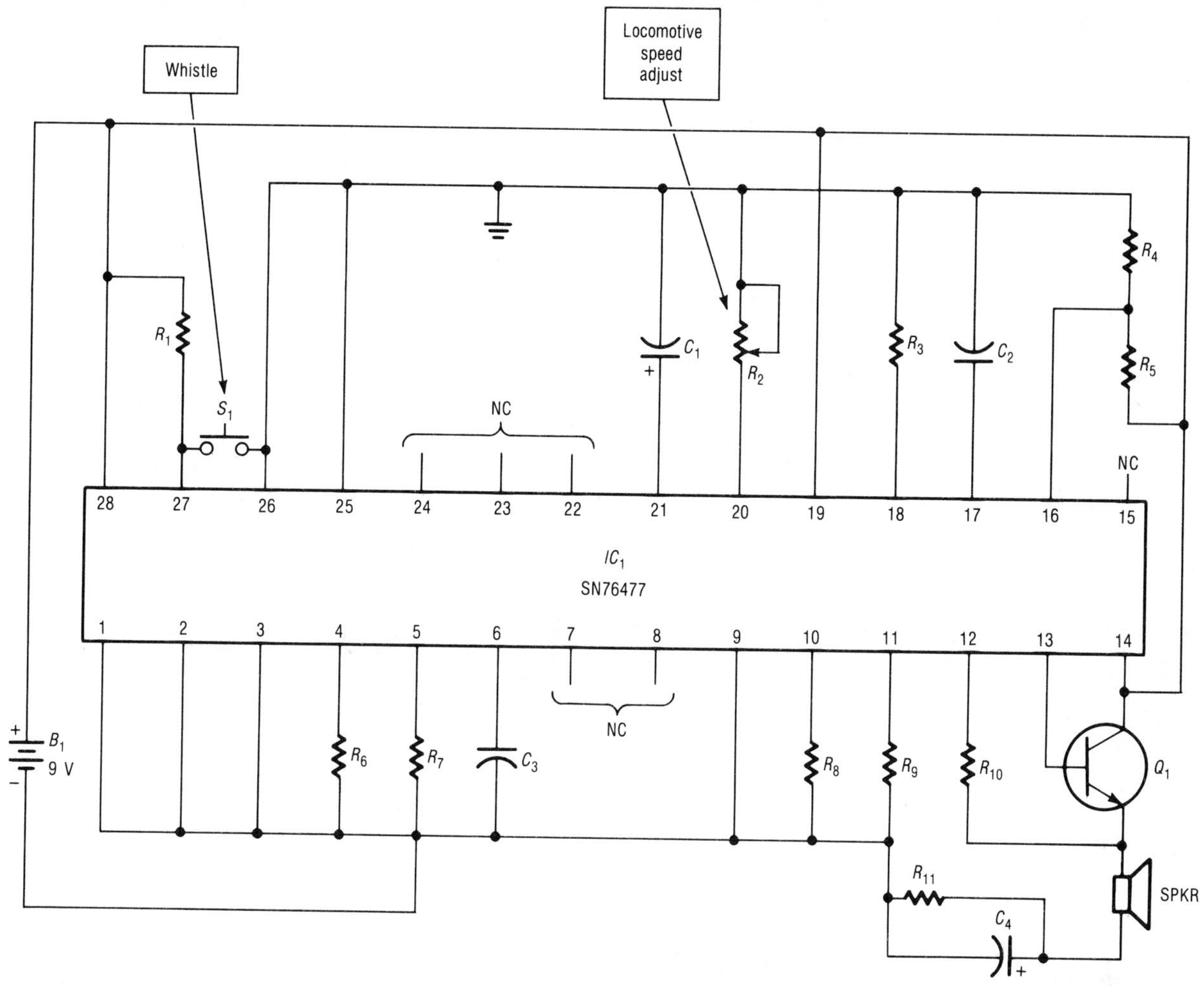

PARTS LIST

- B_1 9-V battery
- C_1 2.2-μF, 16-V electrolytic capacitor
- C_2 0.0047-μF, 25-V disk capacitor
- C_3 470-pF, 100-V disk capacitor
- C_4 10-μF, 16-V electrolytic capacitor
- IC_1 SN76477 complex sound generator IC (Texas Instruments)
- Q_1 2N2222 NPN transistor (or similar)
- R_1 4.7-kΩ, ¼-W resistor
- R_2 1-MΩ potentiometer
- R_3,R_7,R_{10} 47-kΩ, ¼-W resistor
- R_4 27-kΩ, ¼-W resistor
- R_5 68-kΩ, ¼-W resistor
- R_6 39-kΩ, ¼-W resistor
- R_8,R_9 100-kΩ, ¼-W resistor
- R_{11} 100-Ω, ¼-W resistor
- S_1 Normally open push-button switch
- SPKR 8-Ω speaker

FIG. 13-9 Steam engine and whistle sound synthesizer. (Forrest M. Mims, *The Forrest Mims Circuit Scrapbook,* McGraw-Hill, New York, 1983, p. 136. Used by permission of Forrest M. Mims, III.)

of this circuit. Custom sounds can be created by changing the external components connected to the chip. For the locomotive puffing sound, the IC noise generator is turned on and off by its SLF oscillator. Potentiometer R_2 changes the frequency of the SLF oscillator, changing the frequency of the locomotive puffing. Depression of switch S_1 causes the IC to generate the locomotive's whistle sound. This is generated by the voltage-controlled oscillator (VCO) section of the SN76477. The output (from pin 13) of the IC is amplified by transistor Q_1 driving the speaker.

2-MINUTE BEEPER

The 2-minute beeper in Fig. 13-10 is a favorite among students who want to torment friends, teachers, or parents. The beeper emits a short beep for several seconds followed by silence for about 2 minutes before emitting another beep. When hidden, the short periodic beep is difficult to locate.

The circuit in Fig. 13-10 consists of two oscillators: the tone oscillator and the time delay unit. The tone oscillator consists of Q_3 and Q_4. Capacitor C_2 and resistor R_5 feed back part of the output to the base of Q_3 to keep Q_3 and Q_4 oscillating. The rest of the energy at the collector of Q_4 drives the speaker. The tone oscillator is turned on when the collector of Q_2 goes positive and off when it goes more negative. The long time delay in the Q_1–Q_2 oscillator is caused by the longer time constant of the R_1–C_1 feedback circuit. The time between beeps can be changed by adjusting the value of C_1. The beep can be shortened by adjusting the values of R_1 and R_2. Changing the values of R_5 and C_2 shifts the pitch of the tone oscillator.

WATERFALL AND OCEAN BACKGROUND SOUNDS

The circuit in Fig. 13-11 electronically generates signals that simulate several restful natural sounds. The rotary switch S_1 selects one of the following sounds of nature:

1. Rain on tin roof
2. Rainstorm

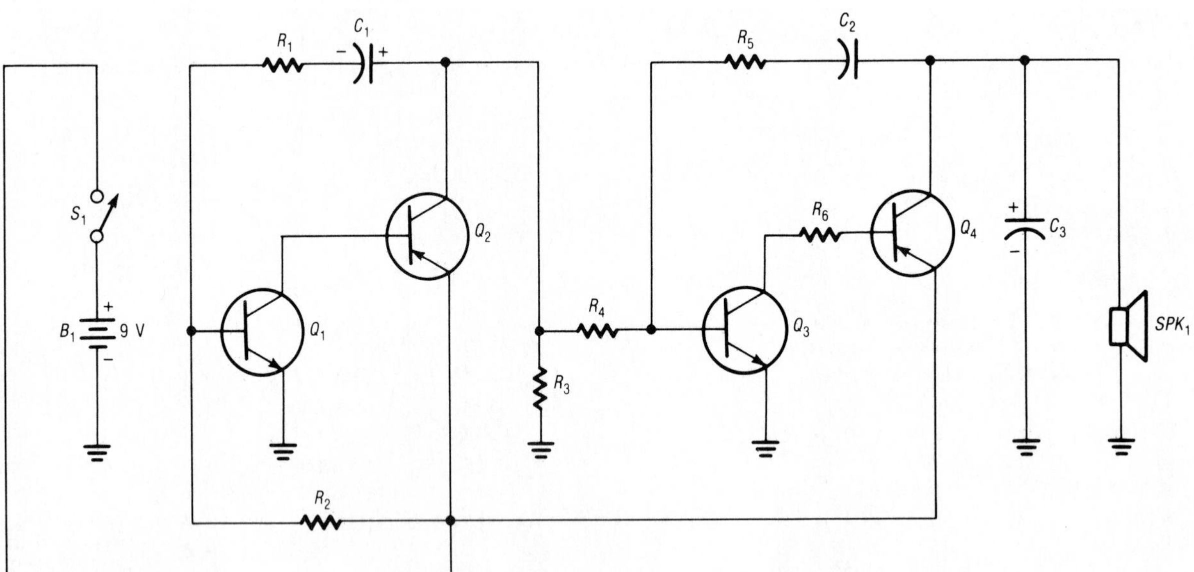

PARTS LIST

B_1	9-V battery	R_2	2.7-MΩ, ½-W resistor
C_1	47-μF, 16-V electrolytic capacitor	R_3	1.5-kΩ, ½-W resistor
C_2	0.02-μF, 25-V disk capacitor	R_4	100-kΩ, ½-W resistor
C_3	4.7-μF, 16-V electrolytic capacitor	R_5	1-kΩ, ½-W resistor
Q_1, Q_3	2N3904 NPN transistor	R_6	470-Ω, ½-W resistor
Q_2, Q_4	2N3906 PNP transistor	S_1	SPST switch
R_1	22-kΩ, ½-W resistor	SPK_1	8-Ω speaker

FIG. 13-10 2-minute beeper. (Used by permission of Robert Delp Electronics.)

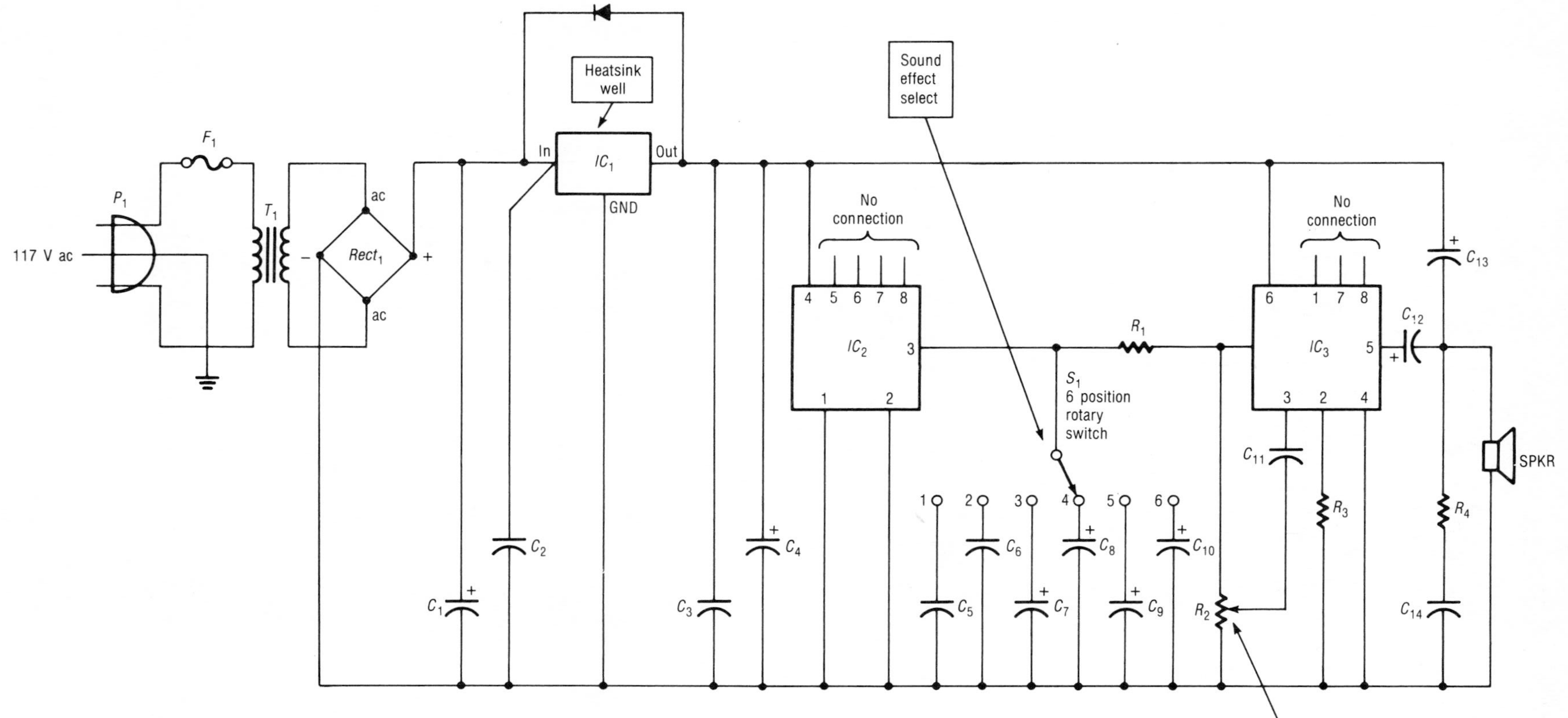

PARTS LIST

C_1	1000-μF, 50-V electrolytic capacitor
C_2	0.22-μF, 50-V disk capacitor
C_3	0.1-μF, 50-V disk capacitor
C_4	100-μF, 25-V electrolytic capacitor
C_5	0.01-μF, 50-V disk capacitor
C_6, C_{11}, C_{14}	0.1-μF, 50-V disk capacitor
C_7, C_{13}	1-μF, 25-V tantalum capacitor
C_8	2.2-μF, 25-V tantalum capacitor
C_9	4.7-μF, 25-V tantalum capacitor
C_{10}	10-μF, 25-V tantalum capacitor
C_{12}	220-μF, 25-V electrolytic capacitor
D_1	1N4002 silicon diode, 1 A, 100 PIV
F_1	0.3-A, fast-blow fuse
IC_1	7815 voltage regulator IC (15 V) (Radio Shack 276-1772)
IC_2	MM5837 digital noise generator (National) (or S2688 by American Microsystems)
IC_3	386 audio power amplifier (Radio Shack 276-1731)
P_1	Three-pronged ac plug
R_1	5.1-kΩ, ¼-W resistor
R_2	10-kΩ audio-taper potentiometer
R_3	6.8-kΩ, ¼-W resistor
R_4	10-Ω, ¼-W resistor
$Rect_1$	Bridge rectifier module, 1 A, 100 PIV (similar to Radio Shack 276-1152)
S_1	6 position (make-before-break) rotary switch (similar to Radio Shack 275-1385)
SPKR	8-Ω speaker
T_1	25-V, 300-mA power transformer (similar to Radio Shack 273-1386)

FIG. 13-11 Waterfall and ocean background sounds. (Michael Gannon, *Workbench Guide to Semiconductor Circuits and Projects*, Prentice-Hall, New Jersey, 1982, pp. 118–119, 207–208. Used by permission of Prentice-Hall, Inc.)

3. High wind through forest
4. River rapids
5. Waterfall
6. Heavy ocean surf

The background sounds circuit in Fig. 13-11 consists of three sections: the power supply, the noise generator, and the audio amplifier. In the power supply, T_1 drops household voltage to 25 V ac while the rectifier module changes that to dc. Integrated circuit IC_1 regulates the voltage to +15 V dc with capacitors C_1 through C_4 acting as filters. The central component of the noise generator section is IC_2. The MM5837 is a white noise generator whose output is modified by capacitors C_5 through C_{10}. The 386 IC is the audio power amplifier which drives the speaker. Try a better-quality speaker on this project for listening pleasure. Potentiometer R_2 is the volume control.

chapter 14

Power Supply Circuits

0- TO 15-V DC VARIABLE POWER SUPPLY

The variable power supply in Fig. 14-1 is a low-cost unit using just a few easy-to-find parts. The unit will deliver 0 to 15 V dc with a maximum output current of 300 mA. A parts kit and a pc board for this power supply are currently available from Graymark International, Inc.

Transformer T_1 steps down the 117 V ac to about 25 V while diodes D_1 and D_2 act as rectifiers. Capacitors C_1 and C_3 act as filters. Power transistor Q_1 acts as a variable resistor in controlling the voltage across R_3 and the output. Potentiometer R_1 supplies the bias voltage at the base of Q_1. As the voltage at the base increases, the emitter-to-collector resistance drops, allowing a greater voltage drop across R_3 (output voltage increases). As the voltage at the base of Q_1 decreases, the transistor gradually turns off and less voltage drops across R_3 and the parallel load connected to the outputs. The transistor should be mounted on a heat sink. Zener diode Z_1 maintains a constant 16 V across the potentiometer.

The high-voltage wiring must be done carefully to make sure that it is completely isolated from the enclosure. The green ground wire on the three-pronged plug must be attached to the metal chassis. Use proper strain relief techniques where cords enter and exit the case.

5-V DC REGULATED POWER SUPPLY FOR TTL CIRCUITS

The dc power supply circuit in Fig. 14-2 is one of the most useful units to have on an electronics workbench. The circuit will deliver up to 1 A at a regulated 5 V. This unit is ideal for operating popular TTL digital ICs. A parts kit and a pc board for the 5-V dc power supply are currently available from Graymark International, Inc.

Transformer T_1 drops the receptacle voltage to about 8 to 12 V ac, while the bridge rectifier D_1 through D_4 changes the voltage to dc. Capacitors C_1 and C_2 are filter capacitors. The internal circuitry of the 7805 IC regulates the voltage to 5 V at the output of the power sup-

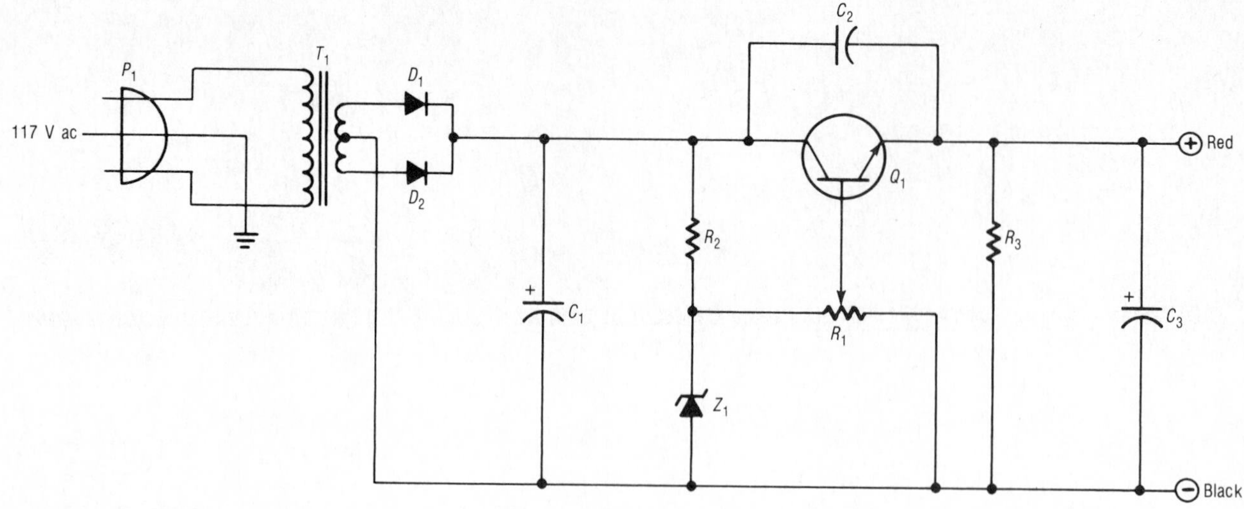

PARTS LIST

C_1	100-μF, 35-V electrolytic capacitor
C_2	0.02-μF, 50-V disk capacitor
C_3	470-μF, 25-V electrolytic capacitor
D_1,D_2	1N4001 silicon diode, 1 A, 50 PIV
P_1	Three-pronged ac plug
Q_1	C1173 NPN power transistor (similar to ECG-152)
R_1	10-kΩ linear potentiometer
R_2	820-Ω, ½-W resistor
R_3	1-kΩ, ½-W resistor
T_1	50-V (center-tapped), 300-mA secondary power transformer (similar to Radio Shack 273-1366)
Z_1	1N5246B zener diode, 16 V, 500 mW

FIG. 14-1 0–15 V variable dc power supply. (Used by permission of Graymark International, Inc.)

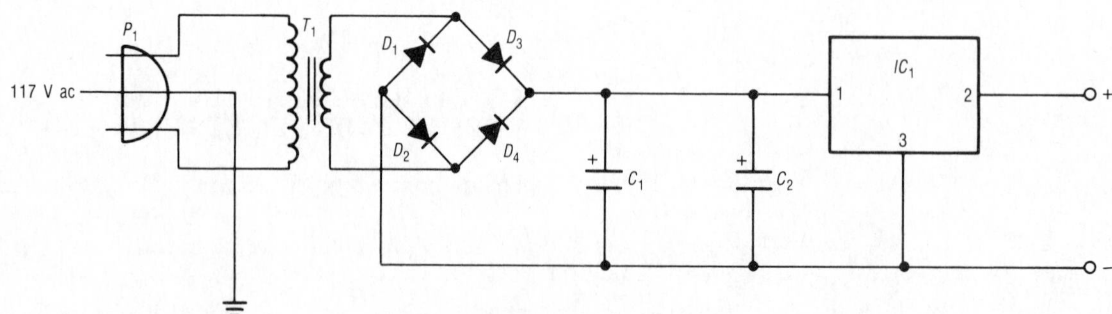

PARTS LIST

C_1,C_2	2200-μF, 16-V electrolytic capacitor
D_1 through D_4	1N5400 silicon diode (similar to Radio Shack 276-1114)
IC_1	7805 voltage regulator (5-V) (Radio Shack 276-1770)
P_1	Three-pronged ac plug
T_1	8- to 12-V, 1.5-A (or more) power transformer (similar to Radio Shack 273-1511)

FIG. 14-2 5-V dc regulated power supply for TTL ICs. (Used by permission of Graymark International, Inc.)

ply. The 7805 IC should be mounted on a heat sink to dissipate extra heat.

The high-voltage wiring must be done carefully to make sure that it is completely isolated from the enclosure. The green ground wire on the three-pronged plug must be attached to the metal chassis. Use proper strain relief techniques where cords enter and exit the case.

CB POWER SUPPLY

The power supply illustrated in Fig. 14-3 is designed to operate citizen's band, marine, and many other mobile transceivers during bench use and testing. The power supply produces 13.6 V dc, which is the customary voltage of the so-called 12-V battery found in most vehicles. The power supply is capable of delivering up to 5 A to a mobile transceiver. Several capacitors C_2, C_3, C_5, and C_6 have been added to the circuit to help bypass radio frequency.

Transformer T_1 steps the voltage down, and the bridge rectifier changes the ac to dc. Capacitors C_1 and C_4 filter out low-frequency variations in voltage, whereas C_2 and C_3 bypass higher frequencies. The LAS-19CB IC regulates the output voltage to 13.6 V and will deliver up to 5 A to the transceiver or other load device. The large-value filter capacitor C_4 at the output helps to smooth out the surges that occur when a transceiver switches to the transmit mode.

Both the voltage regulator IC and the bridge rectifier module should be placed on a good heat sink. The high-voltage wiring must be done carefully to make sure that it is completely isolated from the enclosure. The green ground wire on the three-pronged plug must be attached to the metal chassis. Use proper strain relief techniques where cords enter and exit the case.

15-V DC REGULATED POWER SUPPLY

The power supply circuit in Fig. 14-4 features good voltage regulation at 15 V dc and a low parts count. It is a low-current unit with a limit of about 300 mA.

The 117-V ac household current is stepped down to 25 V ac by T_1 and rectified by $Rect_1$. Capacitors C_1, C_2, and C_3 act as filters. The

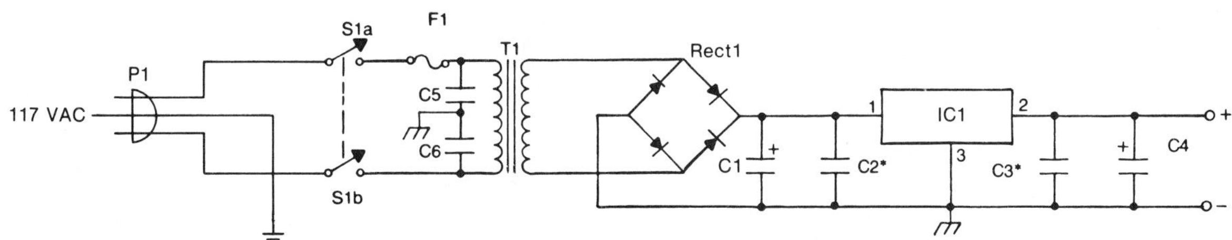

PARTS LIST

C_1	20,000-μF, 35-V electrolytic capacitor
C_2,C_3	(Each two capacitors in parallel) 1-μF, 35-V tantalum capacitor, 0.01-μF, 50-V disk capacitor
C_4	2000 μF or more, 35-V electrolytic capacitor
C_5,C_6	0.01-μF, 2000-V capacitor (these may be left out of circuit)
F_1	1-A fuse
IC_1	LAS-19CB voltage regulator IC (13.6 V, 5 A) (Lambda Electronics) (similar to ECG-934)
P_1	Three-pronged ac plug
$RECT_1$	Bridge rectifier module, 25 A, 50 PIV (Radio Shack 276-1185)
S_1	DPST switch (or use DPDT)
T_1	12.6-V, 5-A power transformer (similar to Stancor P-8642)

FIG. 14-3 13.6-V dc, 5-A CB power supply. (Joseph J. Carr, *104 Weekend Electronic Projects*, TAB Books, Pennsylvania, 1982, pp. 172–176. Used by permission of TAB Books, Inc.)

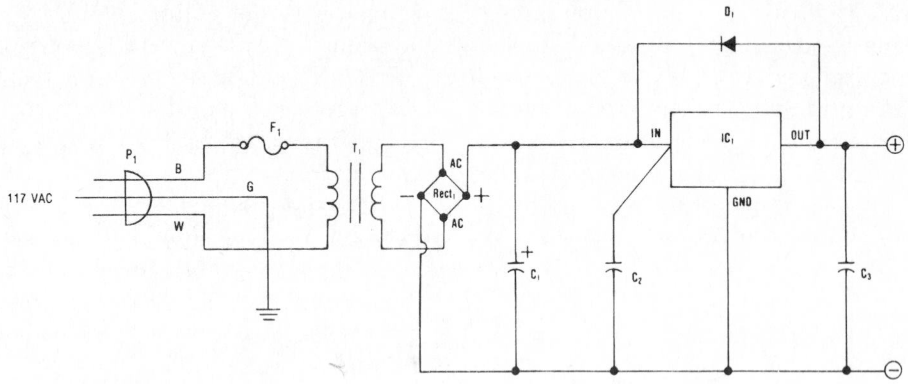

PARTS LIST

C_1	1000-μF, 50-V electrolytic capacitor
C_2	0.22-μF, 50-V disk capacitor
C_3	0.1-μF, 50-V disk capacitor
D_1	1N4002 silicon diode, 1 A, 100 PIV
F_1	0.3-A fuse
IC_1	7815 voltage regulator IC (15 V) (Radio Shack 276-1772)
P_1	Three-pronged ac plug
$Rect_1$	Bridge rectifier module, 1 A, 100 PIV (Radio Shack 276-1152)
T_1	25-V, 300-mA power transformer (similar to Radio Shack 273-1386)

FIG. 14-4 15-V dc regulated power supply. (Michael Gannon, *Workbench Guide to Semiconductor Circuits and Projects,* Prentice-Hall, New Jersey, 1982, pp. 207–208. Used by permission of Prentice-Hall, Inc.)

7815 IC regulates the voltage to +15 V. The voltage regulator IC should be mounted on a heat sink.

The high-voltage wiring must be done carefully to make sure that it is completely isolated from the enclosure. The green ground wire on the three-pronged plug must be attached to the metal chassis. Use proper strain relief techniques where cords enter and exit the case.

1.25- TO 30-V DC REGULATED POWER SUPPLY

The power supply in Fig. 14-5 produces highly regulated dc which can be varied from 1.25 to about 30 V. This unit also features a voltmeter to monitor output voltage, and the power supply can deliver up to 1.5 A. The voltage regulator IC_1 has the built-in ability to protect itself against most overload conditions.

Transformer T_1 drops the voltage to 25 V ac, while the bridge rectifier $RECT_1$ converts the ac to dc. Capacitors C_1 and C_2 act as filters before the regulator. The LM317K maintains a constant 1.25-V difference between the adjustment and the output terminals. The output terminal is 1.25 V more positive than the adjustment terminal. Therefore, when the voltage adjust potentiometer is at its lowest resistance, the output voltage is set at 1.25 V. As the resistance of R_3 increases, so does the output voltage. Capacitor C_3 acts as a filter at the adjustment terminal of the regulator. Resistor R_2 maintains a minimum current between the adjustment and output terminals of the IC. Capacitor C_4 filters out transient voltages. Diode D_2 protects the circuit in case an external reverse voltage is accidentally applied to the output terminals with the reverse polarity. Diode D_1 is a discharge path for C_3 in case a short circuit occurs at the output of the power supply.

The high-voltage wiring must be done carefully to make sure that it is completely isolated from the enclosure. The green ground wire on the three-pronged plug must be attached to the metal chassis. Use proper strain relief tech-

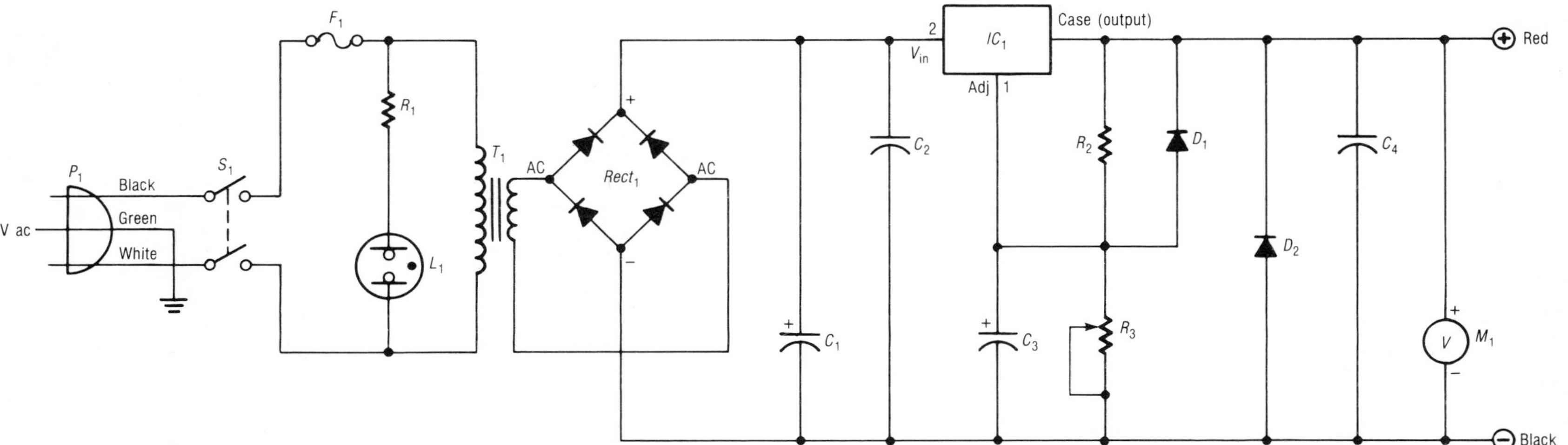

PARTS LIST

C_1	3900-μF, 50-V electrolytic capacitor
C_2	0.1-μF, 50-V disk capacitor
C_3	22-μF, 50-V electrolytic capacitor
C_4	0.1-μF, 50-V disk capacitor
D_1,D_2	1N4002 silicon diode, 1 A, 100 PIV
F_1	0.3-A fuse
IC_1	LM317K voltage regulator IC
L_1	NE-2 neon lamp
M_1	0- to 30-V (or 0- to 50-V) panel meter (Mouser Electronics has panel meters)
P_1	Three-pronged ac plug
R_1	47-kΩ, ½-W resistor
R_2	240-Ω, 5 percent, ½-W resistor
R_3	5-kΩ linear potentiometer
$Rect_1$	Bridge rectifier module, 6 A, 200 PIV (see Mouser Electronics)
S_1	DPST switch (or use DPDT)
T_1	25-V, 2-A power transformer (similar to Radio Shack 273-1512)

FIG. 14-5 1.25- to 30-V dc regulated power supply.

niques where cords enter and exit the case. The voltage regulator IC_1 and the bridge rectifier module $Rect_1$ should be mounted on a good heat sink. Insulate the case (output terminal) of IC_1 from the heat sink, using a mica washer. Use heat sink compound on both sides of the mica washer.

DUAL-POLARITY +/−12-V DC REGULATED POWER SUPPLY

The power supply circuit in Fig. 14-6 is the type used with many operational amplifiers and other linear ICs that require dual-polarity supplies. The outputs of this unit are +12 V compared to ground and −12 V compared to ground. The power supply will deliver up to 1 A of well-filtered and -regulated dc to the outputs. A single 12-V supply or both supplies may be used at once.

Household voltage is stepped down by transformer T_1 to about 13 V ac on each side of the center tap of the transformer secondary. This is rectified by $Rect_1$ and sent to each of the voltage regulators (IC_1 and IC_2). All the capacitors (C_1 through C_8) act as filters to smooth out the ripple in the pulsating dc and handle transient voltages. Voltage regulator IC_1 regulates the output voltage to +12 V compared to ground (GND). In like manner, IC_2 regulates the output voltage to −12 V compared to ground.

The high-voltage wiring must be done carefully to make sure that it is completely isolated from the enclosure. The green ground wire on the three-pronged plug must be attached to the metal chassis. Use proper strain relief techniques where cords enter and exit the case. Mount both voltage regulators (IC_1 and IC_2) on heat sinks.

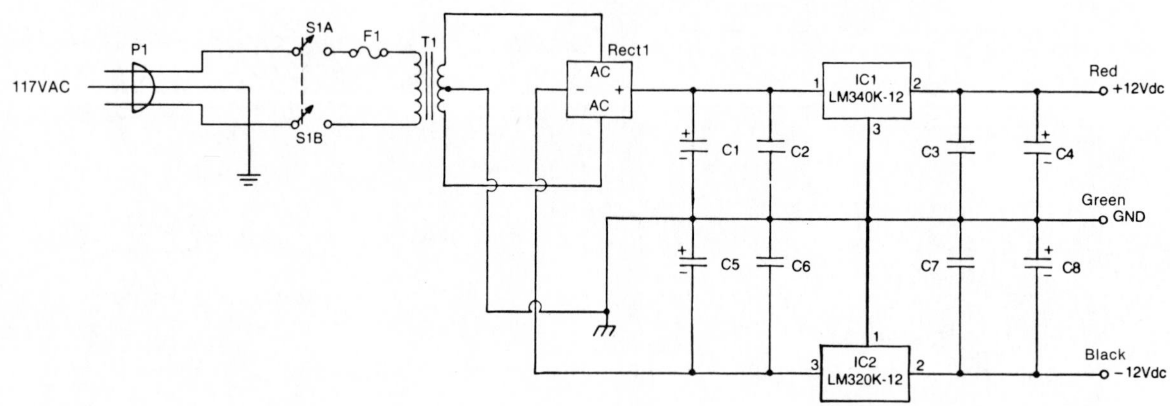

PARTS LIST

C_1, C_5	3300-μF, 50-V electrolytic capacitor
C_2, C_3, C_6, C_7	0.33-μF, 50-V capacitor
C_4, C_8	100-μF, 50-V electrolytic capacitor
F_1	½-A fuse
IC_1	LM340K-12 voltage regulator (+12 V) (replace with 7812 IC—Radio Shack 276-1771)
IC_2	LM320K-12 voltage regulator (−12 V) (replace with 7912 IC—Radio Shack 276-1774)
P_1	Three-pronged ac plug
$Rect_1$	Bridge rectifier module, 4 A, 100 PIV (Radio Shack 276-1171)
S_1	DPST switch (or use DPDT)
T_1	25.2-V, 2-A, center-tapped secondary power transformer (Radio Shack 273-1512)

FIG. 14-6 Dual-polarity +/−12-V dc regulated power supply. (Joseph J. Carr, *104 Weekend Electronic Projects*, TAB Books, Pennsylvania, 1982, pp. 205–208. Used by permission of TAB Books, Inc.)

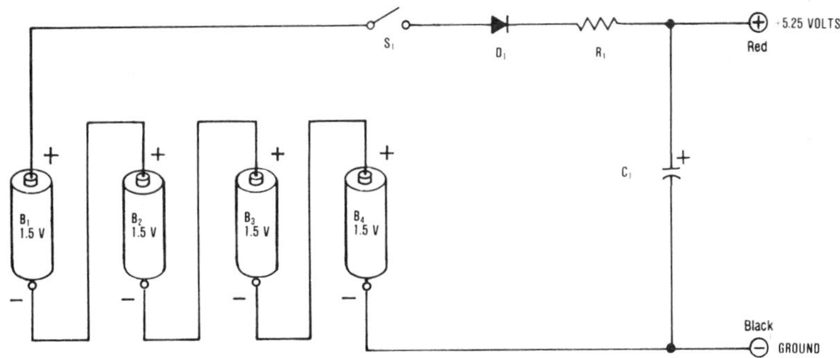

PARTS LIST

B_1 through B_4	1.5-V batteries
C_1	10-μF, 25-V tantalum capacitor
D_1	1N4001 silicon diode, 1 A, 50 PIV
R_1	0.56-Ω, ¼-W resistor
S_1	SPST switch

FIG. 14-7 5-V battery-powered supply for TTL ICs. (Michael Gannon, *Workbench Guide to Semiconductor Circuits and Projects,* Prentice-Hall, New Jersey, 1982, pp. 225–226. Used by permission of Prentice-Hall, Inc.)

5-V BATTERY-POWERED SUPPLY FOR TTL ICs

The circuit in Fig. 14-7 is an inexpensive solution to powering TTL ICs while experimenting. The output voltage is about 5 to 5.25 V dc as need by the popular TTL family of digital integrated circuits.

Four alkaline or four inexpensive carbon-zinc AA, C, or D cells mounted in an appropriate battery holder serve as the power source. Diode D_1 drops the voltage somewhat to about 5 V dc. Capacitor C_1 and resistor R_1 help to suppress transient voltages.

DUAL-POLARITY +/−9-V DC BATTERY-POWERED SUPPLY

The circuit in Fig. 14-8 is an inexpensive solution to powering operational amplifiers and some other linear ICs during experimentation.

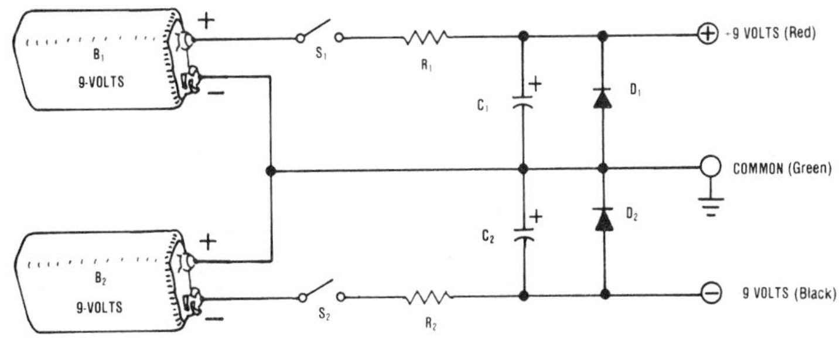

PARTS LIST

B_1,B_2	9-V battery (alkaline or ni-cad)
C_1,C_2	10-μF, 25-V tantalum capacitor
D_1,D_2	1N4001 silicon diode, 1 A, 50 PIV
R_1,R_2	0.56-Ω, ¼-W resistor
S_1,S_2	SPST switch

FIG. 14-8 Dual-polarity +/−9-V dc battery-powered supply. (Michael Gannon, *Workbench Guide to Semiconductor Circuits and Projects,* Prentice-Hall, New Jersey, 1982, p. 227. Used by permission of Prentice-Hall, Inc.)

The unit is a dual-polarity supply and will deliver +9 and −9 V compared to ground at low currents.

Alkaline batteries work the best and will deliver higher currents. The diodes, capacitors, and resistors in this circuit help to suppress transient voltages.

DC POWER SUPPLIES FROM RECYCLED AC ADAPTERS

Many discarded ac adapters or battery eliminators can be used to build inexpensive dc power supplies. The ac adapter in the sample circuit in Fig. 14-9 produces a +5-V dc regulated power supply with a maximum output current of about 200 mA. This would be capable of powering many TTL ICs.

It is customary for the input voltage and the output voltage and current to be printed on ac adapters. In the example in Fig. 14-9, the output voltage was 9 V dc with an output current of 200 mA. Capacitors C_1, C_2, and C_3 are all filter capacitors. The 7805 voltage regulator IC_1 drops the output voltage down to +5 V dc. An ac adapter with 6 V dc listed on its label would work with this circuit if the load across the output were not too great.

Unlike the example in Fig. 14-9, some ac adapters list an ac output instead of dc. A bridge rectifier module must then be added between the adapter and the first filter capacitor (C_1) to make a dc power supply. Use of ac adapters to build a dc power supply is an inexpensive and safe method for younger students to attempt such a project.

TRIPLE-OUTPUT, DUAL-TRACKING POWER SUPPLY

The circuit in Fig. 14-10 is three dc power supplies in one, all of which are well filtered and regulated. The +5-V dc supply along the bottom will deliver currents of up to 3 A and is ideal for powering TTL digital ICs. The other two supplies, driven by T_1 and T_2, form a dual-polarity unit that is useful in testing many operational amplifier circuits. The top supply will deliver from 0 to +20 V dc at 1 A and is controlled by potentiometer R_4. The center supply (driven by T_2) will deliver −5 to −20 V dc at 1 A. Potentiometer R_{10} controls the voltage output from the middle supply. All three supplies are current limited for short-circuit protection.

The dual-polarity supply can be operated in either the independent (N or normal) or the tracking (T) mode as selected by S_2. In the normal mode, R_4 adjusts the positive output from voltage regulator IC_2, whereas R_{10} adjusts the negative voltage output from IC_3. When S_2 is placed in the tracking mode (T), R_{10} becomes the master control and both supplies track

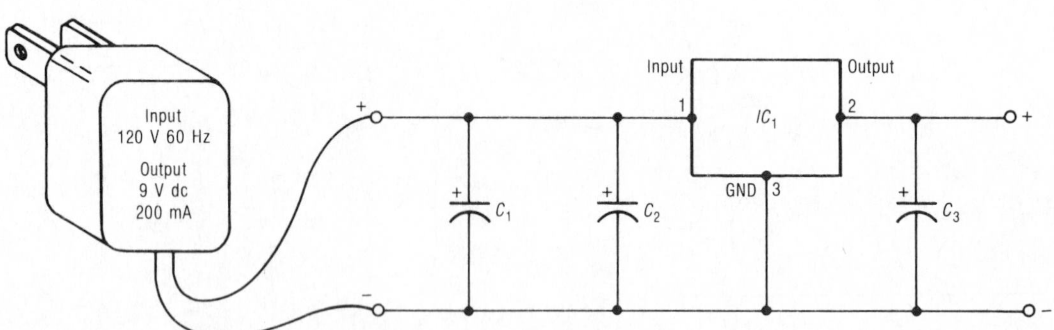

PARTS LIST

C_1	470-μF, 16-V electrolytic capacitor
C_2,C_3	1-μF, 25-V tantalum capacitor
IC_1	7805 voltage regulator IC (+5 V)

FIG. 14-9 Dc power supply from recycled ac adapter. (Reprinted from *Popular Electronics*. Copyright © February 1981, Ziff-Davis Publishing Company.)

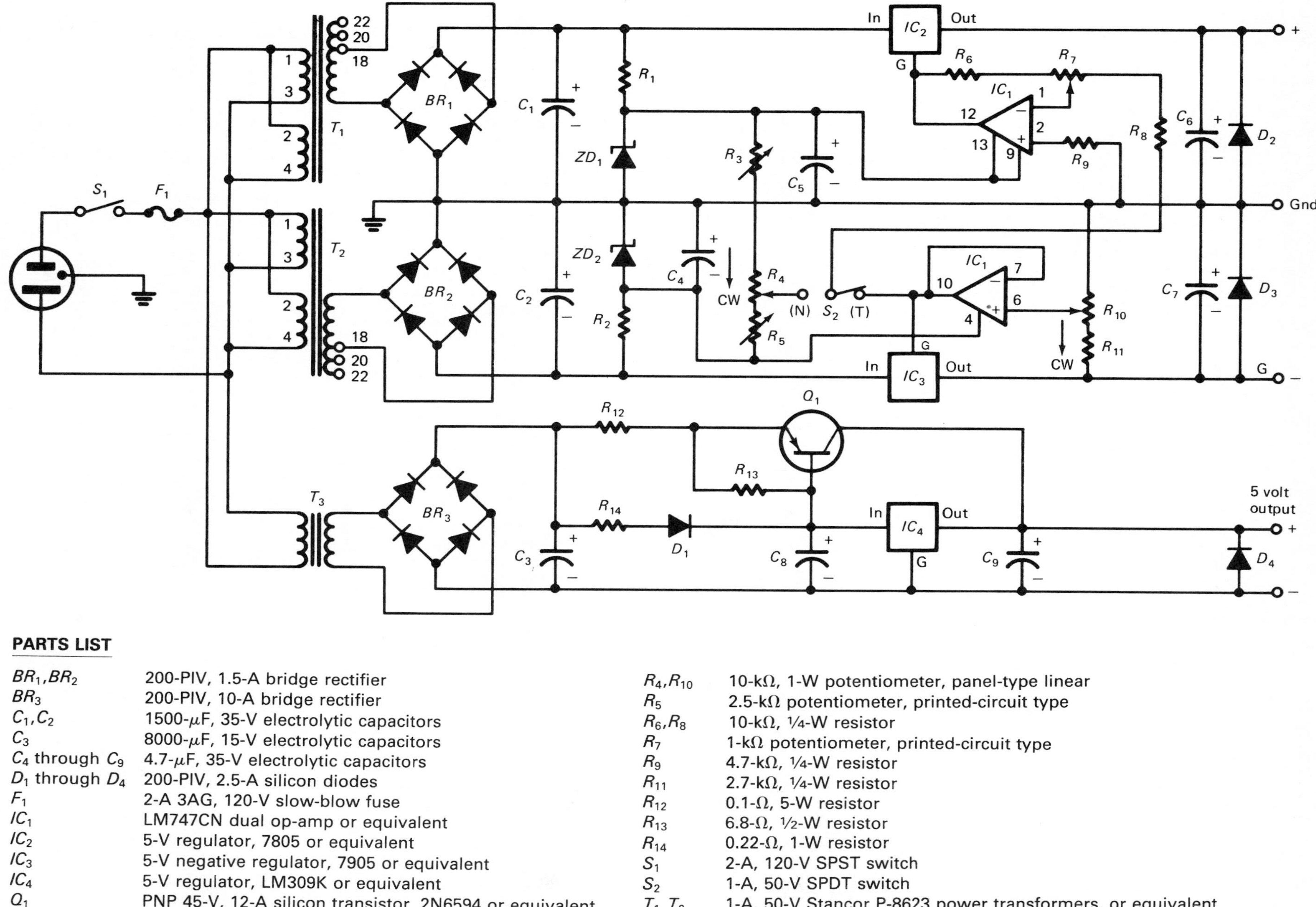

PARTS LIST

BR_1,BR_2	200-PIV, 1.5-A bridge rectifier
BR_3	200-PIV, 10-A bridge rectifier
C_1,C_2	1500-μF, 35-V electrolytic capacitors
C_3	8000-μF, 15-V electrolytic capacitors
C_4 through C_9	4.7-μF, 35-V electrolytic capacitors
D_1 through D_4	200-PIV, 2.5-A silicon diodes
F_1	2-A 3AG, 120-V slow-blow fuse
IC_1	LM747CN dual op-amp or equivalent
IC_2	5-V regulator, 7805 or equivalent
IC_3	5-V negative regulator, 7905 or equivalent
IC_4	5-V regulator, LM309K or equivalent
Q_1	PNP 45-V, 12-A silicon transistor, 2N6594 or equivalent
R_1,R_2	330-Ω, ½-W resistors
R_3	10-kΩ potentiometer, printed-circuit type
R_4,R_{10}	10-kΩ, 1-W potentiometer, panel-type linear
R_5	2.5-kΩ potentiometer, printed-circuit type
R_6,R_8	10-kΩ, ¼-W resistor
R_7	1-kΩ potentiometer, printed-circuit type
R_9	4.7-kΩ, ¼-W resistor
R_{11}	2.7-kΩ, ¼-W resistor
R_{12}	0.1-Ω, 5-W resistor
R_{13}	6.8-Ω, ½-W resistor
R_{14}	0.22-Ω, 1-W resistor
S_1	2-A, 120-V SPST switch
S_2	1-A, 50-V SPDT switch
T_1, T_2	1-A, 50-V Stancor P-8623 power transformers, or equivalent
T_3	7.5-V, 4-A Stancor P-5015 power transformer or equivalent
ZD_1,ZD_2	18-V, 1-W zener diodes 1N4746A or equivalent

FIG. 14-10 Triple-output, dual-tracking power supply. (Charles A. Schuler, *Activities Manual for Electronics: Principles and Applications*, 2d ed., McGraw-Hill, New York, 1984, pp. 165–168. Used by permission of McGraw-Hill Book Company.)

from 5 to 20 V. Trimmer potentiometers R_3, R_5, and R_7 are used for calibrating the voltage range and tracking of the dual-polarity power supplies.

The 5-V regulated supply across the bottom in Fig. 14-10 is powered by step-down transformer T_3. Bridge rectifier BR_3 converts the ac to dc. The filter capacitors and IC_4 will regulate the voltage to +5 V up to 1.5 A. Power transistor Q_1 is gradually turned on by current through R_{14} and raises the current capacity of the power supply to about 3 A.

The transformers, bridge rectifiers, and filter capacitors in the upper two power supplies generate about 25 V dc, which enters each of the regulators (IC_2 and IC_3). About 18 V is tapped off by the zener diodes and associated resistors to power the op-amps (IC_1). With S_2 in the normal (N) position, R_4 taps off voltage from divider R_3, R_4, and R_5, sending it to the top op-amp. The output of this op-amp causes IC_2 to output the correct regulated voltage at the (+) terminal. In like manner, R_{10} sends a voltage to the input of the bottom op-amp, causing IC_3 to output the correct voltage to the (−) terminal. In the tracking mode (T), R_{10} controls both voltage regulator ICs, causing each to output the same voltage (one positive, one negative). Diodes D_2, D_3, and D_4 protect against reverse polarity.

The high-voltage wiring must be done carefully to make sure that it is completely isolated from the enclosure. The green ground wire on the three-pronged plug must be attached to the metal chassis. Use proper strain relief techniques where cords enter and exit the case. Mount the voltage regulators IC_2, IC_3, and IC_4 and transistor Q_1 on heat sinks.

chapter 15

Solar and Optoelectronic Circuits

AUDIBLE LIGHT METER

The light meter circuit in Fig. 15-1 emits a tone when light strikes the photocell. The tone becomes higher as the light striking the photocell becomes greater. This is a fun starter project for young people.

The audible light meter circuit is basically an oscillator circuit. When light strikes the surface of photocell PC_1, its resistance drops, turning on Q_1 and the oscillator. Capacitor C_1 is the feedback capacitor to keep the unit oscillating. The frequency of the oscillator increases as more light strikes the photocell dropping its resistance. This drop in resistance allows C_1 to charge more quickly.

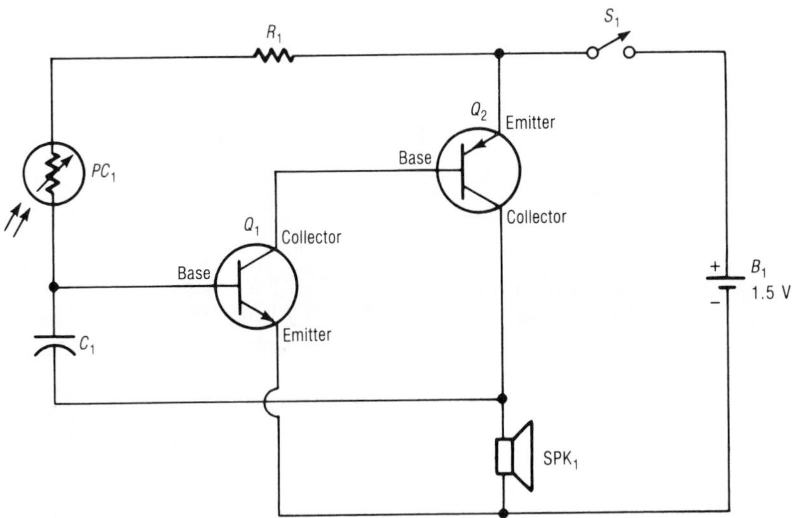

FIG. 15-1 Audible light meter.

PARTS LIST

B_1	1.5-V battery
C_1	0.22-μF, 25-V disk capacitor
PC_1	Cadmium sulfide photocell (Radio Shack 276-116)
Q_1	2N3904 NPN transistor (or similar)
Q_2	2N3906 PNP transistor (or similar)
R_1	4.7-kΩ, ½-W resistor
S_1	SPST switch
SPK_1	8-Ω speaker

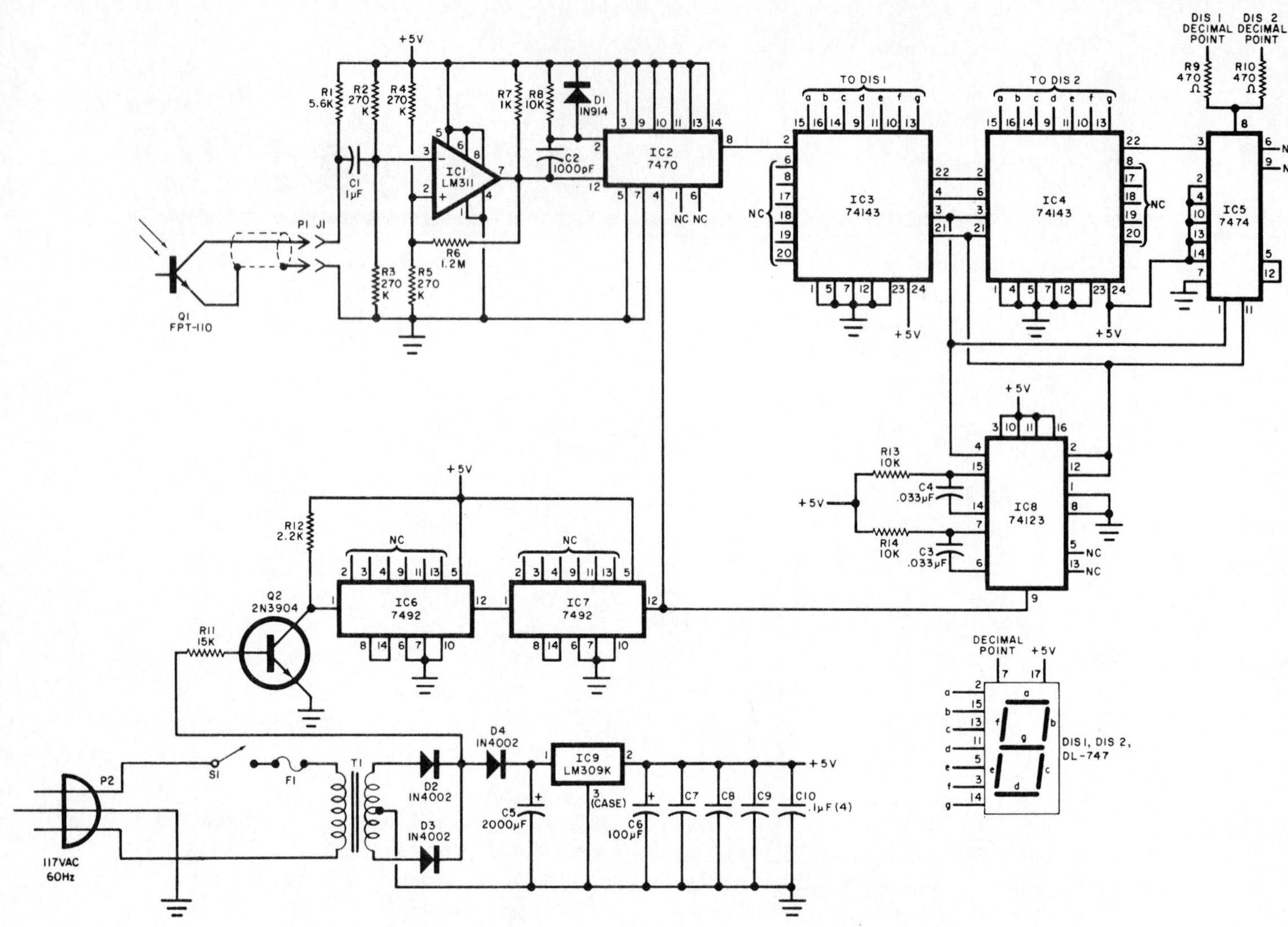

PARTS LIST

C_1	1-μF, 50-V capacitor
C_2	1000-pF, 100-V disk capacitor
C_3,C_4	0.033-μF, 50-V capacitor
C_5	2000-μF, 35-V electrolytic capacitor
C_6	100-μF, 16-V electrolytic capacitor
C_7,C_8,C_9,C_{10}	0.1-μF, 25-V disk capacitor
D_1	1N914 signal diode
D_2,D_3,D_4	1N4002 silicon diode, 1 A, 100 PIV
DIS 1, DIS 2	DL-747 common-anode, seven-segment LED display
F_1	¼-A fuse
IC_1	LM311 voltage comparator IC
IC_2	7470 *J-K* flip-flop IC
IC_3,IC_4	74143 decade counter-decoder-display driver IC
IC_5	7474 dual D flip-flop IC
IC_6,IC_7	7492 divide-by-12 counter IC
IC_8	74123 dual monostable multivibrator IC
IC_9	LM309K voltage regulator IC (5 V)
J_1	RCA phono jack
P_1	RCA phono plug
P_2	Three-pronged ac plug
Q_1	FPT-110 phototransistor (Fairchild)
Q_2	2N3904 NPN transistor (or similar)
R_1	5.6-kΩ, ½-W resistor
R_2 through R_5	270-kΩ, 5 percent, ½-W resistor
R_6	1.2-MΩ, ½-W resistor
R_7	1-kΩ, ½-W resistor
R_8,R_{13},R_{14}	10-kΩ, ½-W resistor
R_9,R_{10}	470-Ω, ½-W resistor
R_{11}	15-kΩ, ½-W resistor
R_{12}	2.2-kΩ, ½-W resistor
S_1	SPST switch
T_1	16-V, 1-A, center-tapped secondary power transformer (Signal No. 241-5-15)

FIG. 15-2 Digital phototachometer. (Reprinted from *Popular Electronics*. Copyright © March 1978, Ziff-Davis Publishing Company.)

DIGITAL PHOTOTACHOMETER

The phototachometer circuit in Fig. 15-2 optically senses the rotation of motor shafts or fans with no connection to the motor. The tachometer features a digital readout calibrated in hundreds and can measure 100 to 9900 rpm. The phototachometer uses the power line frequency as a time base, and accuracy is equal to about ±100 rpm. To use, paint the shaft with a flat black paint and then stick a small piece of white reflective tape on the part that will be rotating. Move the sensor close to the rotating shaft. The sensor (phototransistor Q_1) should be housed in a tube with the light-sensitive section aimed at the open end. This will cut down on false triggering of the phototransistor.

The block diagram in Fig. 15-3 outlines the main functional sections of the phototachometer. Note that the major parts associated with a function are listed on the block diagram. The sensor Q_1 is turned on and off by a piece of reflective tape on a dark shaft. The signal is shaped by IC_1 and fed into the gate (IC_2). For 0.6 second the gate (IC_2) passes the signal to the counters (IC_3 and IC_4), where the count is accumulated. The gate turns off, the count is decoded, latched, and displayed on the seven-segment displays. The counters are then cleared by a pulse from IC_8. The sequence is then repeated, updating the display each time. The power supply produces +5 V for the circuit. It also sends a 120-Hz signal to the divide-by-144 counter section. This section generates the 0.6-second count pulse sent to the gate (IC_2). It also signals the latch-clear block when the count pulse is finished so that the count can be latched on the displays and the counters cleared for the next counting sequence.

The high-voltage wiring must be done carefully to make sure that it is completely isolated from the enclosure. The green ground wire on the three-pronged plug must be attached to the metal chassis. Use proper strain relief techniques where cords enter and exit the case.

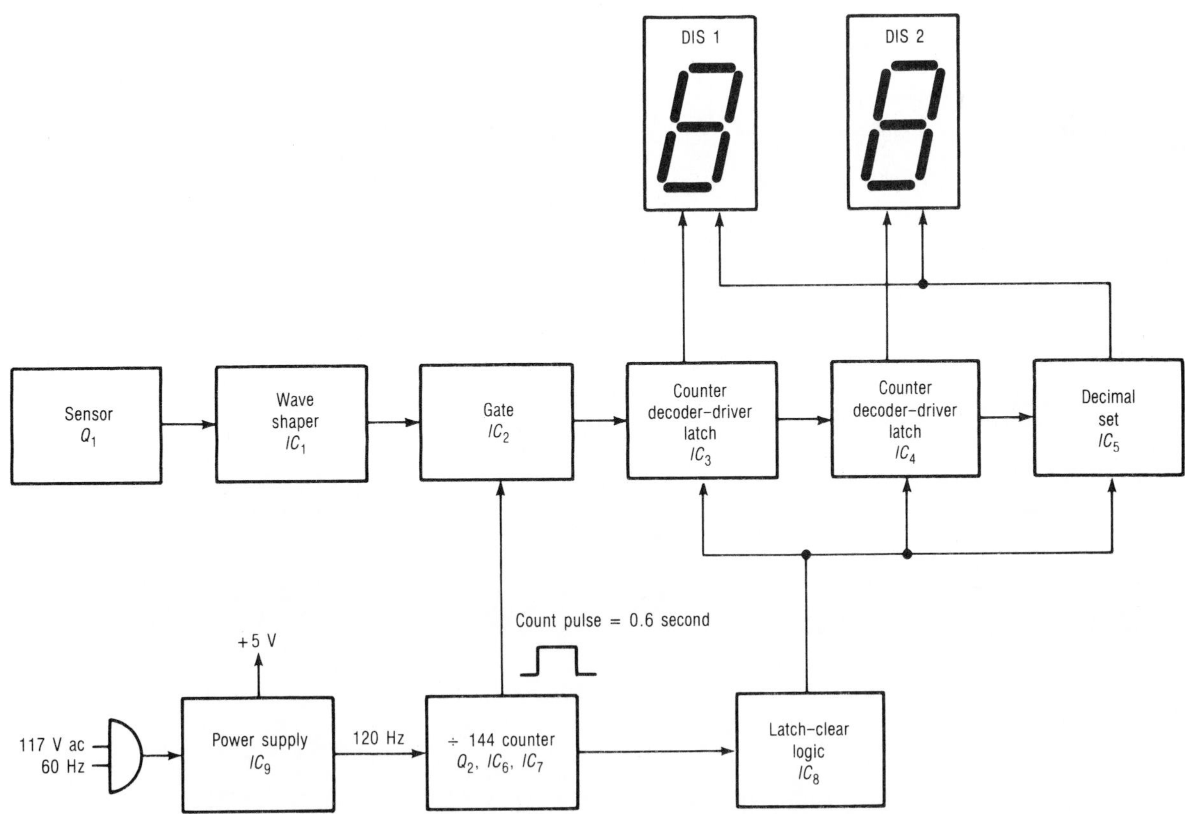

FIG. 15-3 Block diagram of digital phototachometer.

ELECTRIC EYE CONTROLLED OSCILLATOR

The electric eye circuit in Fig. 15-4 will produce a tone when the beam of light striking the photocell is broken. It can be used as a warning device indicating that someone is entering a room or store. This is a favorite among beginning students.

Under normal conditions, light falling on the photocell (PC) causes it to have lower resistance and Q_1 and Q_2 are turned off. When the beam of light aimed at the photocell is interrupted, the photocell's resistance increases and the high-gain darlington amplifier (Q_1 and Q_2) turns on. The increased positive voltage at the gate of the Q_3 turns on the SCR. This turns on the audio oscillator (Q_4 and Q_5) driving the speaker. When the light beam strikes the photocell again, the darlington amplifier, the SCR, and the oscillator are turned off. The electric eye's sensitivity to light can be adjusted using potentiometer P_1.

The photocell may be mounted in a short cardboard tube to shield it from stray light. As

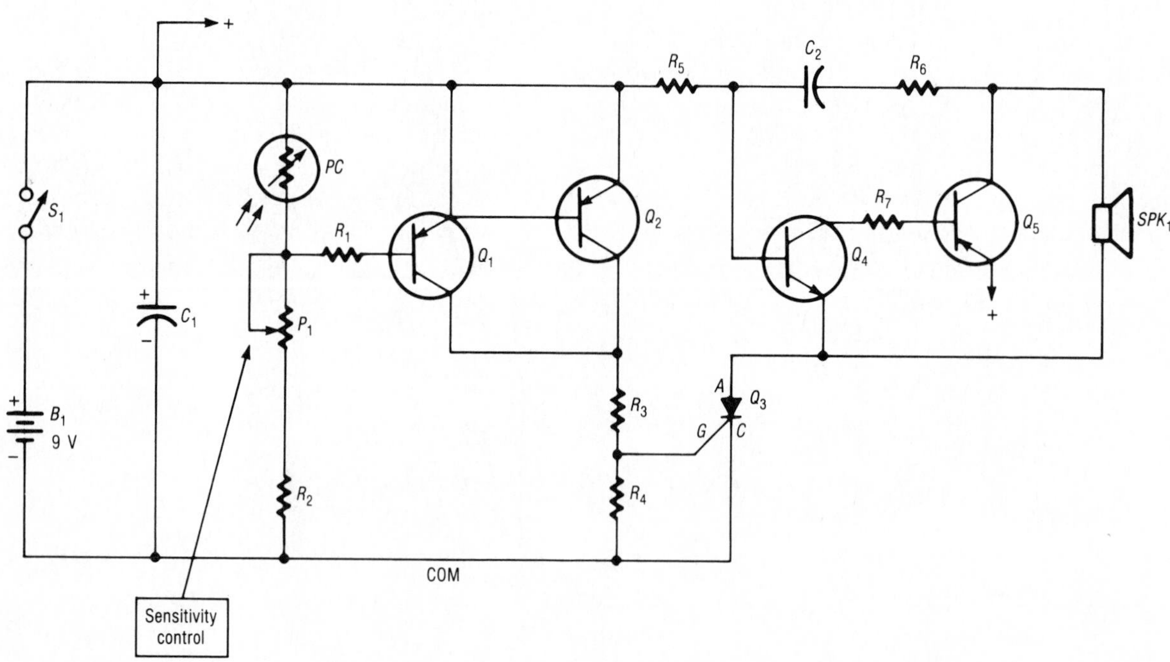

PARTS LIST

B_1	9-V battery
C_1	22-μF, 16-V electrolytic capacitor
C_2	0.02-μF, 50-V disk capacitor
Q_1, Q_2, Q_5	2N3906 PNP transistor
Q_3	C103Y silicon controlled rectifier
Q_4	2N3904 NPN transistor
P_1	10-kΩ linear trimmer potentiometer
PC	Photocell (cadmium sulfide) (similar to Radio Shack 276-116)
R_1, R_3, R_4, R_6	1-kΩ, ½-W resistor
R_2	680-Ω, ½-W resistor
R_5	220-kΩ, ½-W resistor
R_7	330-Ω, ½-W resistor
S_1	SPST switch
SPK_1	8-Ω speaker

FIG. 15-4 Electronic eye controlled oscillator. (Used by permission of Robert Delp Electronics.)

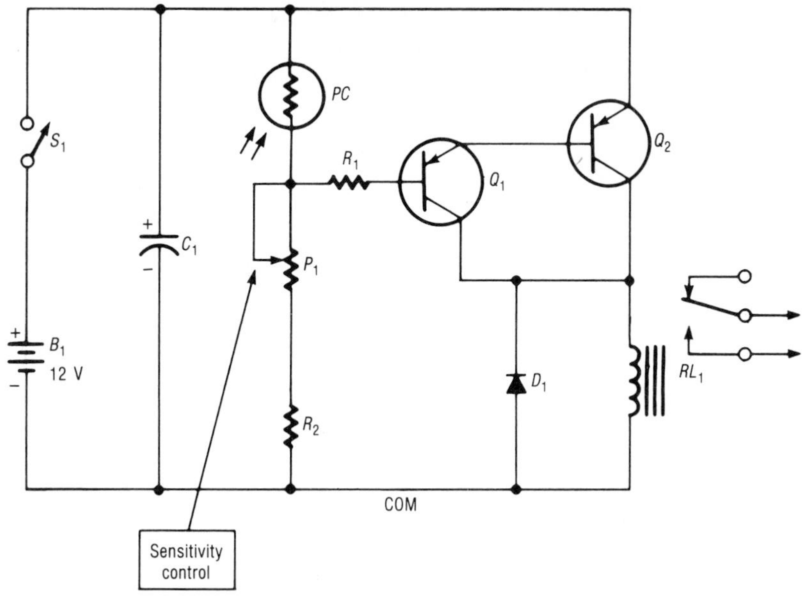

PARTS LIST

B_1	12-V battery
C_1	22-μF, 25-V electrolytic capacitor
D_1	1N4002 silicon diode, 1 A, 100 PIV
Q_1,Q_2	2N3906 PNP transistor
P_1	10-kΩ linear trimmer potentiometer
PC	Photocell (cadmium sulfide) (similar to Radio Shack 276-116)
R_1	1-kΩ, ½-W resistor
R_2	680-Ω, ½-W resistor
RL_1	12-V coil, relay (similar to Radio Shack 275-206)
S_1	SPST switch

FIG. 15-5 Electric eye controlled relay. (Used by permission of Robert Delp Electronics.)

with most oscillator circuits, an alkaline battery works the best.

ELECTRIC EYE CONTROLLED RELAY

The electric eye circuit in Fig. 15-5 will snap a relay closed when the beam of light striking the photocell (PC) is broken. The relay can be used for controlling any number of devices, including alarms and counters.

This circuit operates very much like the previous one, except a relay is used instead of the SCR and oscillator. When a beam of light striking the photocell is broken, the darlington amplifier (Q_1 and Q_2) turns on, causing the relay to snap closed.

LIGHT-SENSITIVE TONE GENERATOR

The tone generator circuit in Fig. 15-6 features two photocells which provide a unique up/down tone as light strikes either one or the other or both. Children and students will love to experiment with this project.

The 741 op-amp IC_1 is wired as an oscillator with the photocells as the tone control elements. Potentiometer R_3 also affects the tone of the oscillator. The LM386 IC_2 is an audio am-

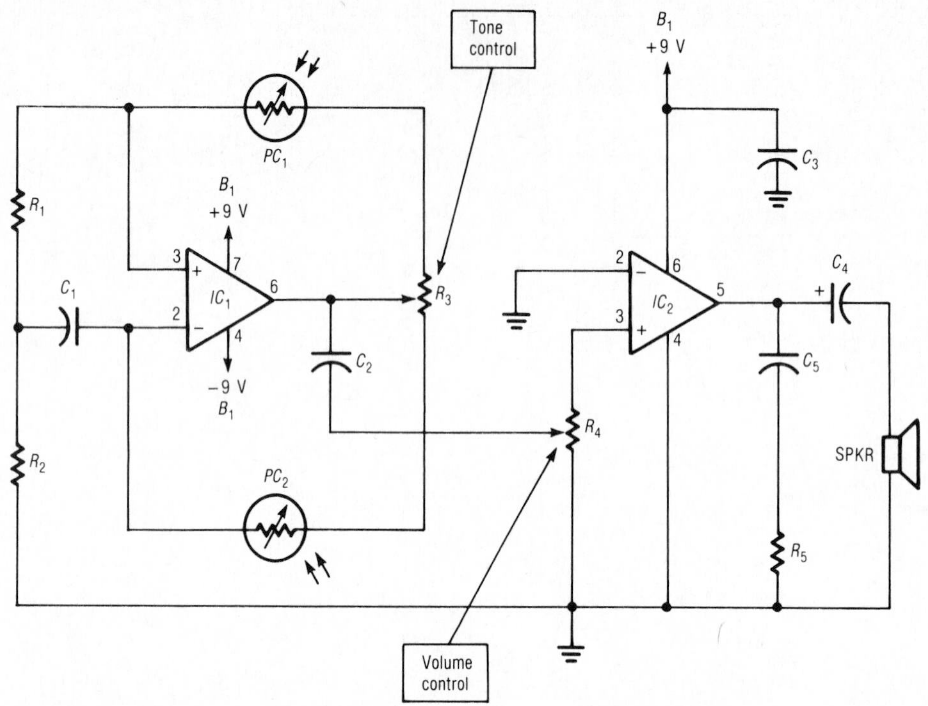

PARTS LIST

B_1	+/−9-V dual-polarity supply (two 9-V batteries)
C_1,C_2,C_3	0.1-μF, 25-V disk capacitor
C_4	100-μF, 25-V electrolytic capacitor
C_5	0.05-μF, 50-V disk capacitor
IC_1	LM741 op-amp IC
IC_2	LM386 audio amplifier IC
PC_1,PC_2	Photocell (cadmium sulfide) (Radio Shack 276-116)
	1-kΩ, ½-W resistor
R_1	1kΩ, ½-W resistor
R_2	100-kΩ, ½-W resistor
R_3	50-kΩ potentiometer
R_4	10-kΩ potentiometer
R_5	10-Ω, ½-W resistor
SPK_1	8-Ω speaker

FIG. 15-6 Light-sensitive tone generator. (Forrest M. Mims, *The Forrest Mims Circuit Scrapbook,* McGraw-Hill, New York, 1983, p. 49. Used by permission of Forrest M. Mims, III.)

plifier which boosts the oscillator's signal and drives the speaker. Potentiometer R_4 is the volume control.

PHOTOELECTRIC NIGHT-LIGHT

The photoelectric night-light circuit in Fig. 15-7 is a simple and inexpensive way to turn on lights at dusk and turn them off at daybreak. The night-light may be used with bulbs up to about 200 W. The lamp will light to half brightness as a result of the action of Q_1, which acts like a rectifier.

During daylight hours, bright light strikes the photocell (PC), causing it to have low resistance. This keeps the voltage at the gate of the SCR low. At dusk, less light strikes the photocell, raising its resistance and the voltage at the gate of the SCR. The SCR conducts, ener-

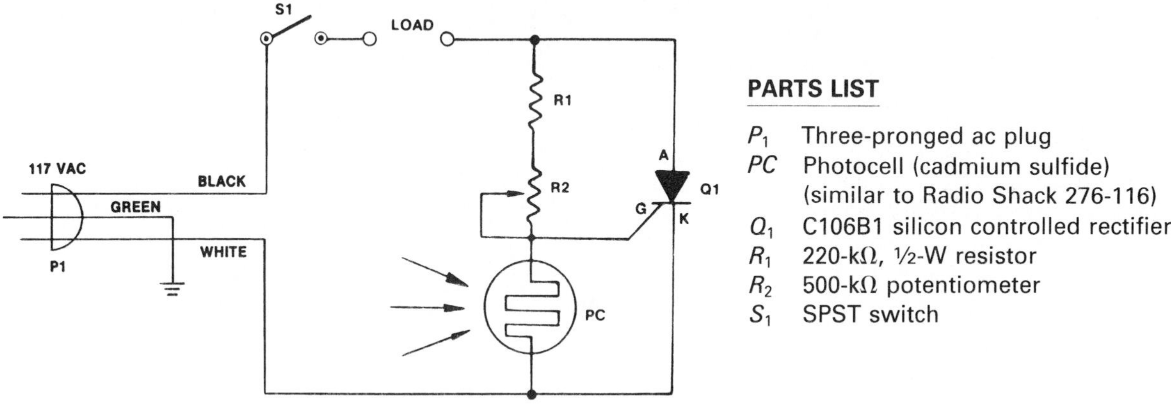

PARTS LIST

P_1	Three-pronged ac plug
PC	Photocell (cadmium sulfide) (similar to Radio Shack 276-116)
Q_1	C106B1 silicon controlled rectifier
R_1	220-kΩ, ½-W resistor
R_2	500-kΩ potentiometer
S_1	SPST switch

FIG. 15-7 Photoelectric night-light. (Used by permission of Mode Electronics.)

gizing the load (lighting the bulb). Potentiometer R_2 is the sensitivity control, while S_1 is a master ON-OFF switch.

The high-voltage wiring must be done carefully to make sure that it is completely isolated from the enclosure. The green ground wire on the three-pronged plug must be attached to the metal chassis. Use proper strain relief techniques where cords enter and exit the case. Note that the "hot" side of the power cord is switched. Heat-sink the SCR, being very careful to isolate it from the chassis. The light cell (PC) must be shielded from the light coming from the load.

OPTICAL PICKUP TACHOMETER

The tachometer circuit in Fig. 15-8 will measure the speed of a rotating object. It will sense as many as 300 light pulses per second. The reading of the tachometer is taken from the analog meter M_1 which may be calibrated with potentiometer R_6. A common method of interrupting the light beam emitted by the LED in PC_1 is illustrated in Fig. 15-9. The circuit counts only the leading edge of light pulses and ignores normal ambient light levels.

TTL AND CMOS LOGIC TO AC POWER COUPLER

The circuit in Fig. 15-10 uses a TTL logic gate to turn on and off high-voltage ac. The H74C2

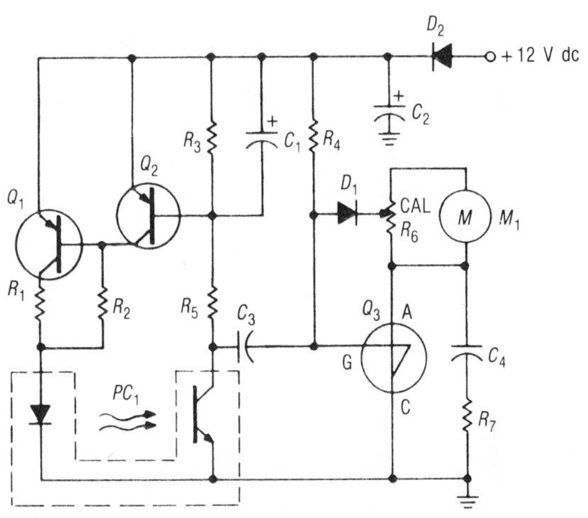

PARTS LIST

C_1, C_2	22-μF, 16-V electrolytic capacitor
C_3	0.0047-μF, 50-V disk capacitor
C_4	0.47-μF, 50-V capacitor
D_1	DZ806 silicon signal diode (10 mA)
D_2	DT230F silicon signal diode (250 mA)
M_1	1-mA dc panel meter
PC_1	H21A1 interrupter module (General Electric)
Q_1	D29E1 PNP transistor
Q_2	2N6076 PNP transistor
Q_3	2N4987 unilateral switch (ECG-6404)
R_1	240-Ω, ¼-W resistor
R_2	7.5-kΩ, ¼-W resistor
R_3	4.7-kΩ, ¼-W resistor
R_4	1.8-kΩ, ¼-W resistor
R_5	10-kΩ, ¼-W resistor
R_6	250-Ω potentiometer
R_7	100-Ω, ¼-W resistor

FIG. 15-8 Optical pickup tachometer. (Used by permission of General Electric Company.)

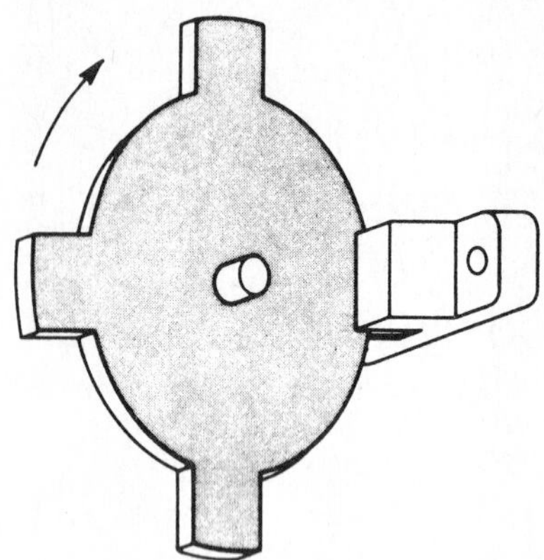

FIG. 15-9 Typical mechanical shaft encoder used with interrupter module. (Used by permission of General Electric Company.)

PC_1 optically couples the standard TTL gate to ac and completely isolates the two circuits. The light-activated SCR in the optocoupler has current limitations of about 300 mA. This circuit will work for standard TTL gates, but not for low-power TTL or CMOS outputs.

When the output of the logic gate in Fig. 15-10 is high, no current flows through the gallium arsenide infrared- (IR-) emitting diode in the optocoupler and the LASCR is off. When the output of the TTL logic gate drops low, the LED lights, causing the LASCR to turn on. This provides high-voltage ac power to the load device.

The circuit in Fig. 15-11 shows how low-power TTL gates could be used to drive the same optocoupler. When the output of the logic gate is high, the transistor is turned off, as is the LED in the optocoupler. When the output of the gate drops low, the transistor is turned on and the LED lights, turning on the rest of the circuit. This circuit would also be used with most NMOS outputs found on many calculators and microcomputers.

The circuit in Fig. 15-12 shows how the output of a CMOS logic gate can drive the LED in the optocoupler. The logic is reversed from the previous examples. When the output of the CMOS gate is high, the transistor is turned on and the LED lights, energizing the rest of the circuit. When the output of the CMOS gate goes low, the transistor is turned off and the LED in the optocoupler is off.

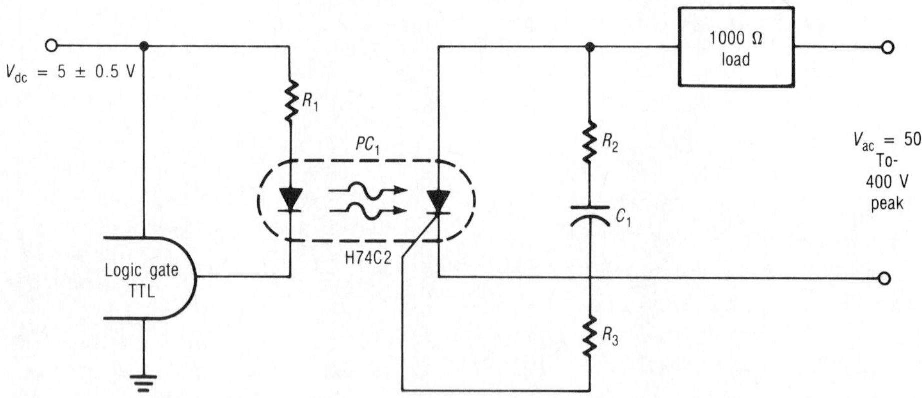

PARTS LIST

- C_1 0.1-μF, 250-V capacitor
- PC_1 H74C2 optocoupler (General Electric)
- R_1 390-Ω, 5 percent, ½-W resistor
- R_2 47-Ω, 5 percent, ½-W resistor
- R_3 56-kΩ, 5 percent, ½-W resistor

FIG. 15-10 TTL logic to ac power-coupling circuit. (Used by permission of General Electric Company.)

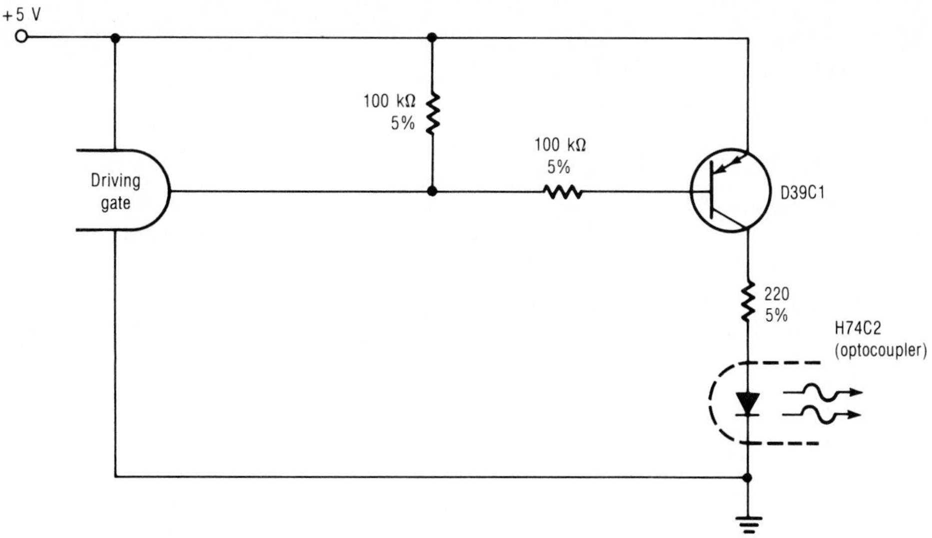

FIG. 15-11 Driving an optocoupler's LED with low-power TTL, MSI and LSI TTL, and NMOS outputs.

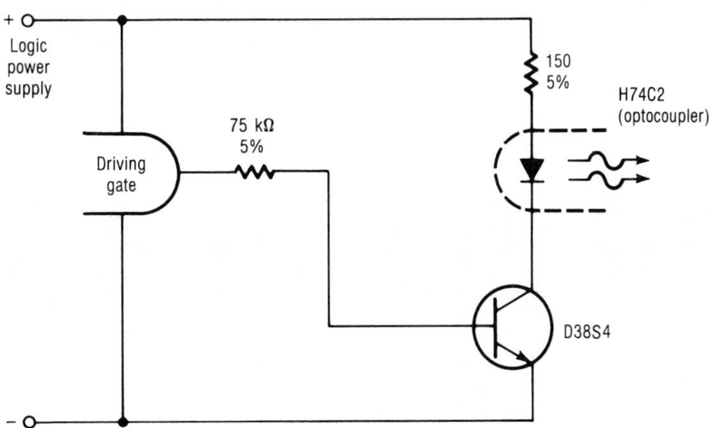

FIG. 15-12 Driving an optocoupler's LED with CMOS outputs.

chapter 16

Stereo, Radio, and TV Circuits

TWO-IC RADIO

The AM radio circuit in Fig. 16-1 has a very low parts count, with only two ICs and associated parts. The two-IC radio will receive many local stations without an external antenna. Variable capacitor C_2 is used for tuning stations, and R_3 is the volume control. A parts kit and a pc board for the two-IC radio are currently available from Graymark International, Inc.

The ZN414 chip IC_1 is a compact unit by Ferranti Electric that contains circuitry equal to a 10-transistor tuned radio frequency (TRF) circuit. Tuning is accomplished with the tank circuit consisting of L_1 and C_2. Resistor R_2 sets the automatic gain control (AGC) on IC_1. Components R_4, R_5, D_1, and D_2 form a voltage divider feeding the supply voltage to IC_1. Capacitors C_4 and C_5 are coupling capacitors between IC_1, the volume control, and IC_2. Potentiometer R_3 controls the amount of audio signal that enters the audio power amplifier. The LM380 audio power amplifier IC_2 drives the speaker.

SIX-TRANSISTOR AM RADIO

The AM radio in Fig. 16-2 uses a "traditional" superheterodyne circuit. It will pick up local stations using its internal antenna L_1. A parts kit, a pc board, and a radio theory course for this six-transistor radio are currently available from Electronic Kits International, Inc.

A block diagram of the superheterodyne AM radio circuit is shown in Fig. 16-3. The antenna coil L_1 and variable capacitor VC_1 form the tuning section. In the block diagram, the tuner has selected a station at 1000 kHz. The local oscillator (VC_2 and L_2) generates a frequency that is 455 kHz higher than that selected by the tuner. In this example, the local oscillator produces a frequency of 1455 kHz. The two frequencies are mixed by transistor Q_1. The output (difference) signal of 455 kHz from the mixer is allowed to pass through the first frequency-sensitive transformer T_1. The 455 kHz is called the *intermediate frequency* (IF) and is the carrier frequency for the intelligence through the first and second IF amplifier

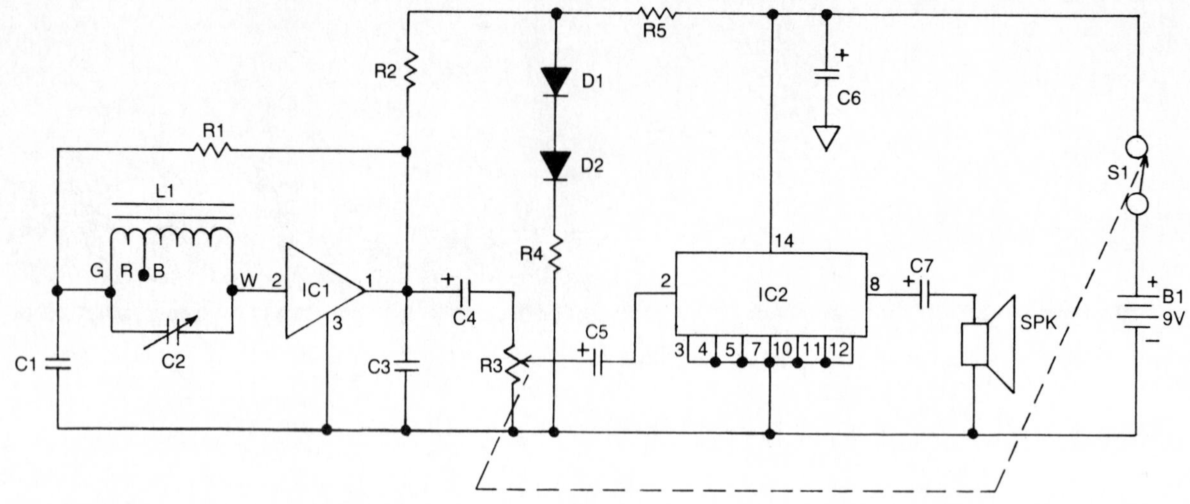

PARTS LIST

B_1	9-V battery	IC_2	LM380 audio power amplifier IC
C_1	0.01-μF, 50-V disk capacitor	L_1	Antenna coil with ferrite core
C_2	Variable-tuning capacitor (365 pF)	R_1	100-kΩ, ¼-W resistor
C_3	0.047-μF, 50-V disk capacitor	R_2	3.3-kΩ, ¼-W resistor
C_4,C_5	4.7-μF, 16-V electrolytic capacitor	R_3	50-kΩ potentiometer with switch
C_6,C_7	470-μF, 16-V electrolytic capacitor	R_4	150-Ω, ¼-W resistor
D_1,D_2	1N4148 silicon diode	R_5	8.2-kΩ, ¼-W resistor
IC_1	ZN414 TRF AM radio IC	*SPK*	8-Ω speaker

FIG. 16-1 AM radio using two ICs. (Used by permission of Graymark International, Inc.)

PARTS LIST FOR FIG. 16-2

B_1	9-V battery	R_2,R_{11}	1-kΩ, ¼-W resistor
C_1,C_3,C_5,C_6,C_7,C_9	0.02-μF, 25-V disk capacitor	R_3	10-kΩ, ¼-W resistor
C_2,C_8	0.01-μF, 25-V disk capacitor	R_4	2.7-kΩ, ¼-W resistor
C_4,C_{11}	10-μF, 16-V electrolytic capacitor	R_5	120-kΩ, ¼-W resistor
C_{10},C_{13}	47-μF, 16-V electrolytic capacitor	R_6	100-kΩ, ¼-W resistor
C_{12}	0.03-μF, 25-V disk capacitor	R_7	470-Ω, ¼-W resistor
C_{14}	330-μF, 16-V electrolytic capacitor	R_8	820-kΩ, ¼-W resistor
D_1	1N60 germanium diode	R_9	1.5-kΩ, ¼-W resistor
D_2,D_3	1N4148 silicon diode	R_{10}	15-kΩ, ¼-W resistor
E/J	Earphone jack	R_{12}	560-Ω, ¼-W resistor
L_1	Antenna coil with ferrite core	R_{13}	680-Ω, ¼-W resistor
L_2	Oscillator coil	R_{14},R_{15}	0.5-Ω, ½-W resistor
PVC	Variable tuning capacitor (TC_1, TC_2, VC_1, VC_2)	R_{16}	150-kΩ, ¼-W resistor
		SP	8-Ω speaker
Q_1	2SC1390(G) NPN transistor	T_1	IF transformer
Q_2,Q_3,Q_4	2SC1390(I) NPN transistor	T_2	IF transformer
Q_5	2SC735(O) NPN transistor	T_3	IF transformer
Q_6	2SCA1048G PNP transistor	VR	10-kΩ potentiometer with switch (volume control)
R_1	22-kΩ, ¼-W resistor		

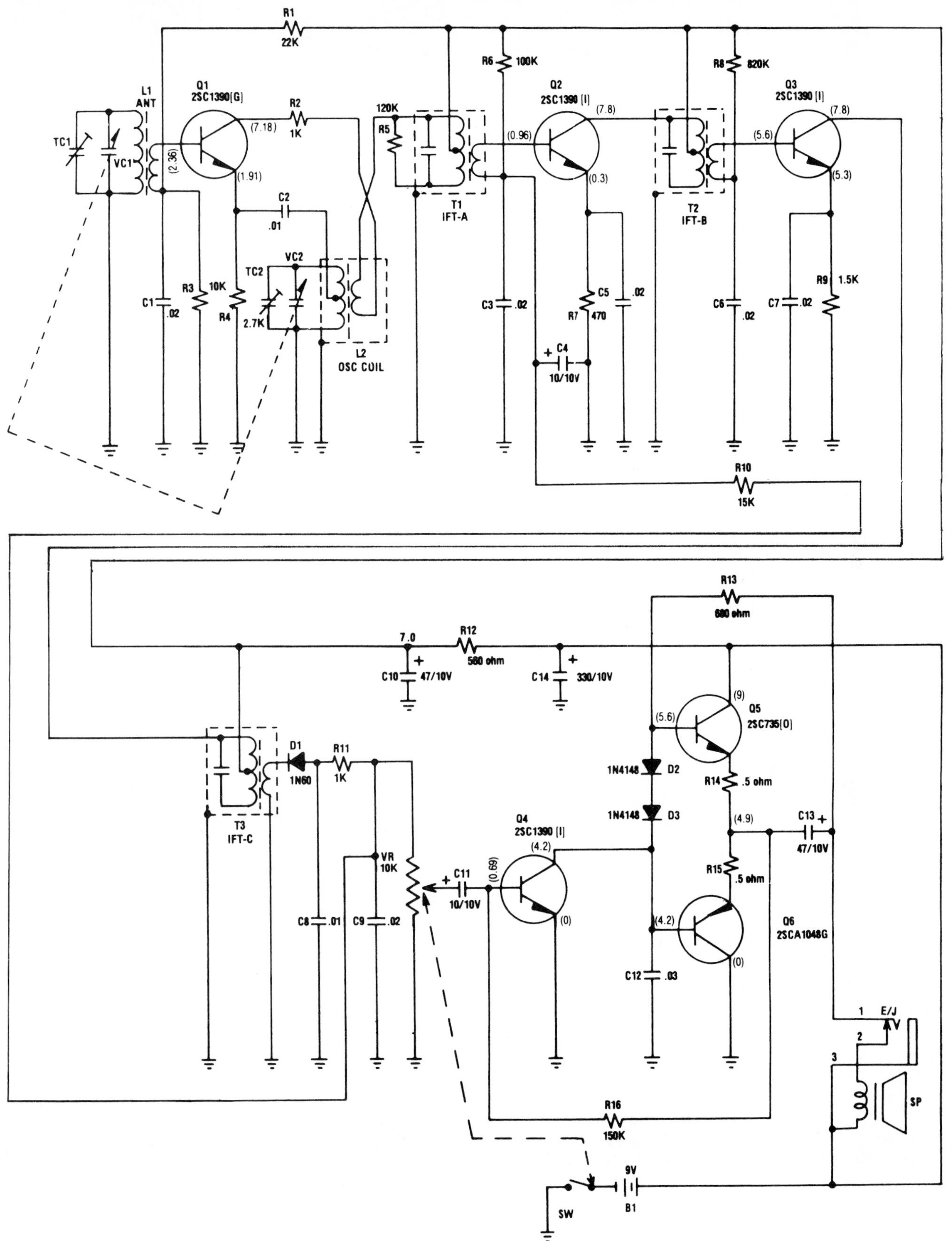

FIG. 16-2 Six-transistor AM radio. See parts list on facing page. (Used by permission of Electronic Kits International, Inc.)

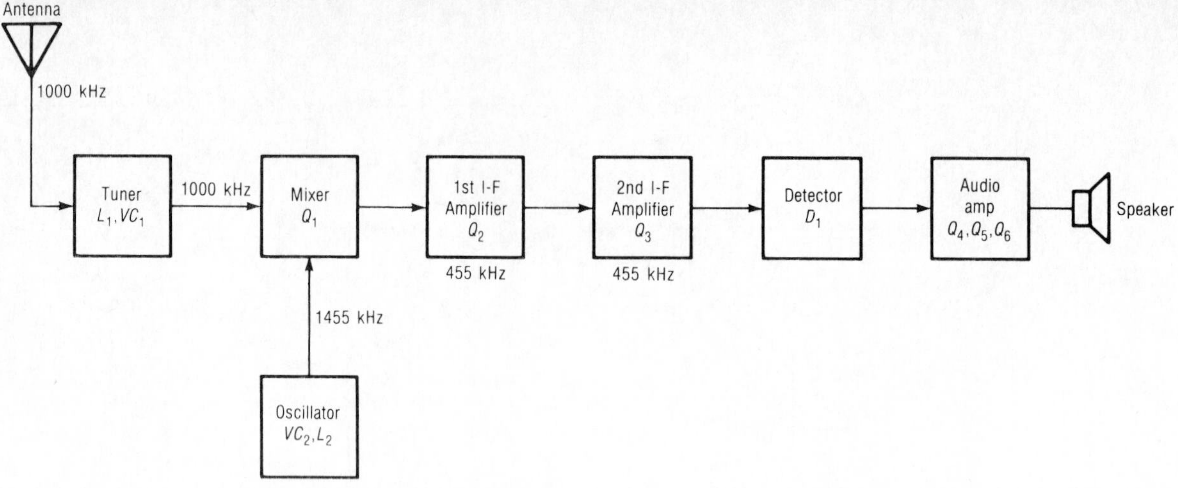

FIG. 16-3 Block diagram of a superheterodyne AM radio circuit.

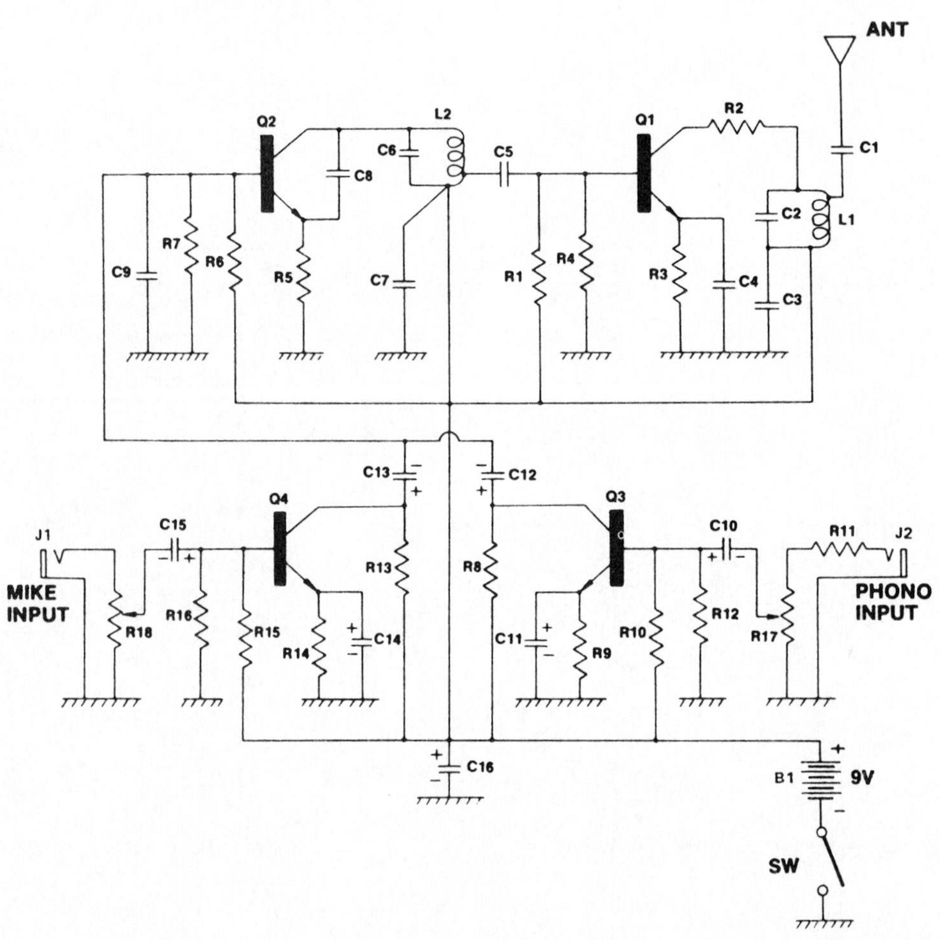

PARTS LIST

B_1	9-V battery	Q_3, Q_4	2SC373 NPN transistor
C_1, C_2, C_5, C_6, C_8	10-pF, 100-V disk capacitor	R_1, R_6, R_{10}, R_{15}	22-kΩ, ¼-W resistor
C_3, C_4, C_7, C_9	0.01-μF, 50-V disk capacitor	R_2, R_3	100-Ω, ¼-W resistor
$C_{10}, C_{12}, C_{13}, C_{14}, C_{15}$	10-μF, 16-V electrolytic capacitor	$R_4, R_8, R_{12}, R_{13}, R_{16}$	4.7-kΩ, ¼-W resistor
C_{11}	47-μF, 16-V electrolytic capacitor	R_5	220-Ω, ¼-W resistor
C_{16}	220-μF, 16-V electrolytic capacitor	R_7	3.9-kΩ, ¼-W resistor
J_1, J_2	Miniature phono jacks (3.5 mm)	R_9, R_{14}	1-kΩ, ¼-W resistor
L_1	Coil (Mode Electronics #29L1)	R_{11}	100-kΩ, ¼-W resistor
L_2	Coil (Mode Electronics #29L2)	R_{17}	10-kΩ potentiometer with switch
Q_1	2SC394 NPN transistor		
Q_2	2SC380 NPN transistor	R_{18}	10-kΩ potentiometer

FIG. 16-4 FM minibroadcaster. (Used by permission of Mode Electronics.)

sections of the radio. The IF amplifiers have selectivity and gain to boost the signal considerably. Diode D_1 detects the signal and the radio frequencies are filtered out, leaving the audio frequency signal (the song or the voice). The audio signal is amplified by voltage amplifier Q_4 and then by power amplifier Q_5 and Q_6, which drive the speaker or earphone.

FM MINIBROADCASTER

The minibroadcaster circuit in Fig. 16-4 will transmit FM radio signals which can be detected by standard FM radio receivers. The range of the minibroadcaster is limited to about 100 feet (ft). Music or speech can be fed into the transmitter by either the microphone input J_1 or the phono input J_2. A crystal microphone or a crystal or ceramic cartridge on a record player will work. The antenna wire should be limited to a wire about 1.5 meters (m) long. A parts kit and a pc board for the FM minibroadcaster are currently available from Mode Electronics.

Assume that a microphone is sending a signal through the level control R_{18} to audio amplifier Q_4. The output of the microphone amplifier is sent to Q_2, where the audio signals from both the microphone and phono inputs are mixed and then used to vary the frequency of the oscillator. Transistor Q_1 supplies the FM signal to the antenna. Oscillator coil L_2 is set for a clear frequency from 90 to 100 MHz; then the output coil L_1 is adjusted to match this frequency by listening to the broadcast on an FM radio. Note that if both the microphone and phono inputs are used at the same time, the signals will be mixed.

AUDIO POWER METER

The circuit in Fig. 16-5 will display relative output from a speaker by lighting LEDs in as-

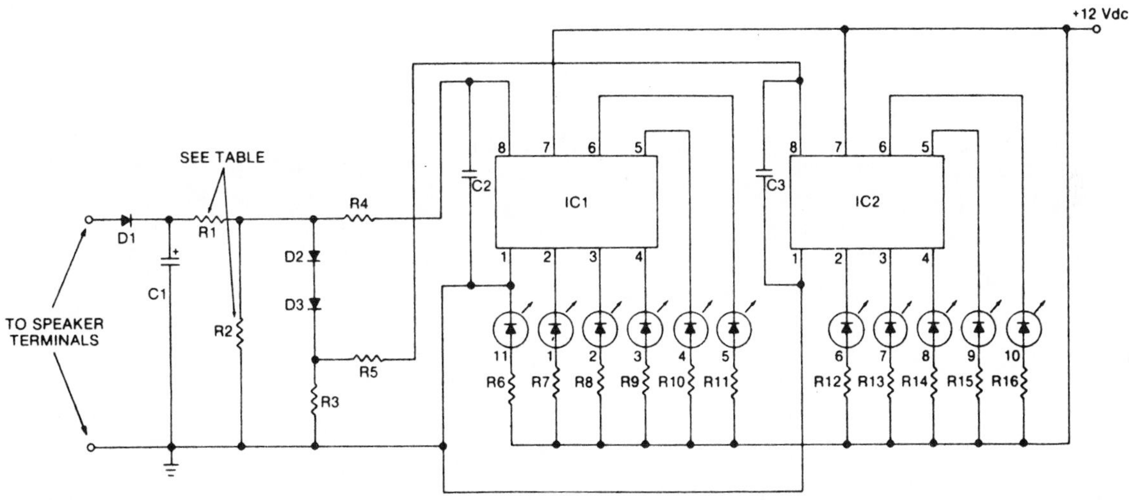

PARTS LIST

C_1	22-μF, 50-V electrolytic capacitor
C_2,C_3	0.01-μF, 50-V disk capacitor
D_1	1N4005 silicon diode, 1 A, 600 PIV
D_2,D_3	1N4148 signal diode
IC_1,IC_2	TL489C IC (Texas Instruments)
LED_1 through LED_{11}	Light-emitting diode
R_1,R_2	(See Table 16-1 for values)
R_3	620-Ω, ¼-W resistor
R_4,R_5	100-kΩ, ¼-W resistor
R_6 through R_{16}	470-Ω, ¼-W resistor

FIG. 16-5 Audio power meter with LED display. (Used by permission of Mode Electronics.)

cending order as the audio power increases. The power meter may be connected across 4-, 8-, or 16-Ω speakers with no loading. Table 16-1 shows the values for resistors R_1 and R_2 for each speaker and for the expected maximum power the meter can measure. With a 4-Ω speaker, for example, Table 16-1 shows that for measurements up to 10 W of root-mean-square (rms) power, calibration resistor R_1 must equal 1.5 kΩ while R_2 equals 470 Ω. A parts kit and a pc board for the power meter are currently available from Mode Electronics.

TABLE 16-1 Values of Calibration Resistors R_1 and R_2 for Audio Power Meter

	R_1 R_2	470 Ω 10 kΩ	1.5 kΩ 470 Ω	10 kΩ 1.5 kΩ
At 4 Ω	Peak rms	3 W 1 W	30 W 10 W	120 W 40 W
At 8 Ω	Peak rms	1.5 W 0.5 W	15 W 5 W	60 W 20 W
At 16 Ω	Peak rms	0.75 W 0.25 W	7.5 W 2.5 W	30 W 10 W

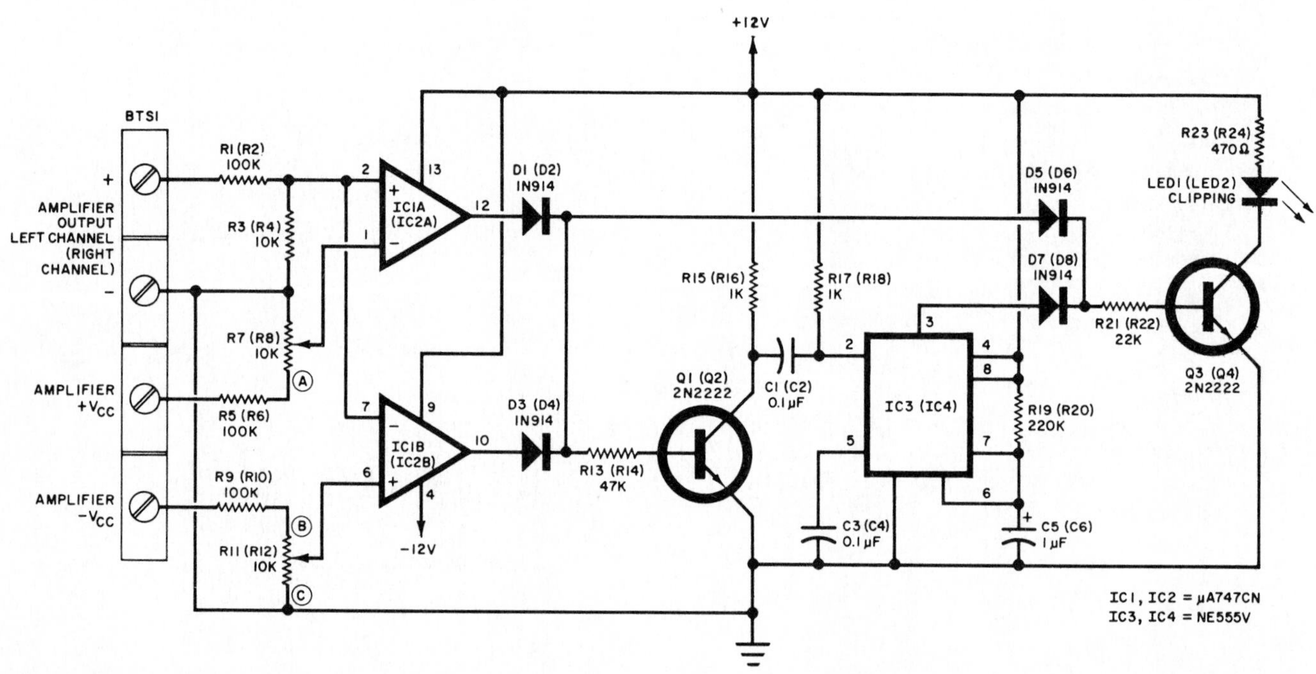

PARTS LIST

BTS_1	4-position barrier terminal strip
C_1, C_3	0.1-µF, 50-V disk capacitor
C_5	1-µF, 50-V tantalum capacitor
D_1, D_3, D_5, D_7	1N914 silicon diode
LED_1	Light-emitting diode
Q_1, Q_3	2N2222 NPN transistor
R_1, R_5, R_9	100-kΩ, 1/4-W resistor
R_3	10-kΩ, 1/4-W resistor
R_7, R_{11}	10-kΩ linear trimmer potentiometer
R_{13}	47-kΩ, 1/4-W resistor
R_{15}, R_{17}	1-kΩ, 1/4-W resistor
R_{19}	220-kΩ, 1/4-W resistor
R_{21}	22-kΩ, 1/4-W resistor
R_{23}	470-Ω, 1/4-W resistor

FIG. 16-6 Clipping indicator for one channel of stereo amplifier. (Reprinted from *Popular Electronics*. Copyright © November 1979, Ziff-Davis Publishing Company.)

CLIPPING INDICATOR FOR STEREO AMPLIFIER

The clipping indicator circuit in Fig. 16-6 will light LED_1 if the amplifier is being overdriven and therefore clipping the output signal. This overdriving can cause distortion of the sound or even damage to the speakers. The schematic diagram and parts list in Fig. 16-6 apply for a single channel, and two must be constructed for stereo. Diagrams for connecting to your power amplifier are shown in Fig. 16-7(*b*).

The block diagram in Fig. 16-7(*a*) outlines the most important functional sections of the clipping indicator. The op-amps in IC_1 are wired as voltage comparators. They compare the peak signal voltages to the positive and negative voltage rails ($+V_{CC}$ and $-V_{CC}$). If the peak voltages of the signal reach these voltages, a "clipping condition" is detected and the clipping indicator (LED_1) is turned on. Transistors Q_1 and Q_3 are turned on when the clipping condition is detected. Transistor Q_3 turns on the LED while Q_1 triggers the one-shot multivibrator IC_3. The one-shot's output also turns on Q_3 and the clipping indicator (LED_1) for at least 0.25 second. The one-shot is the pulse stretching circuit so that even a very short clipping condition will show on the LED. At the end of the clipping condition and

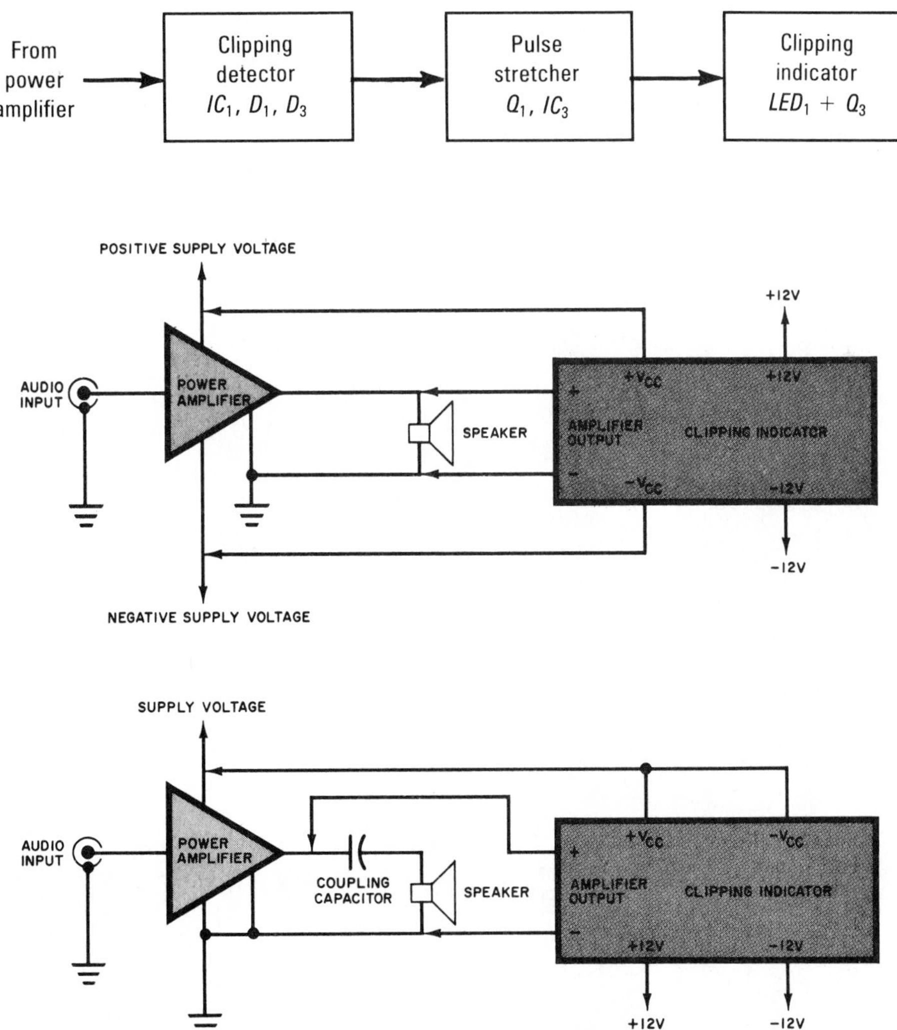

FIG. 16-7 Clipping indicator for stereo amplifier. (Reprinted from *Popular Electronics*. Copyright © November 1979, Ziff-Davis Publishing Company.)

when pulse stretcher turns off, the LED indicator turns off until another clipping condition is detected. Note that circuit uses a +/−12-V power supply. In some cases this power may be available from the stereo amplifier system; otherwise, a separate dual-polarity supply must be constructed. Trimmer potentiometers R_7 and R_{11} adjust the sensitivity of the detector.

TV REMOTE SOUND CONTROL

The remote sound control circuit in Fig. 16-8 allows the viewer to turn the TV sound on or off or use headphones to privately listen to the TV. The remote sound control unit has monophonic and stereophonic phone jacks to accommodate the type of headphone or earphone available. The remote sound control unit also has a volume control for the headphone/earphone jacks.

CAUTION: This project could be dangerous if connected to a TV receiver that has a "hot" chassis and no audio output transformer. Use only on TV receivers with audio output transformers.

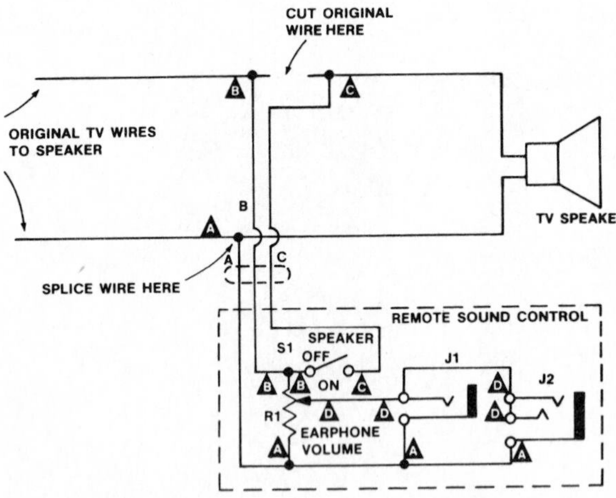

PARTS LIST

J_1 Phone jack (mono)
J_2 Phone jack (stereo)
R_1 Audio taper potentiometer
S_1 SPST switch

FIG. 16-8 TV remote sound control. (Carl G. Grolle and Michael B. Girosky, *Workbench Guide to Electronic Projects You Can Build in Your Spare Time,* Parker, New York, 1981, pp. 139–142. Used by permission of Parker Publishing Company, Inc.)

To install, cut one wire leading to the speaker of the TV and connect *B* and *C* remote switch wires as shown in Fig. 16-8. The third wire from the remote sound control circuit connects to the other side of the speaker.

The on-off feature allows the viewer to turn off the sound from the TV speaker from the remote control box. To use the headphone jacks, plug in the headphones and adjust the volume with potentiometer R_1 on the remote sound control box. The headphones may be used when the TV speaker is either on or off.

DISCO LIGHT ORGAN CONTROL

The light organ circuit in Fig. 16-9 is wired to the speakers of an FM radio, stereo, phonograph, or tape player to give a colorful light show. Three strings of lights (each a different color) are plugged into outlets SO_1, SO_2, and SO_3 and respond to the audio input. The lights plugged into SO_3 respond to bass frequencies, SO_2 to midrange frequencies, and SO_1 to high or treble frequencies. The light organ can handle up to 100 W of lights per channel. Colored Christmas tree lights work very well and are especially impressive when displayed behind frosted or translucent plastic.

When the light organ in Fig. 16-9 has light strings plugged into SO_1, SO_2, and SO_3, line voltage appears across each SCR. For the half-cycle when the SCRs are "ready" (cathode negative and anode positive), a positive voltage at the gate will turn on the attached lights. The input transformer T_1 transfers the signal from the speakers to the audio amplifier Q_1. From Q_1 it is sent to each of the frequency-sensitive filters connected to the gates of the low-, medium-, and high-frequency SCRs. If most of the input frequencies are low, the low-pass filter (R_8 and C_7) triggers SCR_3, lighting the lamps plugged into SO_3. If most of the input frequencies are high, the high-pass filter (R_5 and C_4) triggers SCR_1, lighting lamps plugged into SO_1. Midrange frequencies cause the bandpass filter (R_7, C_5, and C_6) to trigger SCR_2, lighting lamps plugged into SO_2. Potentiometer R_{10} is the overall sensitivity control while R_{11} and R_{12} control the sensitivity of the medium- and low-frequency units, respectively.

The high-voltage wiring must be done carefully to make sure that it is completely isolated

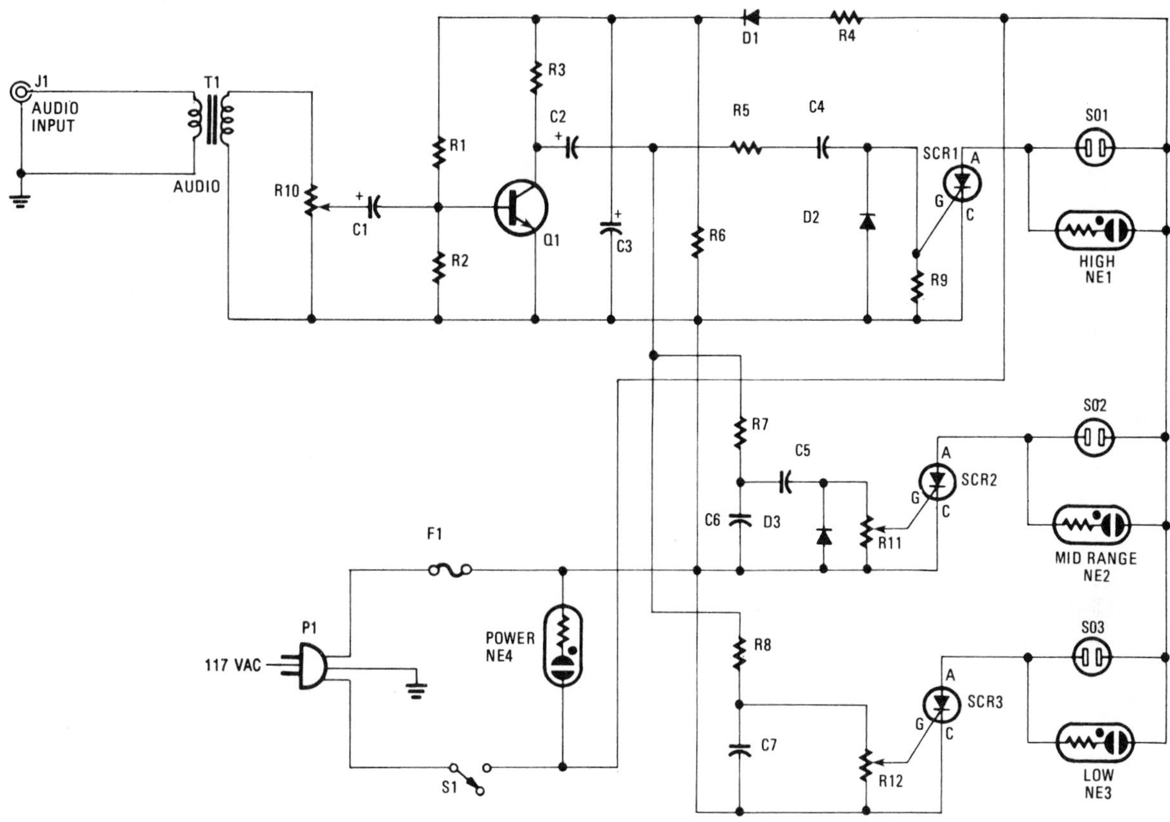

PARTS LIST

C_1,C_2	5-μF, 25-V electrolytic capacitor
C_3	25-μF, 25-V electrolytic capacitor
C_4,C_6	0.001-μF, 50-V disk capacitor
C_5	0.022-μF, 200-V capacitor
C_7	0.2-μF, 200-V capacitor
D_1,D_2,D_3	1N4004 silicon diode, 1 A, 400 PIV
F_1	15-A fuse
J_1	RCA phono jack
NE_1 through NE_4	117-V ac neon pilot lights
P_1	Three-pronged ac plug
Q_1	2N5223 NPN transistor (similar to Radio Shack 276-2009)
R_1	180-kΩ, ½-W resistor
R_2	100-kΩ, ½-W resistor
R_3	1-kΩ, ½-W resistor
R_4	4.7-kΩ, 2-W resistor
R_5 through R_8	10-kΩ, ½-W resistor
R_9	22-kΩ, ½-W resistor
R_{10},R_{11},R_{12}	10-kΩ linear potentiometer
S_1	SPST switch
SCR_1,SCR_2,SCR_3	C106B1 SCR, 4 A, 200 PIV (similar to Radio Shack 276-1067)
SO_1,SO_2,SO_3	AC convenience outlet
T_1	Audio transformer (2 kΩ primary, 10 kΩ secondary)

FIG. 16-9 Disco light organ control. (Barry L. Thompson, "Disco Light Organ Control," *Radio-Electronics Special Projects,* Winter 1983, pp. 20–21, 97. Used by permission of Gernsback Publications, Inc.)

from the enclosure. The green ground wire on the three-pronged plug must be attached to the metal chassis. Use proper strain relief techniques where cords enter and exit the case.

INTERCOM

The schematic diagram and parts list for the 1-W intercom are shown in Fig. 16-10. The LM390 audio power amplifier IC is the central component of this unit. With switch S_1 in the talk position, the master speaker SPK_1 becomes a microphone and the output of the LM390 amplifier IC_1 is fed to the remote speaker. However, with S_1 in the listen position, remote speaker SPK_2 becomes a microphone and the output of the IC is fed to the master speaker. The low-impedance output of the LM380 IC drives the speakers directly through capacitor C_3. Capacitor C_5 and resistor R_3 are used to suppress unwanted oscillations.

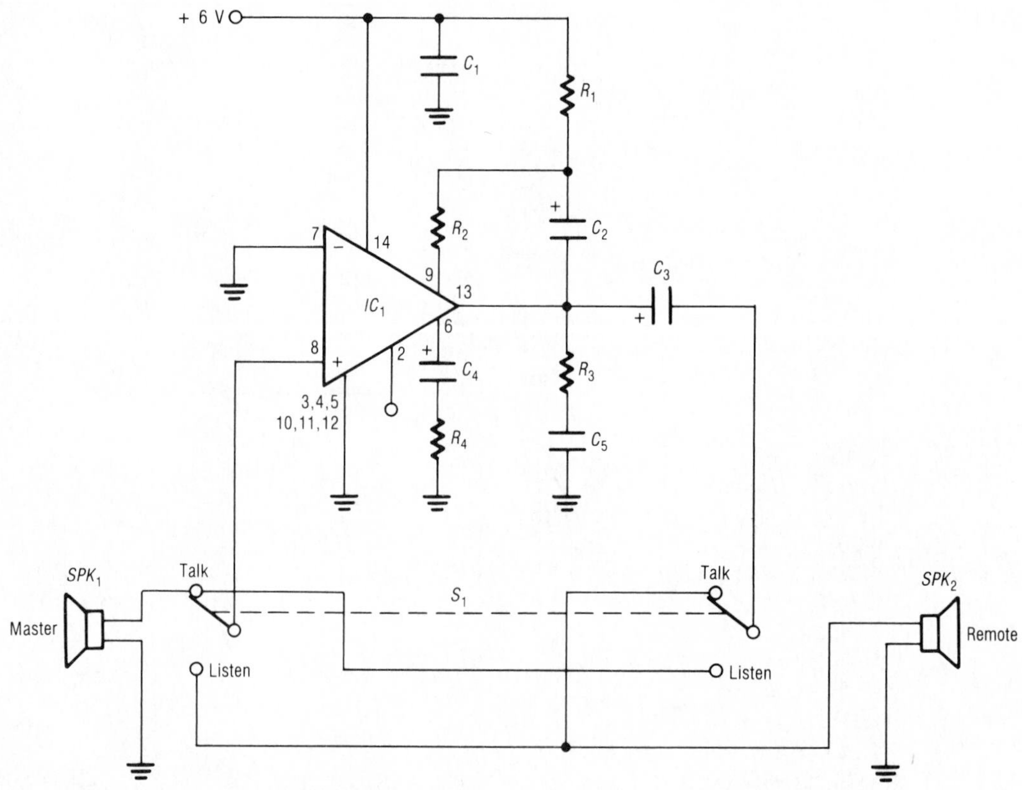

PARTS LIST

C_1	0.1-μF, 50-V disk capacitor
C_2	4-μF, 25-V electrolytic capacitor
C_3	100-μF, 25-V electrolytic capacitor
C_4	25-μF, 25-V electrolytic capacitor
C_5	0.05-μF, 50-V disk capacitor
IC_1	LM390 audio power amplifier IC
R_1	180-Ω, ½-W resistor
R_2	180-Ω, ½-W resistor
R_3	2.7-Ω, ½-W resistor
R_4	51-Ω, ½-W resistor
S_1	DPDT switch
SPK_1,SPK_2	8-Ω speaker

FIG. 16-10 Intercom. (Used by permission of National Semiconductor Corporation.)

1-WATT AMPLIFIER

This 1-W amplifier uses discrete components instead of ICs. The input to the audio amplifier in Fig. 16-11 has high impedance and can be driven with a crystal microphone. Notice that the output stage (Q_4 and Q_5) of the amplifier does not use a transformer. Two 8-Ω speakers wired in series are used in this circuit, or a single 16-Ω speaker could be used as the out-

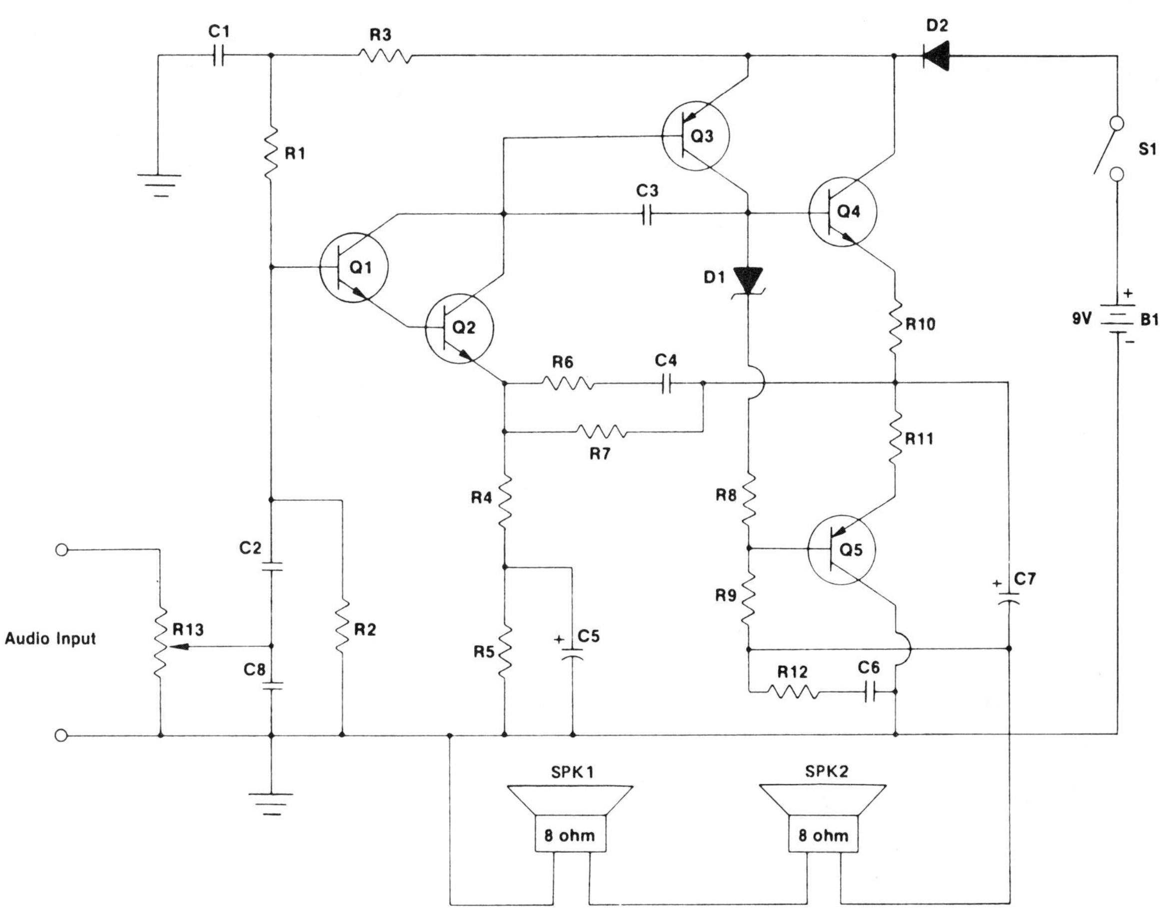

PARTS LIST

B_1	9-V alkaline or ni-cad battery	R_1	5.6-MΩ, ½-W resistor
C_1	0.22-μF, 50-V capacitor	R_2	2.2-MΩ, ½-W resistor
C_2	0.05-μF, 50-V capacitor	R_3	3.9-MΩ, ½-W resistor
C_3	220-pF, 50-V capacitor	R_4	33-Ω, ½-W resistor
C_4	0.02-μF, 50-V capacitor	R_5	1-kΩ, ½-W resistor
C_5	33-μF, 16-V electrolytic capacitor	R_6	1.8-kΩ, ½-W resistor
C_6	0.1-μF, 50-V capacitor	R_7	3.9-kΩ, ½-W resistor
C_7	220-μF, 25-V electrolytic capacitor	R_8	220-Ω, ½-W resistor
C_8	0.001-μF, 50-V capacitor	R_9	560-Ω, ½-W resistor
D_1	Zener diode	R_{10},R_{11}	1-Ω, ½-W resistor
D_2	1N4001 silicon diode, 1 A, 50 PIV	R_{12}	22-Ω, ½-W resistor
Q_1,Q_2	BC 408 or BC 548 NPN transistor (similar to Radio Shack 276-2009)	R_{13}	1-MΩ audio taper potentiometer
Q_3	BC 558 or BC 418 PNP transistor (similar to Radio Shack 276-2023)		
Q_4	BC 338 NPN transistor (similar to Radio Shack 276-2009)	SPK_1,SPK_2	8-Ω speaker
Q_5	BC 328 PNP transistor (similar to Radio Shack 276-2023)	S_1	SPST switch

FIG. 16-11 1-W amplifier. (Used by permission of Mode Electronics.)

put. Output transistors Q_4 and Q_5 should have heat sinks attached if the amplifier is to be used at full power.

A 9-V alkaline battery or 12-V automobile-type battery will operate the amplifier. A parts kit, a pc board, and instructions are available from Mode Electronics.

STEREO AMPLIFIER

This stereo amplifier will produce 6 W per channel and has a very low parts count. The schematic diagram and parts list are detailed in Fig. 16-12. The heart of the circuit is the LM379 dual 6-W audio amplifier IC. The LM379 features a very high input impedance, good channel separation, internal current limiting, thermal shutdown, and a voltage gain A_v of about 50. The LM379 IC is available in a 14-pin power dual-in-line package (DIP) with heat sink. The dc power supply voltage is shown as 28 V on the schematic but can vary from 12 to 30 V.

For low-power applications, either the 2-W LM377 or the 4-W LM378 audio amplifier IC

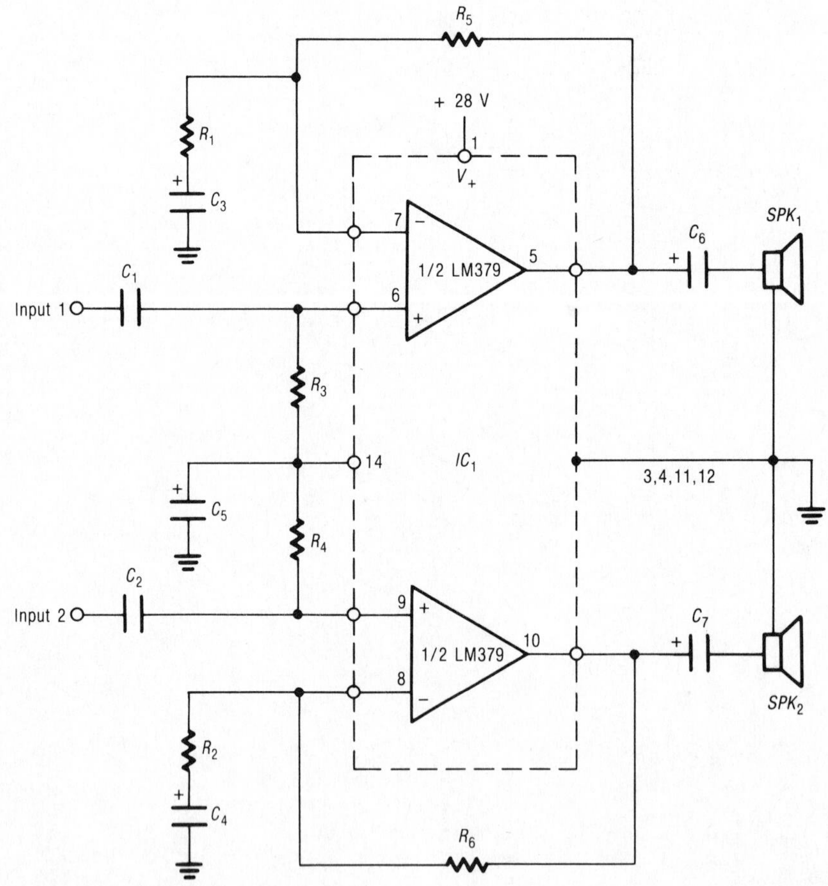

PARTS LIST

C_1,C_2	0.1-μF, 100-V capacitor
C_3,C_4	5-μF, 50-V electrolytic capacitor
C_5	250-μF, 50-V electrolytic capacitor
C_6,C_7	200-μF, 50-V electrolytic capacitor
IC_1	LM379 dual 6-W audio amplifier IC
R_1,R_2	2-kΩ, ½-W resistor
R_3,R_4,R_5,R_6	100-kΩ, ½-W resistor
SPK_1,SPK_2	8Ω speaker

FIG. 16-12 Stereo amplifier. (Used by permission of National Semiconductor Corporation.)

can replace the LM379 IC in the schematic in Fig. 16-12. The LM377 and LM378 audio amplifier ICs operate at lower voltages and are not pin-compatible with the more powerful LM379 6-W audio amplifier IC.

STEREO PREAMPLIFIER

The schematic diagram and parts list for a simple stereo preamplifier are shown in Fig. 16-13. The circuit is based on the 353 wide-bandwidth dual JFET input operational amplifier. The gain of the right and left channels are regulated by the gain adjust controls R_3 and R_{3a}. The circuit features high-input impedance. The stereo preamplifier can drive any commercial amplifier with 2 kΩ or more of input impedance.

Power for the stereo preamplifier may be furnished by a dual −/+12-V dc power supply. To reduce hum, use shielded cable on the input leads to the 353 ICs and place the preamplifier section away from the power supply.

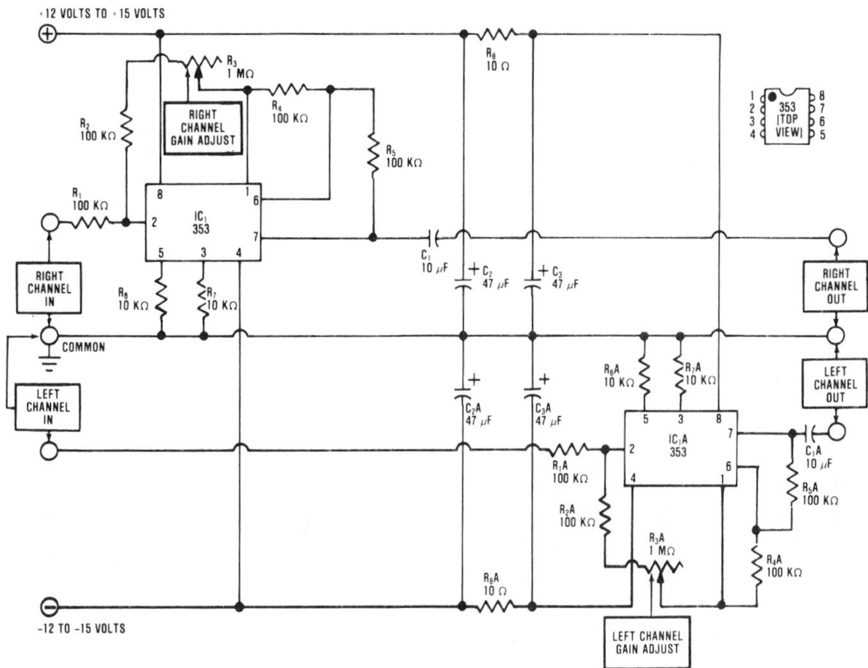

PARTS LIST

C_1, C_{1a}	10-μF, 25-V nonpolarized electrolytic capacitor
C_2, C_{2a}, C_3, C_{3a}	47-μF, 25-V electrolytic capacitor
IC_1, IC_{1a}	353 dual JFET input op-amp IC
$R_1, R_{1a}, R_2, R_{2a}, R_4, R_{4a}, R_5, R_{5a}$	100-kΩ, ¼-W resistor
R_3, R_{3a}	1-MΩ, linear-taper potentiometer
R_6, R_{6a}, R_7, R_{7a}	10-kΩ, ¼-W resistor
R_8, R_{8a}	10-Ω, ¼-W resistor

FIG. 16-13 Stereo preamplifier. (Michael Gannon, *Workbench Guide to Semiconductor Circuits and Projects,* Prentice-Hall, New Jersey, 1982, pp. 126–127. Used by permission of Prentice-Hall, Inc.)

chapter 17

Timer and Thermometer Circuits

STOP-ACTION TIMING TESTER

The timing tester in Fig. 17-1 is a game to test a player's sense of timing. To play, close the ON-OFF switch S_1 and the light-emitting diode LED_1 will flash on every few seconds. The player must try to press the push-button switch S_2 at exactly the right time (when the LED is on). If the player presses the switch when the LED is on, the player wins and the LED is "frozen" in the ON position. A parts kit and a pc board for the stop-action timing tester game are currently available from Electronic Kits International, Inc.

The timing tester circuit uses a 555 timer IC_1, connected as a free-running multivibrator, to turn the LED on and off. The charging time (off time for the LED) is determined by the values of R_1, R_2, and C_1 in the RC circuit. The discharging time (on time for the LED) is determined by the values of just R_2 and C_1. The LEDs on time is much shorter than the off time. The LED is off for about 2 seconds and on for about 1⁄10 second. If the player opens S_2 during the charge time, the LED will remain off and the player does not win. If the player opens S_2 during the discharge time, the LED will remain lit, indicating that the player has won the game. Different timing can be used by changing the values of R_1, R_2, and C_1.

VARIABLE AC TIMER

The timer circuit in Fig. 17-2 will control an ac device up to about 150 W. Depression of switch S_2 starts the timer and turns on the ac device plugged into socket SO_1. After a time, the ac to the device is automatically turned off. The amount of time can vary from a few minutes to about 1 hour. Potentiometer R_1 sets the timer delay. A parts kit and a pc board for the variable ac timer are currently available from Graymark International, Inc.

The central component of the timer section of the circuit in Fig. 17-2 is the 555 timer IC_1. When the start switch (S_2) is pressed, the 555 timer (wired as a one-shot multivibrator) turns

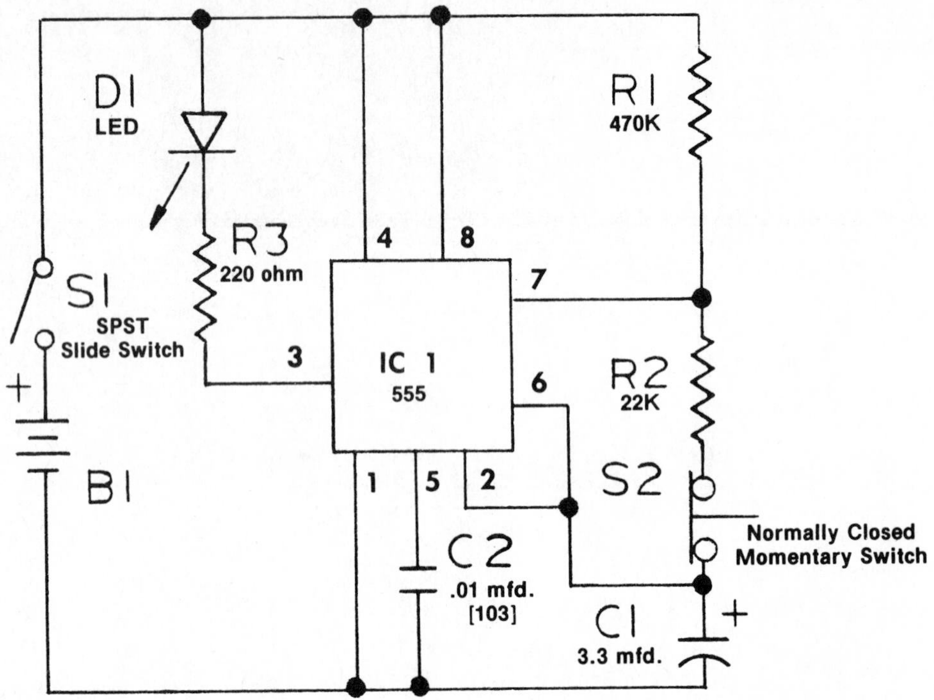

PARTS LIST

B_1	9-V battery
C_1	3.3-μF, 16-V electrolytic capacitor
C_2	0.01-μF, 50-V disk capacitor
IC_1	555 timer IC
LED_1	Light-emitting diode
R_1	470-kΩ, ½-W resistor
R_2	22-kΩ, ½-W resistor
R_3	220-Ω, ½-W resistor
S_1	SPST switch
S_2	Normally closed push-button switch

FIG. 17-1 Stop-action timing tester game. (Used by permission of Electronic Kits International, Inc.)

on the triac Q_1, which activates the load device plugged into the socket SO_1. The capacitor in the RC circuit consisting of R_1, R_2, and C_2 charges. When the capacitor voltage reaches two-thirds of the supply voltage, the output of IC_1 (pin 3) goes low, turning off the triac. Parts T_1, D_1, and C_1 form the dc power supply for the 555 timer IC.

The high-voltage wiring must be done carefully to make sure that it is completely isolated from the enclosure. The green ground wire on the three-pronged plug must be attached to the metal chassis. Use proper strain relief techniques where cords enter and exit the case.

TIME-ON RECORDER

The timer circuit in Fig. 17-3 records the time a television or other appliance is turned on. The time-on recorder circuit automatically turns on when the TV is turned on, and the timer stops when the appliance is turned off, showing the "on time." To use, plug the time-on recorder into an ac receptacle and plug the appliance into the timer. Press the reset push-button S_1 to clear the timer to 00000. As soon as the appliance is turned on, the internal clock chip IC_1 is enabled, causing the on time of the unit to be shown on the digital display

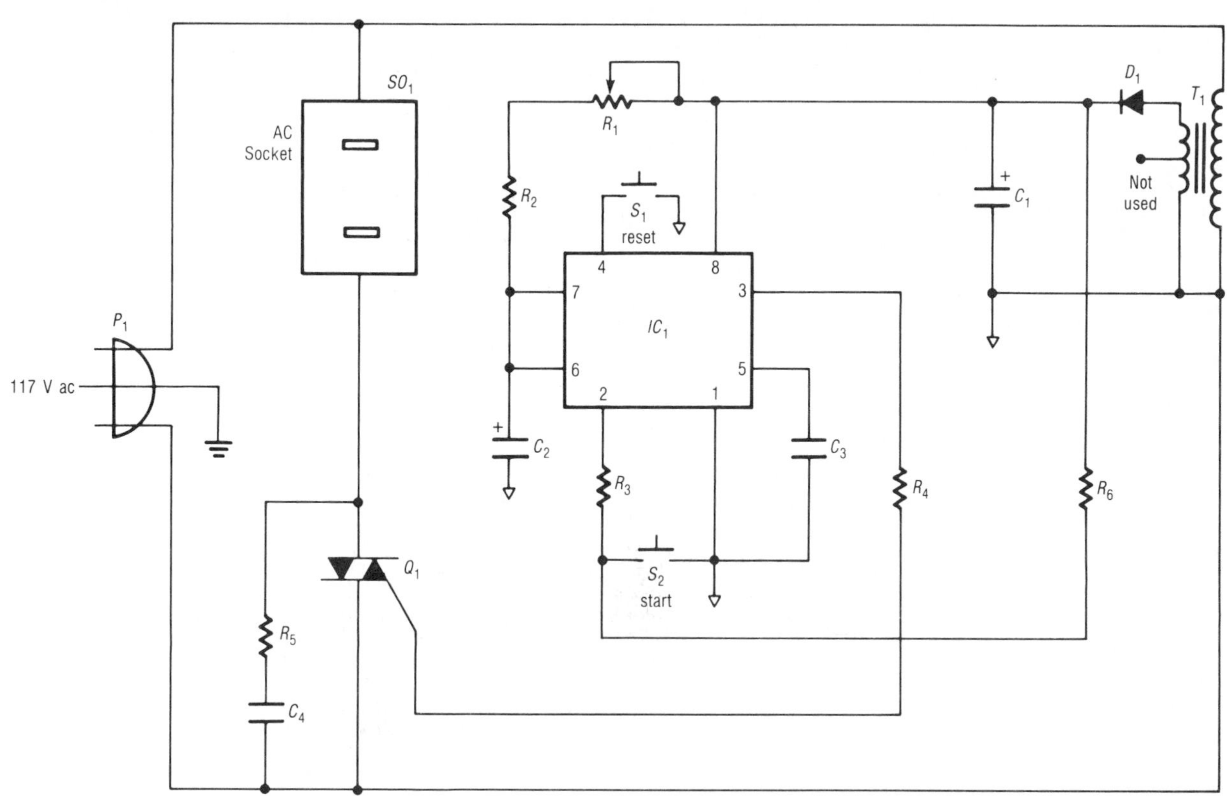

PARTS LIST

C_1	100-μF, 50-V electrolytic capacitor
C_2	500-μF, 25-V electrolytic capacitor
C_3	0.01-μF, 100-V disk capacitor
C_4	0.22-μF, 250-V capacitor
D_1	1N4004 silicon diode, 1 A, 400 PIV
IC_1	555 timer IC
P_1	Three-pronged ac plug
Q_1	Triac, 4 A, 400 V (similar to Radio Shack 276-1000)
R_1	5-MΩ potentiometer
R_2,R_5	100-kΩ, ½-W resistor
R_3	4.7-kΩ, ½-W resistor
R_4	47-Ω, ½-W resistor
R_6	1-kΩ, ½-W resistor
S_1,S_2	Normally open push-button switch
SO_1	AC convenience socket
T_1	12-V, 300-mA power transformer (similar to Radio Shack 273-1385)

FIG. 17-2 Variable ac timer. (Used by permission of Graymark International, Inc.)

($DISP_1$). When the appliance is turned off, the clock chip is disabled and the display freezes the total on time. If the appliance were turned on again, the clock would be enabled, continuing from where it was giving the total on time up to 24 hours.

With an appliance plugged into socket SO_1 in Fig. 17-3, current is flowing through the triac, causing a small voltage drop (about 2 V). This ac voltage is converted to a dc voltage by T_1, D_1, D_2, and C_1. The dc voltage turns on transistor Q_6, dropping the voltage at pin 19 of IC_1. This lets the 60-Hz pulses from T_2 into IC_1, enabling the clock. The clock chip and transistors Q_1 through Q_5 multiplex the five digits. The seven segments are driven through R_8 through R_{14}. Transformer T_2, IC_2, and C_2 form a dc supply to power the IC, transistors,

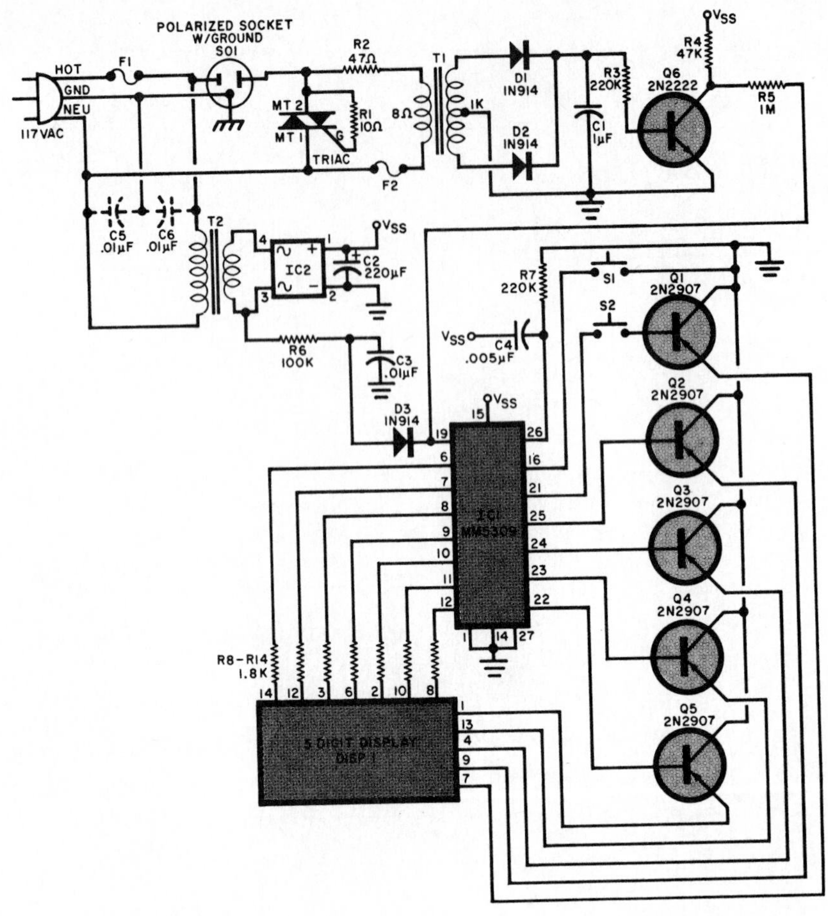

PARTS LIST

C_1	1-μF nonpolarized electrolytic capacitor
C_2	220-μF, 35-V electrolytic capacitor
C_3	0.01-μF, 50-V disk capacitor
C_4	0.005-μF, 50-V disk capacitor
C_5,C_6	0.01-μF, 400-V disk capacitor
D_1,D_2,D_3	1N914 signal diode
$DISP_1$	Five-digit, seven-segment, common-cathode display (Hewlett-Packard 5082 or equivalent)
F_1	6¼-A, 3AG, slow-blow fuse
F_2	¼-A, 3AG, fast-acting fuse
IC_1	MM5309 digital clock IC
IC_2	Full-wave bridge rectifier (similar to Radio Shack 276-1151)
P_1	Three-pronged ac plug
Q_1 through Q_5	2N2907 PNP transistor
Q_6	2N2222 NPN transistor
R_1	10-Ω, ½-W resistor
R_2	47-Ω, ½-W resistor
R_3,R_7	220-kΩ, ¼-W resistor
R_4	47-kΩ, ¼-W resistor
R_5	1-MΩ, ¼-W resistor
R_6	100-kΩ, ¼-W resistor
R_8 through R_{14}	1.8-kΩ, ¼-W resistor
SO_1	Three-pronged ac convenience socket
S_1,S_2	Normally open push-button switch
T_1	Audio output transformer (8 to 1000 Ω) (similar to Radio Shack 273-1380)
T_2	12.6-V, 300-mA power transformer (similar to Radio Shack 273-1385)
Triac	400-V, 6-A triac (similar to Radio Shack 276-1000)

FIG. 17-3 Time-on recorder. (Reprinted from *Popular Electronics*. Copyright © February 1982, Ziff-Davis Publishing Company.)

and displays. Resistor R_7 and capacitor C_4 are the external parts that set the clock chip's multiplex frequency.

Normally the tens and units of hours, the tens and units of minutes, and only the tens of seconds are shown on the five-digit display. Depression of switch S_2 will display the seconds (instead of tens of seconds) and is a test to determine whether the clock is working. Capacitors C_5 and C_6 suppress voltage spikes on the ac line usually associated with inductive loads such as motors. The minimum load to automatically start the time-on recorder is about 10 W, while the maximum load is 600 W.

The high-voltage wiring must be done carefully to make sure that it is completely isolated from the enclosure. The green ground wire on the three-pronged plug must be attached to the metal chassis. Use proper strain relief techniques where cords enter and exit the case. The triac must be mounted on a good heat sink.

FAHRENHEIT THERMOMETER

The simple thermometer circuit in Fig. 17-4 will sense its environment and yield an output voltage proportional to the temperature. The output should equal 1 millivolt per degree Fahrenheit (1 mV/°F).

The temperature sensor is IC_1, while the LM336 (IC_2) is an accurate voltage reference (2.5 V). To calibrate, adjust R_6 so that the voltage across the voltage reference diode IC_2 is exactly 2.554 V. Then adjust trimmer potentiometer R_5 so that the output reads 1 mV/°F. An accurate digital multimeter (DMM) can then read the temperature in degrees Fahrenheit. The LM335 temperature sensor is linear over a range from −40 to +100°C.

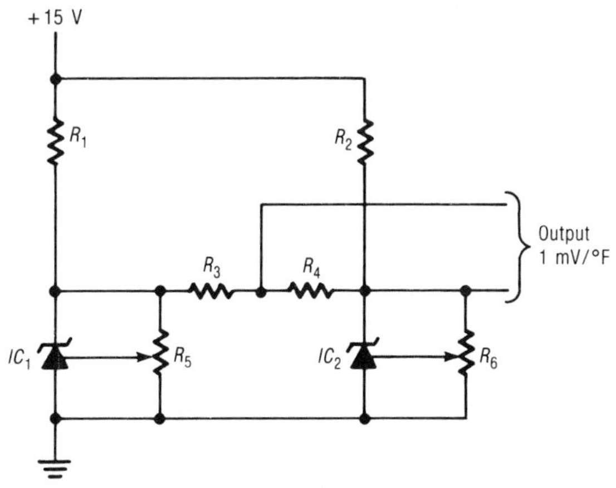

PARTS LIST

IC_1	LM335 precision temperature sensor IC
IC_2	LM336 2.5-V reference diode IC
R_1,R_2	10-kΩ, ¼-W resistor
R_3	4.55-kΩ, ¼-W resistor
R_4	1-kΩ, ¼-W resistor
R_5,R_6	10-kΩ trimmer potentiometer

FIG. 17-4 Fahrenheit thermometer circuit. (Used by permission of National Semiconductor Corporation.)

CELSIUS THERMOMETER

The simple thermometer circuit in Fig. 17-5 will sense its environment and yield an output voltage proportional to the temperature. The output should equal 10 mV/°C.

The temperature sensor is IC_1, while the LM336 (IC_3) is an accurate voltage reference (2.5 V). The LM308 chip (IC_2) is an operational amplifier for scaling the output voltage to 10 mV/°C. Use trimmer potentiometers R_5 and R_6 to give 10 mV/°C. An accurate DMM can read

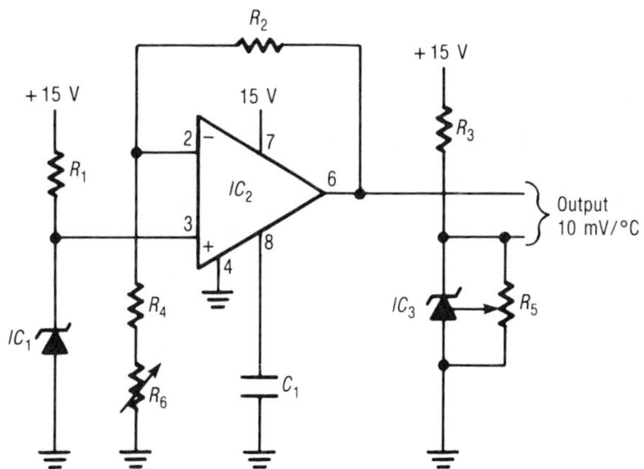

PARTS LIST

C_1	100-pF, 100-V disk capacitor
IC_1	LM335 precision temperature sensor IC
IC_2	LM308 operational amplifier IC
IC_3	LM336 2.5-V reference diode IC
R_1	6-kΩ, ¼-W resistor
R_2	1-kΩ, ¼-W resistor
R_3	12-kΩ, ¼-W resistor
R_4	8.5-kΩ, ¼-W resistor
R_5	10-kΩ trimmer potentiometer
R_6	2-kΩ trimmer potentiometer

FIG. 17-5 Celsius thermometer circuit. (Used by permission of National Semiconductor Corporation.)

the temperature in celsius. The LM335 temperature sensor is linear over a range from −40 to +100°C.

SIMPLE ELECTRONIC THERMOMETER

The thermometer circuit in Fig. 17-6 has a low parts count and is easy to build. The electronic thermometer is a good beginning science project because it takes some testing to determine the calibration of the panel meter in either fahrenheit or celsius. Also, several transistors can be tested to see which makes the most sensitive thermometer.

The current flowing through meter M_1 is the collector current of transistor Q_1. As the temperature of the transistor increases, so does the collector current of Q_1. The builder needs to calibrate the face of the meter based on known temperatures of the transistor. Potentiometer R_1 recalibrates the thermometer while zener diode D_1 maintains a stable bias voltage on the transistor to compensate for battery aging. An ice bath can be used to establish 0°C and boiling water to establish 100°C.

DIGITAL ALARM CLOCK/THERMOMETER

The time/temperature circuit in Fig. 17-7 is a sophisticated clock and digital thermometer. The parts count is held to a minimum by using the MA1026 module by National Semiconductor Corporation. The MA1026 module contains the parts located inside the dashed lines on the schematic diagram in Fig. 17-7, including the LED displays. In addition to the module, only feature select switches, a temperature sensor, a transformer, and a speaker are needed.

Some of the features of the digital alarm clock/thermometer are a large four-digit LED display, a 12- or 24-hour display, a time or temperature display, a fahrenheit or celsius temperature display, an alarm, a multiple snooze alarm, a sleep setting, and a flashing display for power failure. The MA1026 module has six display modes, which are detailed in Table 17-1. The displays are driven directly, eliminating possible radio frequency interference from radio. Other features are bright/dim control, direct alarm output for driving piezo buzzers, an 8-Ω speaker driver, separate inputs for all setting and display modes, and leading zero blanking. The LED display also features PM, colon, degree, and "alarm on" indicators. The clock also has slow- and fast-set features.

Table 17-2 shows the function of the control switches. In the first line, for instance, closing the *time set allow* and *slow set* switches at the same time causes the minutes to advance at a 2-Hz rate and the seconds counter to be reset to zero. A parts kit, a pc board, and an instruction manual for the digital clock/thermometer are currently available from Digi-Key Corporation. The kit from Digi-Key has the added feature of two temperature sensors for indoor and outdoor temperatures.

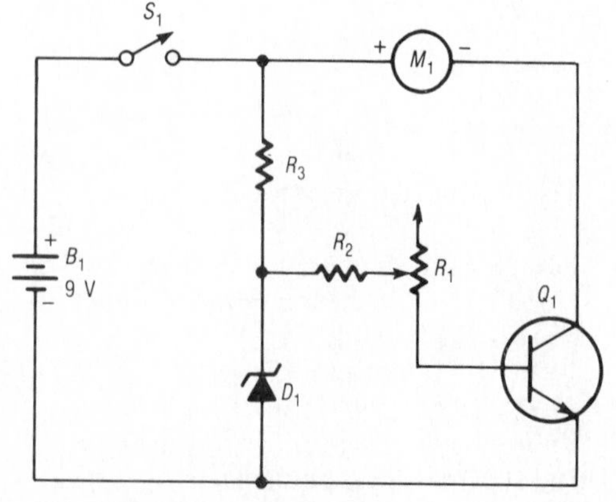

PARTS LIST

- B_1 9-V battery
- D_1 6.8-V, 500-mW zener diode
- M_1 0- to 1-mA panel meter
- Q_1 General-purpose NPN transistor (similar to Radio Shack 276-2009)
- R_1 1-MΩ potentiometer
- R_2 100-kΩ, ¼-W resistor
- R_3 150-Ω, ¼-W resistor
- S_1 SPST switch

FIG. 17-6 Simple electronic thermometer. ("SITRAT- Build This Highly Accurate Electronic Thermometer," *Electronics Hobbyist,* Summer 1984, pp. 22–24. Used by permission of C & E Hobby Handbooks, Inc.)

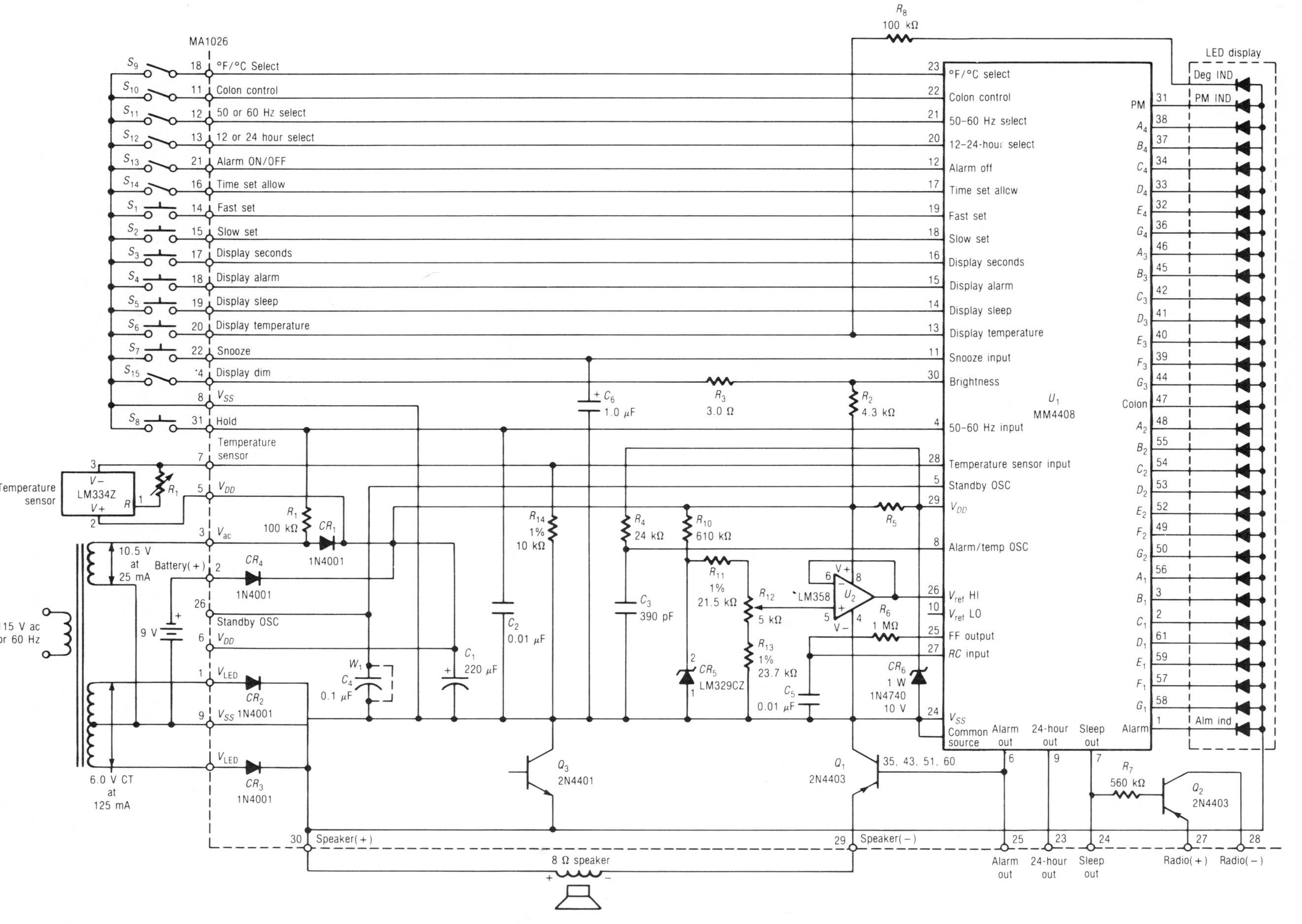

PARTS LIST

IC_1	LM334Z temperature-sensor IC (National Semiconductor)
Module	MA1026 digital alarm clock/thermometer (National Semiconductor)
R_1	300-Ω trimmer potentiometer
S_1 through S_8	Normally open push-button switch
S_9 through S_{15}	SPST switch
SPK	8-Ω speaker
T_1	Power transformer (Digi-Key #MA1026M)

FIG. 17-7 Digital alarm clock/thermometer. (Used by permission of National Semiconductor Corporation.)

TABLE 17-1 Display Modes for the MA1026 Clock/Thermometer Module. (Used by permission of National Semiconductor Corporation.)

Selected Display Modes*	Digit 4	Digit 3	Digit 2	Digit 1
Time display	Time 10s hours, PM indicator	Time Hours	Time 10s minutes	Time Minutes, alarm ON indicator
Seconds display	Blanked	Time Minutes	Time 10s seconds	Time Seconds
Alarm display	Alarm 10s hours, PM indicator	Alarm Hours	Alarm 10s minutes	Alarm Minutes, alarm ON indicator
Sleep display	Blanked	Blanked	Sleep 10s minutes	Sleep Minutes
Temperature display	100s temperature	10s temperature	1s temperature	°C or °F
Alarm and sleep	Lamp test	Lamp test	Lamp test	Lamp test

* If more than one display mode input is applied, the display priorities are in the order of temperature, alarm *or* sleep, seconds, then time. Alarm and sleep have equal priority over seconds; however, when both alarm and sleep are applied, all outputs are ON, providing a lamp test. This display mode has priority above all others.

TABLE 17-2 Control-Setting Functions for the MA1026 Clock/Thermometer Module. (Used by permission of National Semiconductor Corporation.)

Selected Display Mode	Control Input	Control Funciton
Time and seconds display	Time set allow and slow set simultaneously	Minutes advance at a 2-Hz rate and seconds counter is reset to :00
	Time set allow and fast set simultaneously	Minutes advance at a 60-Hz rate; seconds counter not affected
	Time set allow and fast and slow set simultaneously	Hours, minutes, and seconds are reset to: 12:00:00 AM (12-hour mode) 0:00:00 (24-hour mode)
Alarm display	Slow set	Alarm minutes counter advances at a 2-Hz rate
	Fast set	Alarm minutes counter advances at a 60-Hz rate
	Fast and slow set simultaneously	Alarm minutes and hours counters are reset to: 12:00 AM (12-hour mode) 0:00 (24-hour mode)
Sleep display	Slow set	Sleep counter is decremented at a 2-Hz rate
	Fast set	Sleep counter is decremented at a 10-Hz rate
	Fast and slow set simultaneously	Sleep counter is reset to 59 minutes
Sleep display and alarm display	All outputs are driven to provide a lamp test	

appendix A

Schematic Symbols

Electronic schematic symbols are used to represent the *function* or *job* of a component in a circuit. The symbol does not show the outward appearance of the electronic part. Schematic diagrams show the electrical connections in a device by the use of schematic symbols and straight lines.

The collection of schematic symbols on the following pages covers most of the ones used in the 146 circuits in this book. Because the projects originated from various sources, the same function may be represented by a different symbol from project to project.

The physical appearance of two electronic parts that serve the same function may vary considerably. Electronic components vary in appearance because of use, ratings, and the manufacturer's individuality. The best way to find out what components look like is to page through an illustrated electronic parts catalog. For the beginner, an excellent way to learn what parts look like and what they do is to build a few simple battery-powered projects.

Name of device	Circuit symbol
AC plug (3-prong)	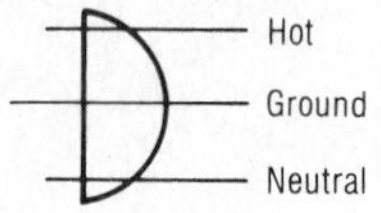 Hot Ground Neutral
AC socket	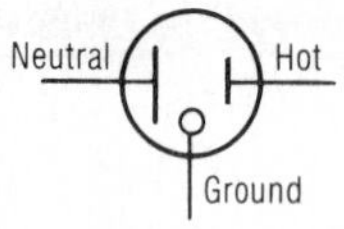 Neutral Hot Ground
Ammeter	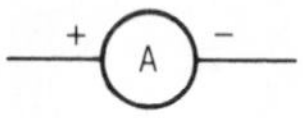 + A −
Antenna	
Battery	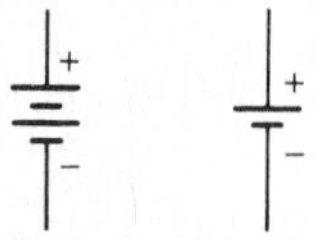 + − + −
Bridge rectifier	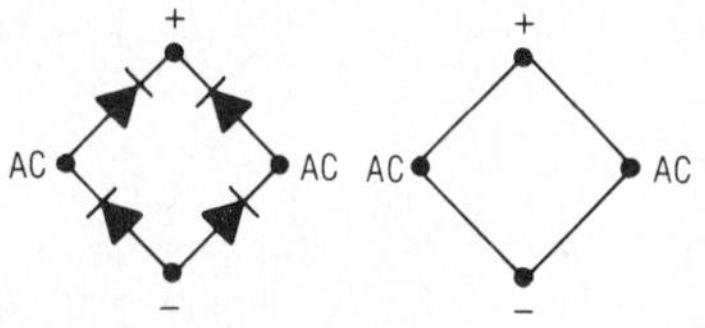 + AC AC − + AC AC −
Capacitor	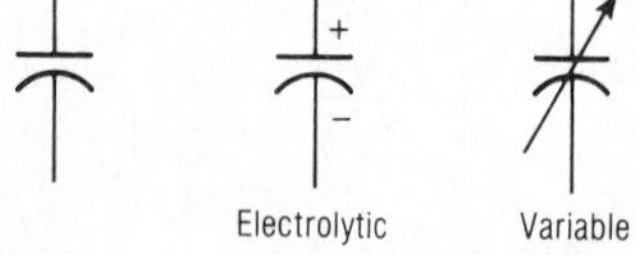 + − Electrolytic Variable
Circuit breaker	

Name of device	Circuit symbol
Coil (inductor)	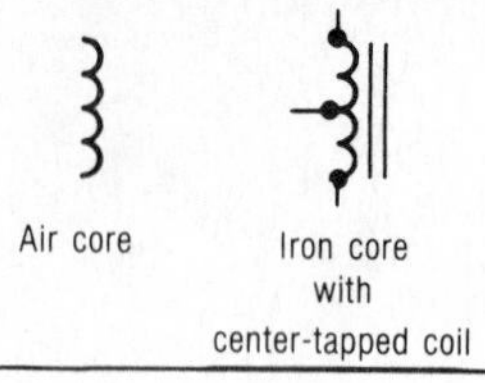 Air core Iron core with center-tapped coil
Crystal (piezoelectric)	
Diode	 Cathode Anode
Fuse	
Ground (common)	
IF transformer	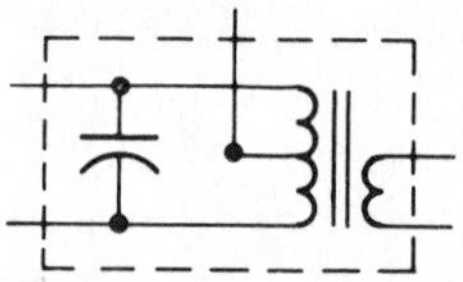
Lamp (incandescent)	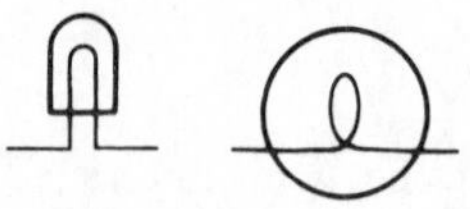
Light-emitting diode (LED)	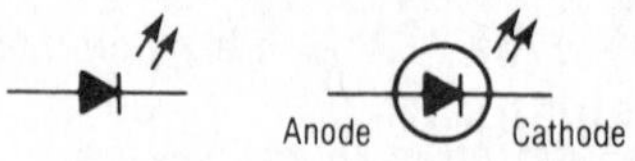 Anode Cathode

Name of device	Circuit symbol
Logic gates	AND gate Inverter NAND gate NOR gate OR gate
Meter movement (panel meter)	+ M −
Motor, dc	Motor
Neon lamp	
Optocoupler (SCR output)	

Name of device	Circuit symbol
Phone jacks	Mono (open circuit) Mono (closed circuit) Stereo
Photoresistor (photocell) (cadmium sulfide)	
Piezo buzzer	+ Piezo buzzer − + −
Potentiometer	
Power connections (typical)	+5 V V_{CC}

Name of device	Circuit symbol
RCA phone jack	
Relay (SPDT)	
Resistor	
SCR (silicon controlled rectifier)	Anode Cathode Gate
Seven-segment LED display	

Name of device	Circuit symbol
Speaker	
Switches	SPST Normally open push button Normally closed push button Rotary
Thermistor	T

Name of device	Circuit symbol
Transistors	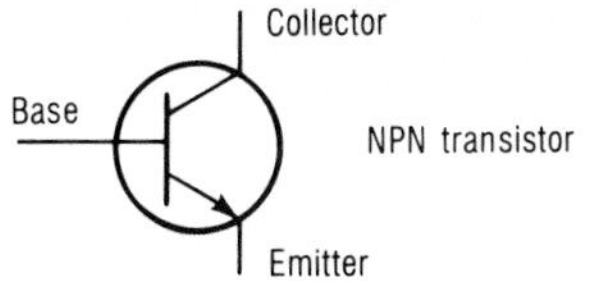 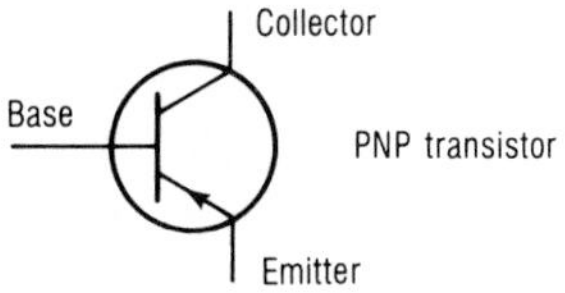 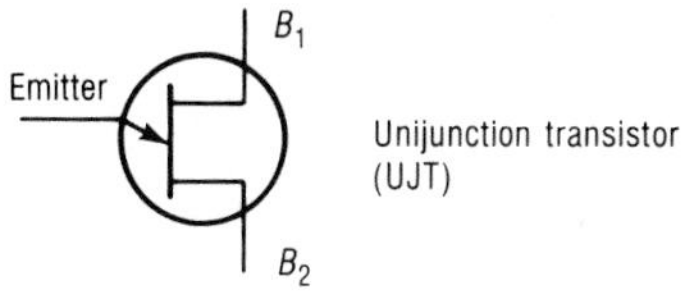 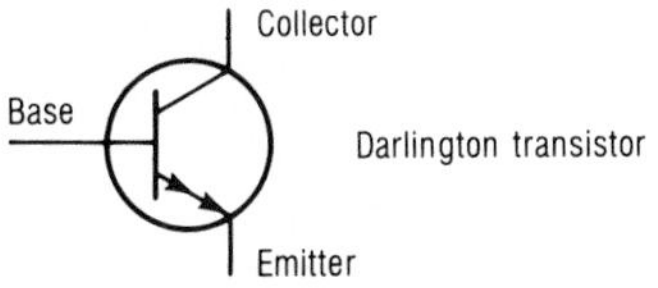 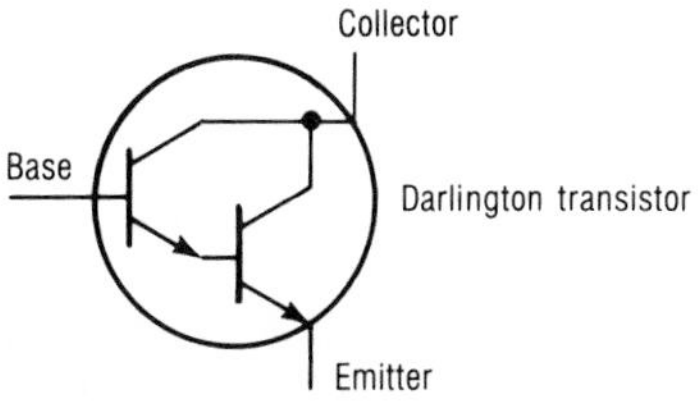 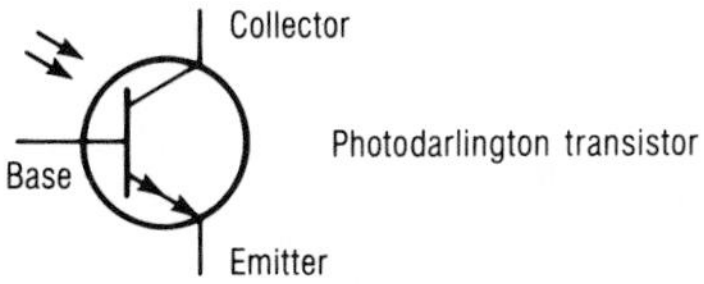 Collector / Base / Emitter — NPN transistor Collector / Base / Emitter — PNP transistor B_1 / Emitter / B_2 — Unijunction transistor (UJT) Collector / Base / Emitter — Darlington transistor Collector / Base / Emitter — Darlington transistor Collector / Base / Emitter — Photodarlington transistor
Triac	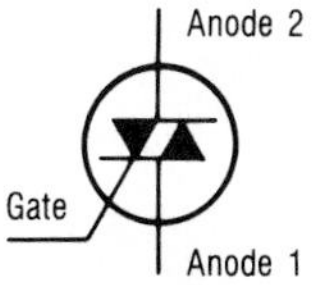 Anode 2 / Gate / Anode 1

Name of device	Circuit symbol
Trigger tube	
Unilateral switch	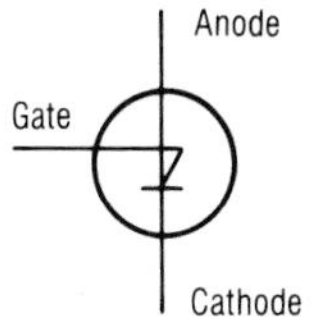 Anode / Gate / Cathode
Varistor (voltage-dependent resistor)	 V
Xenon flashtube	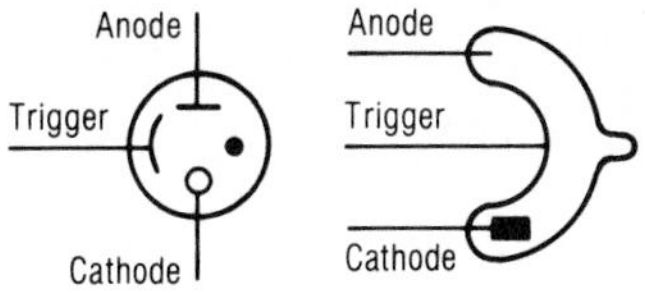 Anode / Trigger / Cathode; Anode / Trigger / Cathode
Zener diode	 Anode / Cathode

appendix B

Circuit Sources

Apple Computer, Inc.
20525 Mariani
Cupertino, CA 95014

C & E Hobby Handbooks, Inc.
300 West 43rd Street
New York, NY 10036

Steve Ciarcia
Byte Magazine
P.O. Box 372
Hancock, NH 03449

Robert Delp Electronics
Box 1026
Fremont, CA 94538

Electronic Kits International, Inc.
23210 Del Lago Drive
Laguna Hills, CA 92653

Electronics Week
1221 Avenue of the Americas
New York, NY 10020

General Electric Company
Semiconductor Products Division
West Genesee Street
Auburn, NY 13021

General Instrument Corporation
Optoelectronics Division
3400 Hillview Avenue
Palo Alto, CA 94304

Gernsback Publications, Inc.
200 Park Avenue South
New York, NY 10003

Graymark International, Inc.
3404 West Castor
Santa Ana, CA 92704

Heath Company
Benton Harbor, MI 49022

Intel Corporations
2565 Walsh Avenue
Santa Clara, CA 95001

David Leithauser
4649 Van Kleeck Drive
New Smyrna Beach, FL 32069

McGraw-Hill Book Company
1221 Avenue of the Americas
New York, NY 10020

Microcomputing
80 Pine Street
Peterborough, NH 03458

Forrest M. Mims, III
Computers & Electronics
309 Laurel Hill
San Marcos, TX 78666

Mode Electronics
500 Norfinch Drive
Downsview, Ontario, Canada
M3N 1Y4

National Semiconductor Corporation
2900 Semiconductor Drive
Santa Clara, CA 95051

OKI Semiconductor
650 North Mary Avenue
Sunnyvale, CA 94086

Parker Publishing Company, Inc.
Route 59 at Brook Hill Drive
West Nyack, NY 10995

Prentice-Hall, Inc.
Englewood Cliffs, NJ 07632

Howard W. Sams and Company, Inc.
4300 West 62nd Street
Indianapolis, IN 46206

TAB Books, Inc.
Blue Ridge Summit, PA 17214

Ziff-Davis Publishing Company
One Park Avenue
New York, NY 10016

Index

ABOUT THE AUTHOR

Roger L. Tokheim holds B.S. and M.S. degrees in Industrial Arts Education from St. Cloud State University and an Ed.S. degree in Vocational Education from the University of Wisconsin-Stout. He is the author of *Digital Electronics,* 2d ed. and its companion, *Activities Manual for Digital Electronics,* 2d ed. (McGraw-Hill, 1984), *Schaum's Outline of Microprocessor Fundamentals* (McGraw-Hill, 1983), and *Schaum's Outline of Digital Principles* (McGraw-Hill, 1980), and numerous other instructional materials on science and industry. An experienced educator at the adult and secondary levels, he is presently an instructor of Industrial Education and Computer Science at Henry Sibley High School, Mendota Heights, Minnesota. Recently he served as chairman of the Adult Electronics, Microprocessor, and Computer Curriculum Committee for the state of Minnesota.